A NOTE ON THE AUTHOR

Alice Bell is a researcher, campaigner and writer based in London, specialising in the politics of science, technology, environment and health.

She was a co-director at the climate change charity Possible for several years, working on a range of projects from community tree planting events to solar-powered railways. She has a BSc in history of science from UCL and a PhD in science communication from Imperial College, where she also held a lectureship in science communication and launched an interdisciplinary course on climate change.

Alice has also worked at the University of Sussex's Science Policy Research Unit, City Journalism School, the Science Museum, and as a freelance writer and editor. Alice has written for a host of publications including *The Times*, *The Observer* and *New Humanist*, researched the 1970s radical science movement for *Mosaic* magazine, co-founded the *Guardian*'s science policy blog, and edited the 'magazine for the future', *How We Get to Next*.

@alicebell

T0173616

OUR BIGGEST EXPERIMENT

A HISTORY OF THE CLIMATE CRISIS

Alice Bell

BLOOMSBURY SIGMA
LONDON · OXFORD · NEW YORK · NEW DELHI · SYDNEY

BLOOMSBURY SIGMA
Bloomsbury Publishing Plc
50 Bedford Square, London, WC1B 3DP, UK
29 Earlsfort Terrace, Dublin 2, Ireland

BLOOMSBURY, BLOOMSBURY SIGMA and the Bloomsbury Sigma logo are
trademarks of Bloomsbury Publishing Plc

First published in the United Kingdom in 2021. This edition published 2022.

Copyright © Alice Bell, 2021

Alice Bell has asserted her rights under the Copyright, Designs and Patents Act, 1988,
to be identified as Author of this work

All rights reserved. No part of this publication may be reproduced or transmitted
in any form or by any means, electronic or mechanical, including photocopying,
recording, or any information storage or retrieval system, without prior permission
in writing from the publishers

Bloomsbury Publishing Plc does not have any control over, or responsibility for, any
third-party websites referred to or in this book. All internet addresses given in this
book were correct at the time of going to press. The author and publisher regret
any inconvenience caused if addresses have changed or sites have ceased to exist,
but can accept no responsibility for any such changes

A catalogue record for this book is available from the British Library

Library of Congress Cataloguing-in-Publication data has been applied for

ISBN: PB: 978-1-4729-7478-5; eBook: 978-1-4729-6687-2

2 4 6 8 10 9 7 5 3

Typeset by Deanta Global Publishing Services, Chennai, India
Printed and bound in Great Britain by CPI Group (UK) Ltd, Croydon CR0 4YY

Bloomsbury Sigma, Book Sixty-Six

To find out more about our authors and books visit www.bloomsbury.com
and sign up for our newsletters

'The past is the present, isn't it? It's the future, too.'
Eugene O'Neill, *Long Day's Journey into Night*.

Contents

Introduction: Experiments

It was Eunice Newton Foote, a scientist, inventor and women's rights campaigner living in Seneca Falls, New York, who in 1856 first warned the world that an atmosphere heavy with carbon dioxide could send temperatures soaring. At the time, no one paid much attention.

Her experiment was reasonably simple. She placed two glass cylinders by a window and planted a thermometer in each of them. Using a pump to remove some of the air from one of the cylinders, she found it didn't catch the heat as well as the other. From this, she figured out the density of the air had an impact on the power of the Sun's rays. This made sense – after all, everyone knew it was colder at the top of high mountains. After comparing a cylinder of moist air with one that had been dried, she found the Sun's rays were more powerful in damper conditions. This wasn't surprising either, as she commented in her notes: 'Who has not experienced the burning heat of the Sun that precedes a summer's shower?' Thirdly, and crucially for our story, she tried filling one cylinder with carbon dioxide. This had the biggest impact: the cylinder became noticeably much hotter and took a lot longer to cool down after the experiment had ended. She concluded, almost in passing: 'An atmosphere of that gas would give to our Earth a high temperature.'

Her husband Elisha was a lawyer, but also undertook science experiments at home and would collect weather data for the local area. That summer they travelled together to the annual meeting of the American Association for the Advancement of Science (AAAS), held that year in Albany, New York. The astronomer Maria Mitchell had become the first female member of the AAAS a few years before, in 1850, but the titles of 'professional' or 'fellow' were still usually reserved for men. The dominant idea of what made for an authoritative 'proper' scientist of the time was still very male (just as it was almost exclusively white), and it's striking that although Eunice's paper, 'Circumstances affecting the heat of the Sun's rays', was presented at the meeting, it was read for her by a man. In contrast, Elisha presented his own paper. Eunice's paper was read by none other than Joseph Henry, the first secretary of the Smithsonian Institution, so it's possible he was chosen simply to give the paper more prominence. In his introduction, Henry made what were described in the press at the time as 'gallant remarks in regard to the ladies', describing Eunice's experiments as interesting and valuable. Still, if he was impressed by her work, he seems to have forgotten about it after the AAAS packed up for the year, as there's no evidence of him celebrating it later. Henry, much like everyone else who read Eunice's paper at the time, seems to have been interested at first before letting it drop entirely from his mind.

A few people did take note of Eunice's paper. There's reference to it in the *Scientific American* write-up of the AAAS meeting, albeit under the dismissive heading 'Scientific ladies'; reports in the *New York Daily Tribune*; and mentions in Canadian, Scottish and German journals. Her paper was also published in the *American Journal of Science and Arts*, alongside Elisha's far less significant work on a similar topic. Elisha's paper was republished in the London-based *Philosophical Magazine*, but whoever picked it must have taken a pass on Eunice's. A fire at the Smithsonian in 1865 destroyed much of the couple's work and saw Eunice's research on carbon dioxide reduced to a few scant references, largely forgotten until 2011 when retired petroleum geologist Ray Sorenson stumbled across it. A few years later, climate scientist Katharine Hayhoe dug it up after a colleague asked why there were so few women in the history of the field. This in turn saw it reported in the climate change press, where the story of a forgotten female scientist who had found a link between carbon dioxide and a warming climate back in the 1850s hit a nerve. And yet, for Eunice and her contemporaries, it was all theoretical, a contribution to our burgeoning understanding of gases and heat. It would be another century before anyone started to worry about it.

In 1956, oceanographer Roger Revelle was one of several American scientists looking at the topic of carbon dioxide relative to climate change afresh. In the intervening years, there'd been a little more scientific research on the topic. There'd also been a lot more carbon dioxide emitted: the problem was rather less abstract for Revelle than it had been for Foote. He'd been studying the ways in which oceans absorbed carbon dioxide and realised it wasn't nearly as much as had initially been imagined. Moved by the consequences of his findings, he concluded his paper with a note that humanity was carrying out 'a large scale geophysical experiment.' At first Revelle saw this experiment with the Earth's climate as a bit of an adventure, as just a fleeting moment in time – telling Congress in 1956 that it was 'an experiment which could not be made in the past because we didn't have an industrial civilisation and which will be impossible to make in the future because all the fossil fuels will be gone'. Like many other scientists of his time, Revelle believed nuclear energy would supersede fossil fuels in a few decades, solving the problem. But that was one prediction he was wrong about. As the 1960s and 1970s rolled on, the evidence for global warming mounted. People started to worry too. But they didn't turn down the gas – quite the opposite.

Revelle's 'experiment' line would be repeated many times, including by UK Prime Minister Margaret Thatcher in an autumn 1988 speech to the Royal Society: 'We have unwittingly begun a massive experiment with the system of this planet itself.' By this point, Revelle and his colleagues had studied further (and checked and rechecked each other's work) and there was a strong scientific consensus that if carbon emissions continued at their current rate, global temperatures would get very uncomfortable by the twenty-first century. Today, we're living in that uncomfortable future that people in the 1960s, 1970s and 1980s used to worry about. Although there's been progress when it comes to clean energy technologies and mechanisms for building climate policy have been set up (the UN climate convention, for example), most people living on Earth are a long way from safe.

★ ★ ★

For anyone who needs a quick recap on the basic science, the way in which the Earth's atmosphere traps some of the Sun's energy is usually called the greenhouse effect. Strictly speaking, greenhouse isn't the best metaphor and it's more as if the planet is wrapped in an insulating blanket of gases. Still, somewhere along the way someone said 'greenhouse' and it stuck. The main gases in this imaginary greenhouse are water vapour, carbon dioxide, methane, ozone, nitrous oxide and CFCs. In some respects, the blanket they provide us is a good thing. Or at least life as we know it has developed under a specific mix of greenhouse gases that keep the Earth at a cosy average temperature of 14°C. Lose this blanket entirely and it'd be nearer -18°C. Mess with the delicate chemistry of the atmosphere even a little, and the complex network of life that's grown up inside this particular greenhouse – the complex network we're part of – starts to falter.

Today, when politicians, scientists and campaigners talk about the danger climate change poses they tend to use the relatively heavy milestones of 1°C, 1.5°C or 2°C global warming (or, if they really want to scare you, 4°C, 5°C or 6°C). One or two degrees might not seem very much, but the figure isn't the difference between when you checked the weather forecast this morning and then later that afternoon. Rather, it's a combination of all the temperatures across the world for the whole year. As such, it can mask many other, more extreme weather events. The comparative warmth they're looking at isn't compared with the year before, but a 'pre-industrial' baseline of the years 1850–1900. They use this baseline because the warming we're talking about isn't just the sorts of climate fluctuations that would be happening whether humans lived on this planet or not, but has been caused by the massive influx of greenhouse gases into the atmosphere since the Industrial Revolution.

The biggest perpetrator of this industrial warming is carbon dioxide, mainly from the burning of fossil fuels. Carbon emissions have also risen due to the destruction of natural 'carbon sinks' such as forests cleared to graze livestock which would otherwise breathe in our emissions. This part of the problem began well before the Industrial Revolution. Indeed, there's evidence that the emergence of farming several thousand years ago saved us from another ice age. Industrial activities have released other greenhouse gases too, like methane, CFCs and nitrous oxide. The fossil fuel industry causes methane emissions, for example, along with carbon, as do livestock (cow farts often get the blame here, though it's more the burps we should be worrying about). And in case you were wondering, yes, those silver canisters of nitrous oxide contribute to climate change too, although the nitrous oxide emissions from agricultural fertilisers and manure are a much larger problem.

One of the many slippery things about the climate crisis is that it doesn't hit people with a clearly identifiable thud. It creeps up gradually over time and does so mixed in with all sorts of other aspects of our world; other problems humans have made and hazards that were already waiting for us. This mixing with other problems is partly what makes the impacts of climate change so hard to predict, but it is also what makes them so toxic. Climate change takes a host of other social, economic and environmental issues, and turns up the heat. It adds new hazards to trip over, squeezes already pressurised systems and further exhausts already depleted resources. As climate scientist Myles Allen puts it: 'People ask me whether I'm kept awake at night by the prospect of five degrees of warming. I don't think we'll make it to five degrees. I'm far more worried about geopolitical breakdown as the injustices of climate change emerge as we steam from two to three degrees.'

The American state of California offers a good example of how the climate crisis tightens the grip of other injustices. Teams of prison inmates – many on minor drug offences and including youth offenders – are sent to fight wildfires for a dollar an hour and the promise of credit towards early parole. This has happened since the 1940s, but as wildfires get worse, the state relies more and more on this cheap, captive workforce. It's been estimated the program saves the state nearly a hundred million US dollars a year. And that's just the tip of the speedily melting iceberg. We can't tell for sure if the 2014–16 Ebola breakout in West Africa was caused by climate change shifting bat populations, but it's likely we'll see more of these interactions in the future as the pressures surrounding rising temperatures push people and other animals closer together. The same can be said about mosquito-borne diseases like Zika or malaria. There's no evidence linking climate change to COVID-19, but it could well mean we see more pandemics, deadlier ones, spreading faster. There's also plenty of research showing that as temperatures rise, so do instances of violence, be that rape, domestic violence or civil war. And, in case you were wondering, Harvard researchers reckon climate gentrification has been discernible for a few years already too, as the rich push the poor out to riskier land.

Greenhouse gas emissions can go down as well as up. As Mark Maslin and Simon Lewis stress in their book on the Anthropocene (the geological era characterised by the impact of humans), *The Human Planet*, there is a noticeable dip in atmospheric carbon around the start of the seventeenth century. Maslin and Lewis trace this back to the colonisation of the Americas a century or so before, or more precisely the deaths of 50 million indigenous people. The dead don't farm and so the unmanaged land shifted back into forests, which in turn inhaled enough carbon dioxide for it to be in bubbles of air from the time preserved deep in the polar ice caps. This regrowth was short lived. European settlers in North America soon got to farming for themselves, not to mention coal mining, inventing kerosene and laying railway tracks, highways, and oil and gas pipelines. Still, this temporary drop in carbon dioxide levels might well have played a role in the so-called 'little ice age', a series of cold snaps between, roughly, 1350 and 1850. This little ice age most likely had a mix of causes – dust from volcanoes intercepting sunlight, for example – but the regrowth caused by colonialism of the Americas might well have been one of them; human forces combining with those from other parts of nature to shift climates, just as they do today.

The little ice age wasn't cold enough to be a true ice age, but it was cold. The carnivalesque end of this involved frost fairs, puppet shows, ox roasts and children playing football on the thickly frozen ice. There are stories of frozen birds falling from the sky, Henry VIII sleighing between palaces, New Yorkers walking from Manhattan to Staten Island and even an elephant being led across the Thames. It's one reason Stradivarius violins are so prized; trees during this period took longer to mature in the cold, making denser wood and thus a very particular quality of sound. The darker side of this mini ice age was people shivering to death. Whole villages in Switzerland were destroyed by growing glaciers. Prolonged cold, dry periods had an impact on crops and livestock. People starved. Some environmental historians spin this as a warning from history, tracing the changes in weather to a rise in anti-Semitism and the witch-hunts as well as several wars. There were winners – there are always people who can make an opportunity out of a crisis – but only off the back of a lot more suffering elsewhere. People in the mid-seventeenth century believed they were living in truly awful times. And, unlike pretty much every other generation that's made that complaint, they had a point. Still, that's nothing compared with what could be in store for people born in the twenty-first century.

★ ★ ★

This book tells the story of how we found ourselves in the middle of Revelle's big, geophysical experiment; how we built systems, technologies and deeply embedded cultures for the burning of coal, gas and oil at scale. Our narrative starts in 1851, the start of this 'pre-industrial baseline' on which those 1.5°C and 2°C warming warnings are based. We kick things off at the Great Exhibition, a big show put on by the British government to celebrate its newly minted industrial power. From there we travel back in time to those cold years of the seventeenth century to understand the roots of the steam age, before moving on through the nineteenth and twentieth centuries, tracing the growth of the oil industry first in the US and then Russia. We'll see the first oil wells drilled in Borneo, Iraq and the Niger Delta, and oil cartels move from inter-war chats over pheasant shooting in the Scottish Highlands to more complex geopolitical deals leading to a crisis at American gas stations in the 1970s.

We'll see the monster of big oil slain by a plucky investigative journalist back in the 1910s, only to re-emerge even more powerful. We'll follow the growth of electricity networks, how the sparks saw off oil and gas in the lighting industry, before going on to market a plethora of electrical devices to further wire up our homes and offices. We'll also see electricity lose out to oil in the battle for transport, at least for the twentieth century. We'll see excitement over solar and wind power start in the 1870s, only to be forgotten about but then rediscovered in the 1970s and finally come of age at the start of the twenty-first century. Throughout, we'll watch an environmental movement grow to fight the dangers of this industrialisation. As we'll see, this movement would be a mixed bunch, folding a variety of ideological takes into environmentalist concerns, from anti-capitalist revolution to white supremacy (as well as a desire to simply breathe more easily).

At the same time, we'll trace the intersecting story of how we discovered the climate crisis was happening in the first place. In some respects, this is the more hopeful end of the story, reflecting humanity's ability to understand itself and the world around it. This strand starts around the same time, rooted in the mid-nineteenth century, with the odd look back to see how we got there. As we'll see, the discovery of anthropogenic global warming didn't arrive in a single 'eureka' moment (or even a single exclamation of 'oh, shiiiiit') any more than the fossil fuel age started with a singular bang. No one woke up one day, looked out of the window, slapped their forehead and exclaimed that burning fossil fuels makes the weather dangerous. As with most science, understanding of the climate crisis unfolded reasonably slowly, with each generation adding their own take.

It took time for people to process what they'd found – emotionally as much as anything else – to appreciate its impacts and causes, to question it, interrogate the gaps in their knowledge, check it was true and link it up with other bits of research that might tell us more. It also took time for this new science to be understood and absorbed by the rest of society, making its way, like any new bit of knowledge, from one laboratory to another, to newspapers, political speeches, chatter over dinner, protests, poems, playgrounds and, eventually, people's everyday way of seeing the world. Some of the slow pace of this gradual unfolding is understandable – annoying, frustrating, losing us valuable time, but also the way science, technology and political systems were set up to run – but some of it was deliberately, mendaciously kept slow too. The oil industry didn't start deliberately spreading doubt about climate change until the late 1980s, but it did spread doubt. We can lay the blame at its feet for at least a chunk of lost time.

★ ★ ★

I'm not going to offer you villains and heroes. This is not a simple story with evil exploitative fossil-fuel baddies on one side and the goodies of renewable energy, environmentalism and climate science on the other. It's more complex than that. What's more, although individual characters played roles that we might, more or less, count as either villainous or heroic, none of them worked alone. The climate crisis is a social project – one that's always been more about the impact of groups of people than individuals.

Spencer Weart puts it well in his 2003 book *The Discovery of Global Warming*, noting that a statement as simple as 'last year was the warmest year on record' is the work of a massive, multigenerational, international effort. Weart means in terms of the many people involved in spotting that shift in global temperature, in building the science that lets us see that far, but we should be aware of the massive effort behind the cause of that warming too. People have only managed to heat the planet to the point they have because they work together. You can play with the idea of a personal carbon footprint if you want, but nothing especially 'high carbon' is done alone. You can drive an SUV on your own, for example, but you still need to buy it from a company and buy petrol from another. Moreover, it was built by multiple hands, using materials mined by others, drawing on the knowledge of generations of engineers, and that's without tracing through networks of advertising, design or the road infrastructure.

In all this, it's vital to remember some people had more of a role in creating the climate crisis than others, and some people are more able to insulate themselves from the dangers too. As Tim Gore, Oxfam's former policy lead on climate change, points out, the poorest half of the global population are responsible for only around 10 per cent of global emissions and yet live overwhelmingly in countries most vulnerable to climate change. So, I invite you to explore 'our' biggest experiment, but to do so critically. We should be aware of our shared humanity and shared planet, as well as the ways many people have worked together over time to create this problem (and how many people will have to work together to undo it). But we must also be mindful of how weighted our social systems are and the inequalities at play; how many people have been excluded, not just in the past but in the present and future too.

It's also worth giving the health warning that this is a story about a lot of white men, many of them rich, and that much of the activity of the book happens in the US and UK. The climate crisis has been and remains a problem of the elite's making, and so it's the powerful we follow to understand how it happened. As the story develops we'll see everything become more globalised. We'll see bigger and more complex trading routes emerge, all chugging out new reasons to burn through fossil fuels in the process. With increasing globalisation, we'll also see an emergence of thinking about the world as a whole, rather than just small bits of it. But that doesn't mean the whole world is working together as equals. Today the idea of thinking about the planet as one is often associated with the sort of hippie ideals of world peace, love and understanding. There's a big difference between a whole- planet approach based on people working together through cooperation and in harmony with nature – the happy, utopian one used by fizzy drinks' ads – and one rooted in more militaristic traditions of control. Both shape our modern conception of the climate crisis and both are likely to continue to be part of how we weave through our warmed future, so it's worth being attuned to them.

Writing this book has, at times, been painful. I would come home from my day job working for a climate charity, supporting my colleagues fighting for a liveable future, and then bury myself in stories of people in the 1770s thinking burning more coal was simply a great way to make more money; others expanding oil drilling in the 1890s; or scientists in the 1970s dismissing the year 2000 as far enough in the future that we didn't need to worry about carbon emissions yet. Sometimes it was hard not to simply shout 'WELL, FUCK YOU VERY MUCH' at whatever source I was taking notes from. Still, it's also been an uplifting experience on occasion too, not least the parts about the history of climate science. And it's certainly helped me understand the climate crisis more fully.

The story of the climate crisis is, undoubtedly, the great tragedy of our time, but it's a story of a lot more than that too. It's the making of our modern world, for good as well as bad. For those of us who live in rich countries, it's easy to take the flicking of a light switch for granted, but we have access to illumination (along with heat, food and transport) that our ancestors could only dream of, access that everyone should be able to enjoy. It's a story of great minds, the pursuit of truth and courageous attempts to make the world better (as well as a dose of eccentricity and whimsy). It's also a story steeped in colonialism, full of inequality, spin, snobbery and hubris. It showcases some of the best of humanity as well as the worst, and may well be the end of us. I've found researching this book a rip-roaring ride and hope you enjoy reading it, even if you find living through the climate crisis a rather less pleasurable experience.

CHAPTER ONE

A Steam-Powered Greenhouse

It's only apposite to start our story inside a giant, overambitious Victorian greenhouse. The Crystal Palace must have been quite dazzling to see up close. A vision in cast iron and plate glass three times the size of St Paul's Cathedral, it covered almost a million square feet of Hyde Park, enclosing four mature elm trees. Created for the Great Exhibition of 1851, the whole thing had been built in rather a rush, with ambition much larger than the deadline or budget allowed. Led by Queen Victoria's husband, Prince Albert, and innovator Henry Cole, the idea for a great, British exhibition had been inspired by similar, though smaller, events that had been running in Paris since the start of the century. The idea hadn't been universally popular at first, but once it opened the critics were, on the whole, proved wrong. Some 25,000 people flocked to the opening on 1 May 1851. By the time the cast-iron doors closed again five months later, 6.5 million visitors had passed through the crystal halls. Allowing for foreign and repeated visits, historian Jeffrey Auerbach estimates a fifth of the British population would have attended the exhibition. Up-and-coming travel agent Thomas Cook arranged special excursion trains, school groups poured in and there's even a story of one woman walking all the way from Penzance.

Based on greenhouses designer Joseph Paxton had previously built for the Duke of Devonshire, the palace's distinctive fan-shaped facade was said to have been inspired by the large ribbed leaves of Amazonian water lilies. Paxton had won fame and a knighthood when he pioneered the growing of these lilies in the UK, replicating their natural warm, swampy habitat with manufactured heat from coal-powered boilers in his greenhouses. He had also noticed the lily pad's seemingly delicate leaves were strong enough to hold the weight of his young daughter (inspiring a brief craze for balancing children on the plants) and put that knowledge to use in the palace design. The 'crystal' walls had been made possible by a new process for producing sheet glass developed in the West Midlands a few decades before, with nearly 300,000 planes of glass shipped along the canal to the building site in central London. Once the various pieces were on site – the glass, as well as iron and wood guttering – a fleet of 75 specially built glazing wagons fitted it all together, with giant lanterns and bonfires of scrap timber allowing workers to keep going well after sundown.

A stained-glass window in the upper galleries filtered the sun in all the colours of the rainbow, and at the centre of the excitement was an iconic crystal fountain, 27ft high, made of 4 tonnes of pink glass. J. J. Schweppe & Co won the catering contract, supplying 2 million Bath buns and more than a million bottles of their relatively new product, artificially fizzing water. When it came to the exhibits, *The Times* calculated you'd need to spend at least 200 hours inside the palace to see each and every one. It contained a diverse perfusion of delights, but all these exhibits had one uniting theme: the awesome power of technology. Although there was a section on fine arts, it was something of an afterthought and the focus was very squarely on the new machines of the age and the raw materials that fed them. For the exhibition's developers, the relationship between science, technology, the Empire and the Earth seemed so simple. 'Science discovers these laws of power, motion and transformation,' Prince Albert told a banquet in March 1850, and 'industry applies them to the raw matter which the Earth yields us in abundance' (no questions to be asked about how the British might have come across raw materials not found within their own islands).

There was iron from Ireland, tin from Cornwall, cedar wood from Cuba, cocoa from Trinidad, tobacco from America, cinnamon from Ceylon and whale oil from the 'South Seas'. Visitors could learn about the history of the steam engine via a special display in the stand run by the Birmingham-based firm Boulton & Watt. There were also displays on the production of steel and cotton, a device for folding paper, a cigarette-rolling machine that produced 100 cigarettes a minute and a printing machine that turned out 5,000 copies of the *Illustrated London News* an hour. You could buy one of the very first weather maps, produced by Greenwich Royal Observatory's superintendent of meteorology James Glaisher, which utilised data that came via a network of amateur weather watchers who sent it to

London via the new electric telegraph. Some of the exhibits were more mundane, featuring the sorts of things middle-class visitors to the exhibition might see in shops or have in their own home: cutlery, chairs, mirrors, clothes and curtains. Other displays showcased ideas for new products; items their designers hoped would become commonplace in years to come. There was a carriage drawn by kites, an early version of the fax machine, false teeth that didn't fall out when you yawned and special furniture designed for steamships that combined a bed with a toilet and could be repurposed as a raft.

When the Great Exhibition is remembered today, it's often as a celebration of science and technology, but really it was about a whole host of things: from giving the Queen's husband something to do, to selling forks, fur coats and fizzy water. Perhaps above all the whole exercise was an expression of imperial power at a time when the English ruling classes were increasingly worried their privileged position in the world might be about to shift. The East India Company (EIC)[*] was provided space in the centre of the galleries devoted to the British Empire, with the commodities of each colonised country shown off for the wealth they provided. This included the infamous Koh-i-Noor diamond, which inspired the detective novel *The Moonstone* and now sits in the Tower of London as part of the crown jewels (despite the governments of India, Pakistan and Afghanistan all claiming ownership). There were also domestic concerns, with an increasingly rebellious British working class pushing for greater share of national prosperity. Tied up in this vision of British supremacy was a celebration of how much British science and technology had developed in the past 50 years or so, and what might come next. It was an indication that life had changed in the UK and was on course to change elsewhere too, with a clear message that this change was unquestionably for the good and those currently in control of the new machines should stay exactly where they were.

---

[*] A history of the East India Company is outside the confines of this book, but it's worth noting that they didn't just buy and sell goods, they were at the forefront of British colonisation, with their own army and navy. For more on the EIC, see William Dalrymple's *The Anarchy* (2019).

And at the heart of all of this was coal. Coal was there in the raw materials' section of the palace, with 18 large lumps of the stuff drawn from fields across the UK, including a chunk from South Wales whose slow trip down to the exhibition site had been reported in detail by the newspapers. Coal also powered the machinery section, via a set of coal-fired steam boilers in the north-west corner. Coal smoke was therefore presumably perfuming the air inside the exhibition too, and the increasingly thick smog of London would have certainly been noticeable to visitors from out of town. Moreover, coal-powered steam would have produced many objects on display at the exhibition, and at least half the visitors would have come by coal-powered trains or on coal-powered boats along the river. Coal was also symbolically front and centre, with that lily-pad design of the frontispiece inspired by Paxton's use of coal-heated greenhouses to grow plants taken back from far-flung parts of the Empire.

Through its various articulations of coal, the Crystal Palace and its contents reflected the start of something new: an age of prosperity for some, built on the burning of fossil fuels. And this is why we start with the story of Joseph Paxton's glass creation; a swollen greenhouse in the centre of London, filled to the brim with machines of the steam age, bourgeois trinkets and resources from around the world often seized by colonisers powered by coal. It reflects a pattern of progress that the world was to follow deep into the environmental crisis we find ourselves in today. The Great Exhibition didn't invent consumer culture or the burning of fossil fuels on a mass scale any more than it invented colonialism. Still, it reflected the establishment of a way of life that would become our modern climate crisis.

* * *

Britain may have led the world into the climate crisis via coal, but they weren't the first to burn it in an organised way. Modern radiocarbon studies suggest people in Inner Mongolia and the Shanxi provinces of northern China were using the fuel more than 4,000 years ago. Then, a few thousand years later, the Northern Song Dynasty in eastern China developed what might be seen as the first fossil fuel-based economy, burning coal to produce iron, gunpowder and ceramics, as well as heat homes. But the power of Northern Song ended when its capital was invaded in 1127, its use of fossil fuels falling by the wayside at the same time. In Europe, the Romans made use of British coal reserves soon after they invaded in the first few centuries AD, setting up coal-trading routes across their empire, using it to heat their homes and garrisons, work iron, process salt and to keep the perpetual fire alive at the Temple of Minerva in Bath. But the British themselves weren't exactly enamoured with the stuff. Coal's polluting effects are pretty obvious even without any awareness of the greenhouse effect. It leaves sooty marks and gets into your chest. This is especially true of English coal, which tends to be the soft and sooty bituminous type. In 1285, Edward I set up the world's first air pollution commission, banning the burning of coal in London. The ban was often flouted, but coal was far from mainstream. When people could afford it, they burnt wood.*

But as the years went on, wood became scarcer. The forests that had once thickly coated the British Isles were sucked up to clear land for agriculture, with the wood used to build houses, roads, bridges, ships and barrels to help in the production of all the salt, lead, glass and beer people were consuming, as well as to simply keep homes warm. This was a political worry – the navy needed wood for ships to fight Spain and explore 'the New World' – and Elizabeth I set up several official investigations into the loss of England's forests. And as the price of wood rose, coal became more attractive. By the middle of the sixteenth century, coal had become an established part of London life. It would have been known as 'sea coal' back then, as it could be found on the beach having fallen from exposed coal seams on cliffs or washed up from underwater deposits. It would come into London up the River Fleet, a now-buried tributary of the Thames, having travelled around the coast of Britain. You can still find a small alleyway half a mile north of Blackfriars Bridge called Old Seacoal Lane, noting the point it would have come into the Fleet. Coal still wasn't universally popular, but it was burnt in homes, blacksmiths, potteries, bakeries and glassworks. Brewers became great users of coal, as did salt boilers (key

---

* For more on the story of coal, see Barbara Freese's brilliant *Coal: A Human History* (2003).

to preserving food in an era before refrigeration). Soap boilers used it too, as did lime burners and sugar refineries. Although parts of the economy still relied on wood, water or animals for energy, an age of fossil fuels had begun.

This was all still relatively small scale compared with what was to come. Indeed, humanity might have crept along with the relatively slow burn of global warming caused by the impacts of agriculture and the odd coal fire if it wasn't for the development of the steam engine. This piece of kit was first sold commercially to help pump water from mines – a way of making coal mining slightly easier – but gradually found uses elsewhere. It thus opened up a much larger market for coal, paving the way for oil and gas industries too, and supercharging our ability to warm the Earth in the process. The first steam engines date back to antiquity. Around the first century in Roman Egypt, Hero of Alexandria developed an 'aeolipile' – a device attached on top of a cauldron that used steam to make a ball spin – which was used largely as a toy. There are whispers of people having more practical ideas for steam-powered devices after this. Basque engineer Blasco de Garay, for example, had an idea for a steam-powered boat back in 1543, but Carlos I didn't want to invest, so it remained just a sketch.

Then, into late-seventeenth-century London, arrived a refugee – indeed a member of the French Huguenot community for which the word 'refugee' was coined – Denis Papin. Papin had already worked with German mathematician Gottfried Wilhelm Leibniz in Paris, and soon found work at the Royal Society as a laboratory assistant to founder member Robert Boyle. Boyle was an established expert on the pressure of gases and Papin worked with him on experiments involving steam, presenting what he called the 'Steam Digester for softening bones' in 1679 – a pot that utilised steam to cook tough foodstuffs (or to put it another way, he invented the pressure cooker). The Royal Society wasn't exactly enthusiastic about Papin's work, repeatedly treating him more as a servant than an equal and refusing to promote him or fund further steam work. Unable to return to France, Papin moved to Vienna briefly and later Marburg in Germany, continuing his experiments on steam and developing ideas for machines that would use steam power not just to cook, but for movement. He returned to

London in the early eighteenth century, but the Royal Society still wouldn't support his steam research. He's thought to have died in poverty in 1713, buried in an unmarked grave in St Bride's Church, just the other side of the road from Old Seacoal Lane.

Military engineer Thomas Savery had more luck showing off his version of a steam engine to London's scientific elite, demonstrating it to Royal Society members in 1699. Savery promoted his device as a way to drain mines, describing it as 'the miner's friend', but it wasn't very efficient, requiring almost as much coal as you'd get out of any mine that used it. Plus, mine owners were worried about the danger of bringing fire to their site, concerned it'd ignite gases found in their tunnels. One thing Savery was really good at, however, was securing intellectual property rights. He patented an invention for 'raising water by the implement force of fire', and speedily secured an Act of Parliament that extended this patent all the way through until 1733. This rather aggressive patent strategy arguably stalled the development of steam power as it kept others out of the game. Still, some developers managed to work with Savery's patent, notably the lay-preacher and ironmonger Thomas Newcomen. His more efficient version of a fire engine borrowed from Savery and Papin, as well as incorporating ideas of his own. Newcomen partnered with Savery to accommodate the patent and by the 1760s there were hundreds of Newcomen engines dotted around mining areas of the UK, soon shipping to other parts of Europe and America too.

If draining a colliery using horses cost £900 a year, a Newcomen engine could do it for £150. Still, they were the size of a house and required a lot of coal to keep running, all overseen by a 'fire man' whose job it was to feed this hungry machine. It was only worth it if you had a healthy supply of coal nearby: fine if you were using it in a coal mine itself, but the costs could soon rack up if you had to transport it anywhere else. Again, we might have stayed there; a medium-sized coal industry helped along by a few Newcomen engines fuelling a slightly more intense burn of global warming. But a new era of steam engines from Boulton & Watt was to change that, offering a cost-effective replacement for water- and horse-powered wheels that could be put to a range of tasks, opening up a massive new market for the burning of fossil fuels. Before long, coal-powered steam was not only powering factories, cotton and flour mills, but also trains, boats and cars, inspiring a host of technologies and infrastructure the oil and gas industries would later build on.

★ ★ ★

James Watt's story starts in a British port in the middle of the eighteenth century – and like many such stories it owes a debt to the slave trade. Watt's family didn't have the wealth of the British upper classes of the time, but they had enough to ensure young Watt had a comfortable childhood and education, with books at home, and even a small forge specially built for him in his father's workshop to allow him to learn wood and metalwork skills. Watt's father worked in shipping, trading in rum, sugar and cotton out of Greenock, a port 20 miles to the west of Glasgow. In 1755, when Watt was in his late teens, he moved to Glasgow to train to be a maker of nautical and scientific instruments.* He had good contacts at the university, where his cousin was a professor of Latin, and he knew the professor of natural philosophy, John Anderson, from school. Soon after Watt arrived in Glasgow, a large collection of scientific equipment was donated to the university by Alexander MacFarlane, a slave-owner in Jamaica. Watt was given the job of fixing the salted-up loot once it had crossed the Atlantic. Although this was piecemeal work, he could make a living as an instrument maker and fixer for the university. He eventually had an apartment and workshop on site, supplementing university work selling scales, balances and compasses to his father's customers in shipping, and eventually his own clientele too. Watt opened a shop in the city centre, employing 16 men, selling and fixing everything from shoe buckles to pistons, nutcrackers, violins, flutes and bagpipes. This story is often told as a geeky Watt finally finding business acumen when he teamed up with the more

---

* Watt's brother John joined their father in shipping and later branched out to directly trading in people too. There's evidence that James Watt himself was involved in the trafficking of a small boy at one point. For an entertaining and clear biography of James Watt that avoids the hagiography all too often applied to him, see Ben Marsden's *Watt's Perfect Engine* (2002). For recent research that pulls out the links with the slave trade, see Stephen Mullen and Simon Newman's 2018 report 'Slavery, abolition and the University of Glasgow'.

outgoing Matthew Boulton, but Watt was doing pretty well on his own before that partnership.

Glasgow at the time was growing rapidly, boosted by slavery via the trading of rum, sugar and cotton. Interest in science, engineering and philosophy was growing off the back of these new riches, and the city was busy with intellectual, industrious chatter. Anderson ran what he called 'anti-toga lectures' in the evenings, where local skilled workers could study. Along with Edinburgh, Glasgow also boasted several 'irregular clubs' where men like Watt could swap ideas over food and drink with other thinkers of the time, and it's through one of these that he met economist Adam Smith and scientist Joseph Black. Fresh from his discovery of carbon dioxide, Black had been appointed professor of chemistry at Glasgow in 1757. After local whisky distillers asked his advice on cost-cutting, Black had gone about investigating how chemicals changed state from solid to liquid or gas (and vice versa). How did water absorb heat to become gas? Why doesn't ice melt into water straight away? He developed the idea of latent heat, arguing that a certain amount of heat would be needed by any material before it could transform. Black's questioning of the nature of heat, along with his friendship and support, was key to what Watt would work on next and would help him make steam engines more efficient.

It's often said Watt was introduced to steam engines when Anderson gave him a model Newcomen engine to fix in 1764. In fact, he'd been working with steam for a few years by then, having first been introduced to it by John Robison, another physicist he'd befriended in Glasgow. Decades ahead of his time, Robison was excited by the prospect of replacing horse-drawn carriages with steam engines and had published a design for an improved Newcomen steam engine back in 1757. Watt was equally intrigued by the power of steam and, inspired by Robison, experimented with Papin's Steam Digester designs. The model Newcomen from Anderson's collection gave Watt something physical to tinker with. He worked out he could waste a lot less heat than the Newcomen system by adding a separate condenser. His work was supported further by investment from a John Roebuck, an old student of Black's who was looking for more efficient engines to use in the coal mines he'd bought near Bo'ness.

It was through this project with Roebuck that in the summer of 1767 Watt visited Birmingham and came into contact with a group called the Lunar Society. Like Glasgow, Birmingham was brimming with talk of science, engineering and philosophy, enriched by the growth of local industry and the many nonconformists who had settled there (that is, people who didn't 'conform' to the Church of England). Birmingham didn't have a university, but its intellectual community was none the worse for it – if anything, it had a positive effect. In England, the choice of university was Oxford or Cambridge and you pretty much had to be a practising member of the Church of England to attend either. Any sniff of the sort of radical politics held by people like Anderson and you might find yourself rather unwelcome too. In the new industrial towns like Birmingham, informal intellectual clubs and academies for religious dissenters thrived, free from such constraints.

The first glimmer of what would become the Lunar Society started sometime in the 1750s in Birmingham. Industrialist Matthew Boulton and local physician Erasmus Darwin had met via mutual friends, hit it off, and started meeting regularly to discuss matters of science and invention over dinner.* They gradually picked up other local men or those passing through town who shared their interests in engineering, botany, geology, or the new sciences of gases and electricity. They'd meet once a month for dinner on the night of a full moon, allowing them enough light to get home safely afterwards. The Lunar Society's members weren't aristocrats, but they tended to have some privilege of education, inherited wealth and contacts. Darwin was one of the few to have attended a formal university. After studying at Cambridge he'd trained at the Edinburgh Medical School, but was interested in pretty much anything and everything, from plants to cosmology to women's education. He'd sometimes entertain himself by writing poems about his science, as well as sketching designs for inventions including several for monitoring the weather (he once suggested the transportation of two icebergs to the equator to cool the tropics and so ease northern winters). Boulton was more straightforwardly a man

---

* Jenny Uglow's *The Lunar Men* (2002) is a chocolate box of a book about these characters, their families and ideas; I highly recommend it.

of business, but was no less excited by new science and inventions. He'd inherited his father's firm making small metal goods, then known as 'toys'. Helped by a marriage to a rich heiress (and when she died, her sister) he'd built this up to a large 'Manufactory', which produced a range of precision craft goods from buckles, buttons and intricately decorated vases, to scientific equipment like thermometers and telescopes. Other Lunar regulars included radical preacher and scientist Joseph Priestley; Quaker, gunmaker and banker Samuel Galton; and Irish politician and inventor Richard Edgeworth (whose 22 children included novelist Maria Edgeworth and whose innovations included a proto-telegraph system). There was also the potter Josiah Wedgwood, a master of both chemistry and marketing who pioneered not only several glazing techniques but also money-back guarantees, celebrity endorsement, free delivery, market segmentation and illustrated catalogues.

When Watt first visited Birmingham, he was shown around Boulton's manufactory by Darwin and was amazed by what he saw. Visiting again, he met the man himself and they immediately formed a strong friendship. As well as friendship, Boulton hoped to find a solution to a problem he'd been pondering for a while. For all his manufactory's success, it could get a bit stuck for power. There was a waterwheel in the near by Hockley Brook, but this could dry up in summer or freeze in winter, and they'd have to buy in horsepower to keep everything going. Boulton had been looking into options for a horseless pump, but the coal-guzzling Newcomen was too expensive to run. He'd heard of the improvements Watt had made to steam engines and wondered if it could be the answer, and something he could sell to mills and other manufactories too.

It'd be several years before Boulton and Watt officially formed a partnership though. Still working with Roebuck, in 1769 Watt took out a patent for his version of a steam engine. They still didn't really have the resources to develop it though, or even much interest. As well as the instruments shop, Watt was busy with engineering contracts, fixing and making Newcomen engines, and building canals too. This practical experience would prove invaluable for what was to come – allowing him to get to grips with the guts of real engines rather than just university models – but for now it seemed like enough work in itself. Boulton, however, was excited by the idea of steam-powered industry – impatient to see Watt's improved steam engine up and running, ready to sell it to the world. Boulton and Roebuck squabbled over contracts, then Roebuck's business failed and Boulton pounced. Still, Watt might well have decided to stay in Glasgow at this point, safely working on his established business rather than risking his future with Boulton. But in 1773 his wife died in childbirth and a move to Birmingham with his two children offered him a fresh start. It would mean leaving behind his friends in Glasgow, but the Lunar Society was a fertile test bed for ideas and stimulating company. Darwin had ideas for a coal-powered internal combustion engine running off charcoal gas, for example. Watt thought this was ridiculous, but it was also his idea of fun. There was Priestley's brother-in-law too, ironmonger John Wilkinson, who had the sort of expertise they'd need to pull off a full-sized version of his engine.

In 1775, Boulton and Watt formalised their partnership, but there was still more work to do. By then, the patent was already five years old, and they knew they needed more time before it'd be a marketable product. Savery had extended his patent back in 1699 with an Act of Parliament. Doing the same with Watt's patent would take some persuasion though. Counter-petitions surfaced, with rival engineers complaining this would cause a monopoly and stifle innovation (and arguably it did). Boulton called in favours from Midlands MPs and the patent was extended until 1800. Watt was convinced he'd end up a pauper's grave, his family starving. He contemplated moving again and finding employment elsewhere. Boulton had done such a good job of talking up the Scottish inventor's genius to anyone who'd listen that Watt was offered work from the Russian government. Boulton was appalled when he heard this, and speedily wrote to his partner, trying to convince him to stay. Gradually orders for this new, improved steam engine arrived and they could both relax.

Back when Boulton was arguing over contracts with Roebuck, he'd been offered the chance to develop the steam engine in and around Birmingham but refused, saying he wanted to sell the idea to the world – and he did just that. Despite the anti-slavery stance of many of their Lunar Society friends, plantations in the Caribbean became a key market for Boulton and Watt's new engines. In Europe, they courted aristocratic patrons, sending a team of engineers led by Watt's eldest son, James Jr, dressed in velvet, to the King of Naples. In 1786, Boulton & Watt opened the world's first steam-powered flour mill, Albion Mills. A showpiece designed to pull in the crowds, it was as much a PR stunt as it was an actual factory, built in the heart of London next to Blackfriars Bridge. Watt was, characteristically, still worried, concerned visitors would see the poor condition of the mill and this might do more harm than good. He had a point too. There were a lot of broken parts, with shoddy, sometimes dangerous, work hidden behind closed doors. The mill burnt down in March 1791, amid rumours of arson. The blackened ruin is said to be an inspiration for the 'dark satanic mills' in the poem by William Blake that would later become the hymn *Jerusalem*. The fate of Albion Mills didn't seem to do the firm any lasting damage. Both Boulton and Watt died very wealthy men.

Boulton and Watt's sons took over the firm, and as they continued to improve and sell their engines to the growing industrial market, it made good PR sense to emphasise Watt's role as the great inventor. James Watt Jr was quick to ensure there was a bust of his father on display at the Royal Society, with another sent to the French equivalent in Paris. Watt's old friend from Glasgow, John Robison, had the job of writing the *Encyclopedia Britannica* entry on the steam engine and underlined his old research partner's role, glossing over the many other people who'd been part of the steam engine story. A few years after Watt's death, in the 1820s, a group of scientists, industrialists and politicians met to plan a tribute, one that ostensibly celebrated the genius of the Greenock tinkerer but also cemented their own role in the new, steam-powered economy. Humphry Davy, who had run with the Lunar men in his youth and was now a wealthy member of the London scientific elite, extolled Watt's 'profound science' claiming he was more important than Archimedes. Robert Peel, Conservative MP and son of one of the richest mill owners in Britain, was greeted with loud applause as he argued that Watt's work had led not only to his own personal riches but those of the whole cotton industry, and so was also the 'source of national wealth'.

Several statues were commissioned, including one for Westminster Abbey that was so ludicrously big the floor gave way when it was pulled in, destroying several rows of ancient gilded coffins beneath. If this was some sign from God of the destruction the fossil fuels were to bring to the world, it was lost on the Victorian establishment which, by this point, was eagerly transforming every bit of society they could around coal.

* * *

Boulton and Watt took the steam engine out of the coal mines and into the mills, but for the steam age to really get going, these engines had to be applicable to transport. Watt had chatted to Darwin and Robison about their ideas for steam carriages back when he was developing his static engine, and they weren't the only ones playing with the idea of steam engines for transport. In France, military engineer Nicolas-Joseph Cugnot experimented with steam-powered vehicles as early as the 1760s, as a way to transport cannons. You can still see his steamer car, the *fardier à vapeur*, on display in the Paris Museum of Arts and Crafts. However, after a few tests, including a mishap with a wall of the Paris arsenal in 1771 (arguably the first recorded car accident), the project was abandoned.

The first coal-powered ship set sail as early as June 1783 on the river Saône in eastern France. The Marquis de Jouffroy d'Abbans had taken a Newcomen engine and wooden fan and mounted them on top of a boat; it chugged along at walking pace for 15 minutes before the hull split and the boiler burst in clouds of steam. Wind power was safe when it came to ships, at least for the time being. A few years later, in 1787, American inventor John Fitch attempted to launch a commercial steamship outfit on the Delaware River. So the story goes (and it might well be just a story), Fitch had been captured by Native Americans while exploring the Ohio River valley; haunted by dreams of canoes chasing him, he imagined a steam-powered ship might have helped him escape. Fitch's first steamboat only went at 3mph, but was enough to convince investors it was a good idea and a second ship in 1790 fared slightly better. He started advertising for passengers but, despite rolling up a few thousand miles, never attracted enough custom to warrant the cost of running it. He tried to sell his idea in France and then England, but failed, never building another boat and ending his life in 1797 with a handful of opium pills washed down with a bottle of whisky.

It would be another two decades before anyone managed to really sell the idea of steamships. The man to do this, Robert Fulton, started off working in the arts and spent most of his career in engineering chasing an idea for underwater bombs. Still, he kick-started a steamship industry in the US, which was soon copied elsewhere. After working as an apprentice jeweller in Philadelphia, Fulton moved to the UK in 1787, ostensibly working as a painter but mainly living off the generosity of rich friends. Around 1790, he seems to have changed track. Perhaps infected by the 'canal mania' gripping Britain at the time, or maybe just realising he'd never be a great painter, he started a career in engineering. Suddenly full of inventions for everything from a canal-digging machine to a marble-cutting saw, he was especially excited by his idea of *Nautilus,* a submarine missile that could explode enemy boats from beneath. He also flirted briefly with the idea of a steamship, writing to Boulton & Watt in 1794 to enquire about a three to four horsepower engine and what size of boat it would need (although they never replied).

In 1797, Fulton moved to Paris. Planning to stay just six months, he ended up there for seven years, settling into a relationship with American poet and diplomat Joel Barlow and his wife Ruth.* The Barlows took Fulton on as a sort of protégée, giving him the affectionate nickname 'Toot', tutoring him in languages and maths, and lobbying the French government to fund his *Nautilus* plans. Joel dabbled in engineering himself, he'd even taken out a patent for a boiler to try on

---

* Letters between the Barlows suggest there was a loving and possible sexual relationship between them and Fulton, though it's hard to tell as they were written in a rather obtuse eighteenth-century baby talk. Whatever it was, the three seem to have had a lot of affection for one another, which continued for the rest of their lives. For more on Fulton, Kirkpatrick Sale's *The Fire of His Genius* (2001) pulls out the social and political contexts and impacts of his engineering work, and Cynthia Owen Philip's *Robert Fulton: A Biography* (1985) offers more details of his personal life, including extracts of the Barlows' letters.

steamships, and could introduce Fulton to people like ballooning pioneers the Montgolfier brothers. In Paris Fulton also met American diplomat Robert Livingston. A member of the 'committee of five' who had drafted the Declaration of Independence, Livingston had been sent to Paris by President Thomas Jefferson to negotiate the Louisiana Purchase. With an inherited fortune that allowed him to indulge an interest in engineering on the side, Livingston had been percolating an idea for trying out steamships in New York, and in Fulton saw a potential business partner with the engineering sense and entrepreneurial spirit to make this steamship wheeze work.

Many were sceptical, the Montgolfier brothers were dubious this steamship business would ever really work and Fulton himself was more interested in his underwater missile ideas. But, boosted by Livingston's powerful support, he borrowed a steam engine and started experiments on the Seine. They drew crowds, including the Napoleons, and soon l'Emperor himself demanded Citizen Fulton report on his progress. However, Napoleon was evidentially unimpressed, or possibly just more enamoured by the Montgolfiers' balloons, leaving Fulton and his inventions be. The British noticed Fulton's work too, buying up his plans for underwater mines. They proceeded to sit on the idea, however, seemingly more interested in preventing the French from getting hold of the tech than using it themselves. Still, this deal finally got Fulton an audience with Boulton (Watt had retired by then) and with that a steam engine to try out on a ship in New York.

Finally, on 17 August 1807, Fulton's steamship was ready to attempt a trip from New York City to the state's capital at Albany, 150 miles up the Hudson. As Kirkpatrick Sale describes in his book on Fulton, a crowd had gathered in Greenwich Village, excited by this unusually long boat with weird paddles, which puffed and roared out clouds of dark smoke like some sort of sea monster. Many of the spectators wouldn't have seen any type of a steam engine before and the idea of a steam-powered boat must have seemed quite fantastic. Some, quite reasonably, thought it might explode and were possibly only there in the hope of some fireworks. Still, it didn't explode or sink and, although it wasn't immediately obvious it would be a success, it fared much better than Fitch's attempts, soon carrying 100 passengers a week.

Fulton busied himself adding several more steamboats to his fleet, including *The Paragon*, a 'floating palace' complete with mahogany staircases, silk curtains and a kitchen capable of serving dinner, on china, to 150 guests. He also had his eye on what he had long suspected would be the real prize of American steamships: the Mississippi. He partnered with Nicholas Roosevelt (first cousin once removed of President Theodore) first to research the feasibility of a steamship in these waters and then build a boat to try it out. In September 1811, Roosevelt set off on a 2,000-mile journey from Pittsburgh to New Orleans in a sky-blue steamship, with his wife Lydia, their infant son, a captain, engineer, six hands and a dog named Tiger on board. Fulton had expected the trip would take a month. It took four, during which they not only celebrated Christmas but had to sit through an earthquake, and Lydia – who had been heavily pregnant when they set sail – gave birth. They soon managed to speed things up, and established a regular service to Natchez, a few hundred miles upriver. Fulton died in 1815, aged only 49, surviving just about long enough to see his business start to flourish, and the riches and accolades pour in. A year after his death, a new street running between his two ferry operations on the south-west tip of Manhattan was named 'Fulton Street', the first of several all over the US. By the middle of the century, it was hard to find any stretch of water in America that didn't have a steam service of some sort. Moreover, the tech was key to colonisation of the interior of the US. In the two decades after the first steamboats arrived on the Mississippi, more people moved to the middle of America than the colonies had attracted in the previous two centuries. And when the army beat Native American communities back, they arrived on steamboats, a sad twist on Fitch's nightmares of being captured on the Ohio River.

Steamships quickly popped up around the world. There was a Liverpool to Glasgow steamer in 1815, and the Post Office added steamboats to their runs to Ireland and France in 1820. By the 1830s, there were steamboats crossing between the UK and Spain, and out towards Egypt too. The East India Company (EIC) commissioned two steamships to run from between Suez and Bombay in the mid 1830s, but the EIC's days, by this point, were numbered. The 'free trade' ideas of Watt's old drinking buddy Adam Smith were increasingly influential and in 1833 an Act of Parliament saw the last vestiges of their trading monopoly formally end. Other traders were quick to pounce on the opportunity, not least the Scottish firm Jardine Matheson & Co. Keen to cash in with the disruptive technology *de jour*, the firm sent out a series of steamboats to transport opium from India to China, hoping steam would withstand monsoons better than sails. The Chinese had another way of looking at it, refusing to have this 'smokeship' (and more pertinently, its opium cargo) in their waters. Fights between China and British traders escalated, and the EIC – keen to show it was still the dominant force in the area – sent out secretly commissioned steam warship *Nemesis* complete with rocket launcher. As what became known as the Opium Wars rolled on, Jardine Matheson & Co realised they could simply put opium on a P&O steamer (it helped that James Matheson was on the P&O board).

★ ★ ★

Back on land, the first railways weren't powered by fossil fuels. They were simply carriages on rails that cut down friction, making it easier to transport a heavy load like coal or slate. In Australia, the first railway was human-powered, using convict labour, but most railways were pulled by horses, occasionally helped along with wind-powered sails. When it came to adding coal to the mix, people had played with ideas for steam cars for decades, but the first serious innovator in steam-powered land transport was Richard Trevithick. Having grown up in Cornwall and watched steam engines pump water from local tin and copper mines, he'd dreamt of taking it further. When Watt's patent expired in 1800, Trevithick seized the opportunity, taking steam-powered road locomotive the *Puffing Devil* out for a ride on Christmas Eve 1801. He didn't give it a proper steering mechanism though and it ended up headfirst in a ditch a few days later.*

---

* There's a story that they simply left it there and went off to eat goose in a local pub, forgetting that this engine, even in the ditch, contained a large fire (until the inevitable explosion, that is). For more great stories like these, see Christian Wolmar's history of early steam trains *Fire & Steam* (2007).

Undeterred, Trevithick produced steam locomotives for ironworks, running them on rails so they wouldn't fall into any ditches. With an engine designed by the flamboyant Count Rumford, Trevithick also set up a short-lived steam car service in London in 1803. Run out of a coachmaker's workshop on Leather Lane, this managed several trips down Tottenham Court Road and along the City Road usually at 4–5mph, but sometimes hitting nearer 9mph. For one special outing, Oxford Street was cleared of horses and carriages, the shops closed but upper floors packed with spectators. People were keen to watch the machine puff its way through London, but unfortunately this enthusiasm didn't extend to investment and the scheme folded. In 1808, still trying to drum up some interest, Trevithick set up a steam circus not far from what is now Euston Station. There, a small steam locomotive, *Catch Me Who Can,* moved at 2mph around a 100ft circular track, with passengers paying a shilling to ride. Sadly, not enough people even came to watch. Trevithick turned his back on the locomotive business and boarded a whaling ship to South America, working in mining in Costa Rica and Peru.

Horses remained the default method to power railways, but they could only go so fast before they needed food and rest, plus the work involved meant they often died quickly. For industrialists needing to transport heavy loads, like iron or coal, steam-powered transport seemed worth a try. One of the engineers employed to build steam-powered trains for the mining industry was George Stephenson. Born in 1781 in Northumberland coal country, Stephenson did not have the advantages Watt was born with. Instead, he built them for himself. Young Stephenson started work aged seven, first in farms and then with his father, feeding the mines' steam-powered water pumps with coal. He put himself through night school and learnt to read for the first time in his mid-teens.* Then, in his mid-20s, things took a turn for the tragic. His wife and baby daughter both died. He hired a

---

* Later in life, people fighting the expansion of the railways would use Stephenson's relative lack of education as a slur. Stephenson struggled with spelling and grammar throughout his life, remaining very self-conscious of his writing even when his engineering work had made him internationally famous, only letting close family members see anything he wrote.

housekeeper to look after his infant son, Robert, and – possibly out of grief but more likely necessity – walked 200 miles north to take a job in Scotland, working with a Boulton & Watt steam engine at a spinning mill. When he returned a year later, he found his father had been badly injured in an engine accident, blinded by steam and unable to work. Times were hard. For a while Stephenson thought about moving to America – to see if the new industries popping up around Pittsburgh's coal could make something of his engine skills – but he decided to stay put. On Saturdays, when the engines were idle, he would take them apart and put them back together, examining each piece as he went to understand how it worked. He was also part of a regular discussion group on mechanical matters, including Trevithick's locomotives. He invested in formal education for his son Robert too, on the agreement that he would, in turn, teach him back.

Stephenson started building steam-powered locomotives to move coal around the mine he worked for, but he really drove steam trains into the mainstream in 1824, when he was appointed chief engineer for a new steam railway planned between Liverpool and Manchester. By this point, there was a sense that steam was the future of transport and there were concerns that if a British company didn't get a move on, America or Russia might beat them to it. The developers had started drumming up PR before it opened. You could visit the new tunnel built just outside of Liverpool for a shilling and walk through it to the sound of a band echoing through the chamber. They also, infamously, picked their trains in the style of a sort of beauty contest for prospective locomotives. This attracted a range of somewhat madcap entries, exciting the press and drawing in crowds too, with more than 10,000 people attending the opening day. George and Robert Stephenson improved engines they'd already been running on coal lines to build the *Rocket*. The local press's favourite was the *Novelty*, from Swedish inventor John Ericsson, which managed a few good runs at 28mph, but leaked and was disqualified. In contrast, the *Rocket* ran smoothly back and forth on the track at 14mph, before speeding up on the last leg to nearly 30mph, to the astonishment of spectators who had rarely seen anything travel faster on land. The £500 prize was Stephenson's, as was the commission to build another four engines for the track.

The Liverpool and Manchester Railway line formally opened on 15 September 1830. It was a less than auspicious start, with a fatality on the track and protests against the Duke of Wellington (the deeply unpopular prime minister who'd been invited to ride the first train). It being Manchester, there was also rather a lot of rain. Still, it didn't seem to do the idea of steam transport much damage. Although the line had been developed with freight in mind (mainly cotton), passengers soon became the core of the business. Within a decade there was a mainline through the spine of England, with a further 7,000 miles of railway built in the 1840s. Engineers like Isambard Kingdom Brunel took time to build grand termini for these new lines, with elegant bridges and viaducts to carry the trains through the countryside. There's a story that London's Paddington Station (which opened in 1854) is so intensely beautiful because Brunel was annoyed Paxton had got the Great Exhibition gig and wanted to outshine him.* As railways history writer Christian Wolmar emphasises, developers like Brunel knew the importance of putting on a good show if the British establishment were going to be kept on side with this rapid development. The new railway infrastructure of the era flattered the landscape, rather than appearing to destroy it, with grand stations designed to win over the affluent and influential members of society who otherwise might make the sort of fuss that could stall development.

★ ★ ★

Even before the opening of the Liverpool and Manchester line, other countries were getting in on the steam train game. Stephenson was soon travelling all over Europe advising new rail plans. Construction for the Baltimore and Ohio Railroad started as early as 1828, after the developers visited railway projects in England. In the midst of

---

* One of the many other stories about Brunel is that he aligned the tunnel he built in 1841 through Box Hill (then the longest in the world) so the Sun shone directly through on the 9 April, his birthday. When the tunnel was closed for maintenance in April 2017, GWR staff tested this theory and confirmed that it did seem to have been designed for the sun to rise on that date, at least from the eastern portal.

the usual debates over whether this railway should be powered by steam, wind or muscle, a public race was planned between a horse and the locomotive *Tom Thumb*, designed by Peter Cooper (who later also patented the main ingredient for Jell-O). After a good start, the engine's belt snapped. Cooper tried to repair the belt and did finish the course, but came second, his hands painfully burnt from the heat of the engine. Still, the investors were impressed by the performance of the locomotive, appreciating that it was carrying several people and the everyday running of a railway would be somewhat different from this staged race. Despite the challenge of scale, it wasn't long before Americans had access to the world's most extensive system of railroads. By 1860, 30,000 miles of track had been laid. The first transcontinental line, the Pacific Railroad, linking existing rail networks in Iowa with San Francisco Bay, was completed in May 1869, with railroad financier Leland Stanford (also famous for the university) ceremonially tapping in the final 'golden' spike with a special silver hammer. Although often forgotten, 90 percent of the workers who'd built this railroad were Chinese, tens of thousands of them, mainly from the Guangdong Province, South China.[*] It's important to remember that the whole project also involved taking millions of acres of land from Indigenous nations. Some Indigenous people worked with the railroads, others resisted. When they took this latter course of action the railroad officials and military authorities responded by cutting off their food supply or, in several cases, extremely brutal massacre. The railways, like steamships (and it should be said, wind power for irrigation too), quickly became key technologies of colonialism in mid nineteenth century America.

Over on the other side of the world, the East India Company infamously brought steam railways to India, with a line between Mumbai and Thane opening in 1853. Presented as the embodiment of civilisation, which the British were graciously endowing on India, it was a nakedly imperial project from the start, designed for British

---

[*]For more on Chinese labour, see Gordon Chang's (2019) *Ghosts of Gold Mountain*. Manu Karuka's (2019) book *Empire's Tracks* also covers this story, alongside those of Indigenous nations.

interests. Like the Manchester and Liverpool Railway and Fulton's Mississippi endeavours, at its heart was the movement of cotton. Moreover, once up and running it was speedily used to transport forces against the First Indian War of Independence in 1857. This war was arguably the nail in the EIC's coffin, leading to widespread condemnation back home in England; its powers passed to the Queen before it was formally dissolved in 1874. This was far from the end of British imperialism though, via railways or otherwise. It was just shaped differently. Jardine Matheson & Co bought up land in China for a railway line between Shanghai and Woosung plot by plot, and shipped over a small locomotive in the 1870s, although the railway was later ripped out by the Chinese government, wary of foreign interference. Cecil Rhodes' plan for a railway from Cairo to Cape Town to establish British dominance throughout Africa – first suggested in a *Daily Telegraph* leader in 1876 – was refused funding by the UK government, but he cobbled together investment from various sources, including his diamond company De Beers. Soon, in pretty much any part of the continent in which the British were active, railroads could be found transporting both luxury tourists and valuable minerals.

The growth of railways also led to the next development in steamships. One October evening in 1835 over dinner with directors of the Great Western Railway (GWR), Isambard Kingdom Brunel cracked a joke: why not run the line all the way from London to New York, simply putting the passengers on a steamship at Bristol? His casual remark planted a seed. After dinner, Brunel got chatting to Thomas Guppy, a sugar refiner from one of Bristol's richest merchant families who was one of the GWR directors, and the two discussed the idea long into the night. They knew about earlier attempts to use steam to cross the Atlantic; that they'd mainly relied on wind in their sails and weren't really worth it. But Guppy and Brunel felt engine tech had developed sufficiently that it was worth another go. Within weeks, a handwritten prospectus for a Great Western Steamship Company was being passed around rich and daring Bristolians. Brunel's ship for the Great Western Steamship Company, *Great Western*, was finished in the spring of 1838. A series of mishaps in London meant it lost a transatlantic race to New York to another ship, *Sirius*, but Brunel's creation was generally recognised as the better design (plus it won the race back to Europe). What's more, the competition between the two had attracted crowds of excited New Yorkers. It was fabulous PR. Great Western's ocean ambitions collapsed when the mail contract went to former EIC merchant Samuel Cunard. Still, the business case for transatlantic travel in steamships had been clearly made. The world was getting smaller, and burning through prodigious quantities of coal in the process.

* * *

Steam power grew up drenched in the prosperity that slavery and colonialism offered Europeans in the eighteenth century. Trade in cotton in particular shaped the development of trains in the UK and India and steamships along the Mississippi. As steam power grew, it was soon put to work to further ingrain and spread the power of white elites. There's one story at least that bucks the trend slightly, however, and is worth telling as tales like these are too often 'steamrollered' by more dominant narratives. In 1906 Tamil lawyer V. O. Chidambaram Pillai – known by his initials VOC – registered an Indian-owned steam shipping company out of the busy port town of Thoothukudi (in the southern tip of India, then known as Tuticorin). VOC had been following the Swadeshi movement, which was using boycotts of British goods as a form of resistance against imperial rule. He figured that in port towns like his, freedom-fighting had to mean control of the seas. He raised a million rupees, selling shares door-to-door, for the Swadeshi Steam Navigation Company, which would be owned by and for Indians, and run in competition with the Scottish-owned British India Steam Navigation Company. His ship, painted in the colours of the Swadeshi flag and emblazoned with a deliberately provocative nationalist slogan from a Bengali poem, ran between Thoothukudi and Colombo, irritating the British intensely. The British firm fought hard against the competition, cutting prices and at one point giving out free umbrellas in an attempt to sweeten customers. Sadly, this era of steam-powered anti-colonialism was short-lived. In 1909, VOC was arrested for encouraging workers at a cotton mill to strike, and the Swadeshi

Steam Navigation Company was liquidated in 1911.* Still, as we'll see in later chapters, the new fossil fuel economy would challenge power structures as well as ingrain them, not least through the mobilisation of coal workers. Indeed, the elites knew this a good half century before VOC started selling shares in his steamship company. Arguably the 1851 Great Exhibition put on such an inflated show of Britain's great industrial might precisely because it was worried the old patterns of power were changing.

And what became of that giant greenhouse in Hyde Park? The Great Exhibition was only ever meant to be a temporary event and so closed a few months after it opened. The Crystal Palace was too beautiful to lose entirely and so was rebuilt in 1854 on the top of Sydenham Hill in south-east London, where it acted as a hall for exhibitions, concerts, theatre and the occasional circus. Other countries, keen to emphasise Britain didn't have a monopoly when it came to invention, put on their own shows following the Great Exhibition's model (which, after all, had been borrowed from the French in the first place). There was the 1878 Exposition Universelle in Paris, for example, which included a display of Alexander Graham Bell's telephone and the head of the Statue of Liberty. Or the 1889 equivalent that left the world the Eiffel Tower. We'll drop by the 1893 Chicago World's Fair when we pick up the story of electricity in Chapter Five. The Crystal Palace itself burnt down in 1936, in a fantastic blaze that could be seen for miles around. The ruins remain though. If you ever get the chance to visit on a foggy day, it's a wonderfully spooky sight.

And coal, this chapter's key protagonist? Just as this sooty, incendiary rock was the first fossil fuel to be embedded in modern economies, it may well be the first to go. It is rapidly disappearing from the British electrical grid, so much so that tourist steam railways like the 'Harry Potter' *Jacobite* route through the Highlands have appealed to the UK government to protect their stocks. Still, coal is far from gone. Globally, it accounts for nearly a third of energy used worldwide, playing an especially important role in the production of iron and

---

* There's a great telling of this story in Sunil Khilnani's *Incarnations: India in 50 Lives* (2016) (also available as a BBC podcast). Thanks to Justin Picard for the tip.

steel. Moreover, arguably Britain has simply outsourced its personal coal problem, able to dress itself up as a climate leader for quitting the stuff in ways other countries simply don't have the means to. Manufactories like Boulton's may be long closed, but the desire for the sorts of consumer 'toys' he produced has grown and grown. And as Britain imports rather than makes its goods these days, children in other countries choke on the coal smoke that produces the everyday stuff of British life; all shipped in on the descendants of Fulton's steamships.

CHAPTER TWO

Discovering Our Hothouse Earth

Every summer, tens of thousands of people attempt to climb Mont Blanc, the highest peak in the Alps. This number is increasing as the desire to catch a summit selfie grows, bringing less-experienced climbers with it. At the same time, warming temperatures are making such trips riskier. The average temperature in the resort of Chamonix, at the foot of the mountain, rose by more than 2°C over the course of the twentieth century (more than double the global rate). This is not just a matter of glaciers retreating, but also means climbers have to contend with landslides and steeper, icier and more dangerous routes. Iconic mountains around the world are struggling with similar overcrowding and ever-drippier glaciers but, as the birthplace of modern mountaineering, Mont Blanc finds itself on an especially large number of bucket lists.

It wasn't until 1786 that anyone claimed to have 'conquered' Mont Blanc. In 1760, young, wealthy scientist Horace de Saussure, fresh from university and a thesis on the science of heat, was in the Chamonix Valley to collect plant specimens. While he was there, he offered a reward to anyone who could reach the summit. After two men finally made it to the top one evening in August 1786, de Saussure managed several trips himself, establishing a scientific tradition of research in high altitudes. In the two and a half decades that passed, he'd become professor of philosophy at the Geneva Academy, making a name for himself with extensive studies of the botany, geology and physics of the Alps. As he developed his studies of the area, he also invented and improved many kinds of apparatus, including a 'cyanometer', for estimating the blueness of the sky; and a 'heliothennometer', a wooden box lined with blackened cork and covered with three sheets of glass, which he used to explore how solar radiation increased with altitude.

De Saussure's research would inspire future generations of environmental scientists studying the Alps and further afield, and as we'll see in Chapter Six the pastime of mountaineering he helped kick off would be the bedrock for parts of the modern environmental movement. What's more, his heliothennometer would act as an early analogy for the greenhouse effect, as the earliest climate change scientists wondered if the atmosphere worked a little like a solar oven. This is the story of how, from those first steps up a mountain, we built up an idea of the Earth wrapped in a sort of insulating blanket of gases and realised that if we changed the chemical composition of this blanket we could catastrophically change the climate. We start with stories of chemists, back when the discipline was still transmuting itself from alchemy, unravelling an understanding of temperatures and picking out a series of invisible gases from the air. We then move on to ideas of the Ice Age and from that the greenhouse effect. Finally, we combine those two threads, as researchers gradually realise that the gas we'd now call carbon dioxide was especially good at trapping heat and, if we added enough of it to the atmosphere, global temperatures would soar.

★ ★ ★

Today, the idea that the air around us is made up of a mix of different, invisible chemicals is pretty intuitive. We're used to the idea that we inhale something called oxygen and then breathe out another invisible substance called carbon dioxide. We know other animals do the same and plants do it the other way around. If we paid attention in school, we know that roughly three-quarters of the sea of air we live within is nitrogen, with a chunk of oxygen, and a sliver of other stuff like carbon dioxide and hydrogen. We might well worry about the quantities of carbon dioxide and other greenhouse gases totting up in the atmosphere, or feel increasingly nervous of tiny, naked-to-the-eye particles of pollution, especially in car-choked cities. But we have to be taught before we can imagine all this; we need science to see many of the gases because they're not obvious to the naked eye.

For the ancients, air was simply air, one of four basic elements along with earth, water and fire. It might be wet or noxious. Perhaps the air up there near the gods, aether, was special. It might be smelly, if you lived near a coal burner, it might make you choke. If it was polluted, miasma or 'bad air', it might make you ill. But there wasn't a modern sense of the atmosphere around us being made up of a variety of changeable gases. It was the alchemist Johannes Baptista van Helmont who, in the early-seventeenth century, first coined the term 'gas', and with that opened up a whole new world for scientists to explore and exploit. With an alchemist's curiosity in the invisible and a more modern desire to overthrow orthodox thinking, van Helmont began a study of woodsmoke. Concluding it was different from both air and water vapour, he called it 'gas' – which is probably from a word scientists were already using, 'gaesen', meaning to effervesce or ferment, but might also be a play on the word 'chaos'. He burnt dung and intestines too, exploring what gases they'd produce, looked into the effervescence of fermenting wine, and examined the fizz you get from mixing vinegar and chalk.

Carbon dioxide – the gas that would become such an iconic problem – was first identified in 1754 by James Watt's friend, Joseph Black. As a medical student in Edinburgh, Black was fascinated by the chemistry of our bodies and the materials with which they interacted. For his final thesis, he studied kidney stones and chemicals that might be used to dissolve them. Playing with magnesia alba, a mild laxative made from Epsom salts, Black noticed it became lighter in weight when it was heated. This fascinated him – what was leaving the material when it got hot? He performed a series of experiments, concluding that this weight was being lost into the air. He also noted that when heated the magnesia alba had an obvious smell, although not an especially bad one, and if you put a candle nearby the flame would go out. On further exploration, he realised that if you immersed a piece of burning paper in the gas, the fire would go out as quickly as if it had been dipped in water.

In 1766, the notoriously shy and eccentric aristocrat scientist Henry Cavendish dropped some zinc into hydrochloric acid and watched a gas emerge as it burnt with a pale blue flame. He called this explosive and lighter-than-air gas 'inflammable air', although we know it better as hydrogen. A few decades later, Italian pneumatic scientist Tiberius Cavallo entertained an audience at the Royal Society with soap bubbles filled with this new lighter-than-air gas, showing how it lifted the bubbles up.* In 1772, Black's student, Daniel Rutherford, developed the study of 'fixed air' and isolated something he called 'mephitic' or 'phlogisticated air', which we'd know as nitrogen. In Italy, Alessandro Volta collected gas from rotting plants in marshes by Lake Como and used it to fuel a lamp. What had previously been a uniform, invisible substance was fast turning out to be a cacophony of different gases, all with exciting possibilities.

You didn't need fancy lab equipment or a specialist education to know air came in a variety of different flavours. Coal miners knew all too well the sometimes deadly nature of the gases they'd find in the mine tunnels. They called them 'damps', from the German word *dampf*, meaning vapour (as opposed to something soggy). There was 'chokedamp', a suffocating mix of nitrogen and carbon dioxide, or for any mix of gases that might explode, 'firedamp'. This latter group might be called 'stinkdamp' if you were lucky enough to be warned of its presence with a smell, or 'whitedamp' if you weren't. William Brownrigg, a physician and chemist living in the English coastal mining town of Whitehaven, became interested in the types of airs miners were breathing and how this might expand science's idea of gases. He collected bladders of firedamp for investigation at the Royal Society in 1730 and was elected a fellow in thanks. Local mining agent Carlisle Spedding had the bright idea of distributing this flammable firedamp throughout the town via underground pipes to use it to light the streets. The town refused, but Spedding used the method to light his own office and Brownrigg's lab. As we'll see in Chapter Three, it would be a while longer before gaslighting caught on. That lamp

---

* I don't know if he also set fire to them, but he could have. It's a trick scientist-entertainers have since played for generations. I used to set fire to towers of bubbles filled with methane when I worked in the children's galleries at the London Science Museum. Do not try this at home.

Volta had made to run on marsh gas was very much a scientific instrument – requiring specialist skills to work – and not a mass-market product.

* * *

It wasn't long before Joseph Priestley was playing around with Black's discovery of carbon dioxide. A preacher, teacher, philosopher and writer as well as a chemist, Priestley couldn't help getting pulled into unpicking new ideas. A radical in both his politics and his science, the two often intermingled. His major work on gases, *Experiments and Observations on Different Kinds of Air*, warned that there could be philosophical and social, not just scientific, ramifications of such research, and that the English hierarchy might have reason to 'tremble even at an air pump or an electrical machine'. Science arrived in Priestley's life already wrapped up in politics. Born into a family of religious dissenters, he was excluded from mainstream English education and so had to challenge political orthodoxy simply to do scientific work. He'd studied at a dissenters' academy in Daventry and, after a stint as an assistant pastor in Suffolk that everyone agreed was a disaster, found a living teaching in Nantwich. As his students were, like him, excluded from Oxford and Cambridge, Priestley felt there was no point working from the usual Classical curriculum designed to prepare for admission. Instead, he'd teach maths, science, English and history, and not just limited to the Greeks and Romans either. This was quite innovative for the time, as was the fact he taught girls as well as boys, albeit in different rooms. Word soon spread, Priestley's reputation grew and in 1761 he was offered a post at the dissenters' academy in Warrington; and after his educational materials were published in 1765, he was awarded an honorary doctorate from Edinburgh.

The Warrington job gave Priestley the chance to learn practical lab skills. He'd help the chemistry lecturer produce nitric acid for his classes, bewitched by the ways in which careful heating and distillation could pull all sorts of coloured fumes and crystals from something as mundane as clay. In 1767, he moved back to Yorkshire and took up a post as a minister in Leeds, and it was there that he started researching airs in earnest. The home his family were due to live in was being refurbished, so they were offered temporary accommodation next to a brewery. Priestley was delighted at the prospect of vats of fermenting liquids just next door emitting a steady supply of Black's 'fixed air' and soon got experimenting, eagerly recruiting the slightly bemused brewery workers to help.

Playing around at the Leeds brewery, Priestley soon found that woodsmoke caught up in this fixed air would swirl and cascade over the sides of the vat, as it was heavier than common air. He also experimented on small animals and realised that although fixed air wasn't poisonous it wasn't exactly good to breathe either. Butterflies became sluggish when enclosed in a vial of fixed air, but they could be revived. The same was true for a frog and mouse, although he killed a snail. Trying similar experiments with plants, he started to unravel respiration. We animals breathe out 'fixed air', but plants breathe it in and then give us back the air we need. For Priestley, the science of respiration was wrapped up in his theology. He saw air as 'injured' by our breath and green plants carefully provided by a God who would not allow mankind to be suffocated by its continual exhalation (we can only imagine what he would have made of humanity's later exhausts, air pollution and the climate crisis). Another of Priestley's discoveries was that if he poured water back and forth over the brewers' vats he could suffuse it with the fixed air, providing a pleasant fizz. Keen to share this new discovery, he wrote up instructions for making this fizzing water and sold copies for a shilling each. News spread around Europe of this delightful discovery, for which Priestley won the Royal Society's Copley Medal in 1772, and various entrepreneurs developed devices for mass production – including Jacob Schweppe, who soon cornered the London market with his force pump. There was already a market for naturally sparkling waters from spa towns and people thought these lab-made fizzy drinks might have medical benefits. Priestley hoped it might be a useful treatment for scurvy and lobbied the navy sufficiently that Captain Cook took soda water machines on his 1772 voyage to the Antarctic. Over in the US, Yale chemistry professor Benjamin Silliman started manufacturing and selling lab-made soda waters too, hoping to bring the sparkling waters of spa towns to the masses. Although it would be a few more decades before the soda industry really took off in America, Silliman's students apparently enjoyed it as a hangover cure.

Priestley's most significant gaseous discovery came in 1774. He'd treated himself to a big, new glass convex lens from London. He was enjoying the new toy, using it to focus the Sun's rays and setting fire to pretty much anything that caught his eye, including some rusty dots of calx that would form on mercury if you heated it. Researchers like Robert Boyle had tried heating mercury calx before and noticed air was released, but hadn't bothered to investigate further. Priestley set up his apparatus so he could collect this gas and test it. As he turned the lens on the calx he saw gas rush out, seemingly three or four times as much as there had been calx. What was it, he wondered? Both fixed and mephitic airs would extinguish a candle, so he reached for one. But the flame leapt up, glowing intensely. The gas he'd discovered was oxygen, but he called it 'dephlogisticated air'. He didn't have time to experiment further as he was about to depart on a tour of Europe. Over dinner in Paris with French scientist Antoine Lavoisier and his wife Marie-Anne, Priestley mentioned this new air he'd found from mercury calx, the methods he'd used and how it had made the candle burn so brightly. It was in faltering French, not helped by Priestley's stutter, but got the key point across. The Lavoisiers' faces lit up at the news – what was this new air that helped things burn?

If Priestley often found himself the outsider in British society, Lavoisier didn't have the same problem in France. Things generally came easily to him. Born in 1743 to a wealthy family, he went to one of the most prestigious schools in Paris. In his early 20s Lavoisier became the youngest member elected to the Académie des Sciences and not long after took a share in the Ferme Générale, a private company that collected taxes on products like tobacco, alcohol and playing cards, as well as taking tolls at gates and boarders. Lavoisier's job was to combat smuggling and fraud among tobacco retailers, and he used his scientific skills to spot contraband tobacco that had been smuggled into France. The work took up a fair amount of his time, but it was also extremely profitable, allowing him to subsidise his interest in science. It was through the Ferme that he met his wife Marie-Anne, the daughter of one of his supervisors. Beautiful, charming and rich, she was clearly a catch, but would go on to play the role of lab partner as much as society wife. As well as hosting dinners for local and visiting scientists, Marie-Anne studied languages so she could translate Antoine's work and correspond with scientists around Europe. She attended lectures at one of the Paris colleges so she was well versed in chemistry; she took notes on his experiments, offering critical commentary, and trained herself in engraving to illustrate his papers too.

In 1775, Lavoisier was appointed scientific director of the Royal Gunpowder Administration, tasked with restarting gunpowder production after France had lost valuable Indian saltpetre reserves to the British. He was successful; his formula was so good it could send a cannonball 50 per cent further than any of the French's European competitors. Benjamin Franklin was impressed and negotiated a secret deal to bring hundreds of tonnes of this new, high-grade gunpowder to support the American Revolutionary forces. By 1779, more than 800 tonnes had been sent across the Atlantic and Lavoisier, with characteristic smugness, would claim America owed its liberty to the quality of his chemistry. As part of the gunpowder job, the couple also moved into the Paris arsenal. It might have seemed like a rather grim place to live, but their quarters were large, and came with a library and laboratory.* Once the Lavoisiers were settled in their new home, one of the first experiments they turned their hands to was a replication of Priestley's play with mercury calx. It took 12 days, but finally they'd found Priestley's new gas. Mice happily jumped about in it and candles burned powerfully. Exploring further, Lavoisier found the gas seemed to change everything it touched. Mix with charcoal and you get

---

* Joe Jackson, in his gripping book on Priestley and Lavoisier, *A World on Fire* (2005), paints a vivid picture of this, dubbing it a 'scientific paradise'. Jackson's book is worth checking out if you're interested in these scientists and the ways in which chemistry, philosophy and politics intermingled at the time.

carbonic acid, with sulphur, sulphuric acid; he was uncovering the basis for what we'd now learn as oxidisation. Applying an idea of acids we now know to be slightly flawed, he shunned the name Priestley had given it – 'dephlogisticated air' – and instead named it 'oxygen' from the Greek for 'acid maker'.

In the process of refining his oxygen, Lavoisier had also filled a vial with another strange gas. When he put a mouse in a jar of this, it died. Candles would go out too. He tested it to see if it was fixed air by seeing if it made limewater cloudy, but it didn't. He didn't know it yet, but this gas was nitrogen, the same gas Rutherford had found a few years before. Lavoisier called it 'mofette' and when he mixed it with his oxygen, he realised he'd remade common air. From this, he could clearly state something others had been circulating around for a while; that the atmosphere is not a simple, single substance, but a mix of airs. There is pure air, he argued, which we breath in, 'which Mr Priestley has very wrongly called dephlogisticated air'; fixed air, which we breathe out; and noxious air, this mofette, which he figured we needed to learn more about.

* * *

What was this word 'phlogiston', the one Rutherford and Priestley had named their new airs after? What was it about nitrogen that meant it was phlogisticated, and oxygen dephlogisticated? And why was Lavoisier so against it? The answer is key to the changes in chemistry at the time and also foundational to our understanding of the climate crisis: it was the matter of heat.

Today, a lot of climate chatter is rooted in talk of degrees Celsius. This scale was first put forward in 1742 by Swedish astronomer Anders Celsius. It runs 100 degrees between the freezing and boiling points of water: 0°C at one end, 100°C at the other. Anything colder than the freezing point of water is -0°C and everything hotter than boiling water is above 100°C. It's intuitive for us humans because our fingers tend to come in tens (and we use them to count) plus we spend a lot of time playing with water. But it's important to remember that numbers like these, which might seem neat and significant, are often just numbers for a human-built scale. Nature itself has its own games to play, and in the early days of heat research plenty was up for grabs. The potter Josiah Wedgwood, who needed a decent thermometer to keep his kilns consistent, developed a system based on small bits of clay; measuring their size before and after heating to check how much they had expanded. Edmond Halley (of comet fame) recommended the Royal Society agree to a standardised scale with the boiling point of alcohol at one end and the freezing point of aniseed oil at the other. Other options included a candle flame, snow, when wax congealed or butter melted, the King's Chamber at the centre of the Great Pyramid of Giza or the hottest water a hand could endure while remaining still (which surely suffered from problems of subjective degrees of endurance, as well as being somewhat pointlessly painful). The wine cellars of the Paris Observatory were especially popular among European scientists of the time, appearing in at least three temperature scales.[*]

Depending on when or where you grew up, you might be more used to thinking with a scale first put forward in 1724 by instrument maker Daniel Fahrenheit. Born in Gdańsk to a wealthy merchant family, he moved to Amsterdam when his parents died suddenly in his mid-teens and discovered the small but rapidly growing business of making scientific instruments. This was the era of the so-called Dutch Golden Age when scientists from across Europe were drawn to the Dutch Republic,[†] and Fahrenheit could learn from a range of different scientists and sell to them too. The upper end of Fahrenheit's scale was found by placing the bulb of a thermometer in the mouth or armpit of a healthy person, and the lower point was found from a

---

[*] For a detailed and fascinating trip through the history of measuring heat, see Hasok Chang's *Inventing Temperature* (2004). You'll never look at the bubbles in a pot of boiling water in quite the same way again.

[†] This is often put down to the relative intellectual tolerance of the Dutch Republic attracting scientists to the area. Dagomar Degroot points out in his book, *The Frigid Golden Age* (2018), that the so-called 'little ice age' of that period also played a role, as the change in climate pushed people in other parts of Europe to move. It also sped up Dutch ships bound for Asia and the Americas (or to put it another way, climate change fuelled colonialism a few hundred years ago, as well as the other way around).

mix of ice and water. It was inspired by a system first developed by Danish astronomer Ole Rømer and it is sometimes said Fahrenheit ran with a misunderstanding of Rømer's use of the term 'blood warm' (it's possible Rømer meant boiling water, not body temperature, which would have made Fahrenheit's scale similar to Celsius'). Fahrenheit also popularised the use of mercury for thermometers. Small and reliable, his temperature-measuring devices spread around Europe, partly through doctors who had trained in Amsterdam. After much deliberation and a lot of study of bubbling and hissing pots of water to decide exactly what constituted a boiling point, Fahrenheit became standard for weather, industrial and medical measurements in English-speaking countries up until the 1960s.

This still left the question of what was heat? The idea Priestley was so fond of, phlogiston, had been developed in the early-eighteenth century. A development of old alchemical ideas, it suggested things burnt because they contained a sulphurous principle of combustion. All substances contained some phlogiston it was thought; good burners like whale oil, marsh gas, alcohol or coal were pretty much pure phlogiston. It might seem ridiculous to us, but it had explanatory power and much of the science of the early Industrial Revolution ran if not on phlogiston theory itself, then something a little like it. There were problems though, especially when it came to weight. If you heated metal it should lose weight, because you were realising its phlogiston, but the opposite was true. Believers in phlogiston tried various ways of explaining away the problem – perhaps phlogiston was of negative mass – but scientists like Lavoisier were increasingly sceptical.

In 1780, Priestley and his family had moved to Birmingham. It was a comfortable set-up. He had a job preaching at the New Meeting House, one of the more liberal congregations in England, and socialised with the Lunar set, having swapped science with Boulton and Wedgwood for years. His home was large enough to fit a laboratory as well as a library, along with space for showing magic lantern shows. One of his first experiments in his new lab space involved passing an electric spark through a mix of Cavendish's inflammable air and his own dephlogisticated air. He was interested to see what might happen. He must have expected an explosion; a gas that burns well mixed with one that seems to help things burn, with added electricity. He got one too, but only a small one, and he must have been baffled by what was left over: a few drops of dew. What he had done, although he wouldn't describe it as such himself, was mix hydrogen with oxygen to make water. He mentioned it to Cavendish, who did some further study and in spring 1783, to some controversy and a lot of scientific chatter, announced he'd discovered that he'd split apart yet another of the ancients' elements. Water was, in fact, two parts inflammable air to one of this dephlogisticated/ pure one. It seemed like an incredible claim: how could water, so wet, be made of these two gases? Moreover, if hydrogen was pure phlogiston because it was so flammable and oxygen was dephlogisticated because things burnt in it, why if you combined the two would you get water? Shouldn't it just be a huge ball of fire? For many, this was the nail in the coffin for phlogiston, who dropped it as a theory of heat.

Although Lavoisier rebuffed phlogiston theory, he still saw heat as a substance, he just called it 'caloric'. For the more modern idea of heat as movement, we need Count Rumford. Born Benjamin Thompson in Woburn, Massachusetts, in 1753, Rumford was largely self-taught and there are stories of him walking 10 miles to Cambridge to attend lectures at Harvard. After a stint as a schoolteacher in Concord, New Hampshire, he married a wealthy widow who was 14 years his senior and then speedily made himself unpopular in the run up to the American Revolution with his vocal support for the Tory cause. He fled, leaving his wife and their daughter behind, utilising his scientific skills to act as a spy for the loyalists, writing letters in invisible ink detailing troop movements. After the war, he found work in Munich, Germany, where he had at least two illegitimate children, and his science-based improvements to the Bavarian army were so successful he was made a Count of the Holy Roman Empire. He made a fortune refining and selling fireplaces, chimneys and kitchens, and ploughed it into, among other things, setting up prizes in Boston and London for scientific studies into heat and light. He invented the range cooker (i.e. most modern ovens) and several types of drip-coffee maker.* When asked of where he wanted to be a count, somewhat mischievously, he picked the old name for Concord, Rumford. America had kicked out the aristocracy, but he'd still be a count of a bit of it.

What makes Rumford important to this bit of our story is his work boring cannons. Working in the arsenal in Munich in the late 1790s, he looked at the way cannons were made, using a dull drill to bore the hole. The heat generated in this process seemed limitless. He couldn't square this with the idea that heat was a substance and from this built an idea of heat as vibrations. Chemistry had already by that point established an idea of molecules of matter vibrating in certain positions. Rumford imagined heat as increasing the frequency of these vibrations. He wasn't the first to suggest such an idea, and it would need more work before we'd see the modern science of thermodynamics (notably Sadi Carnot and James Prescott Joule). Caloric would stick around right through the first few decades of the nineteenth century, and for a while some scientists were enamoured with the idea of heat as a sort of wave, like light. But Rumford and the friction of his cannons offered a strong articulation of the idea of heat as movement, and when John Tyndall (more on him later in this chapter) wrote his best-selling book *Heat as a Mode of Motion* in 1863, he'd declare Rumford the star.

★ ★ ★

Before we leave the chemical side of this story, it's worth saying something about what became of Priestley and Lavoisier – less so because of their scientific contributions, but more as studies of scientists' involvement in politics. Antoine and Marie-Anne Lavoisier

---

* He's sometimes credited as inventing the soup kitchen, although it was perhaps better understood as a workhouse. Having studied the thermal qualities of different cloths and furs at length, he wanted to improve the army's uniforms. Looking around for a workforce to produce the new clothes, he gathered up local beggars and put them in what became workhouses. But now he had to feed them. After scientific investigation he settled on a type of soup made from barley, dried peas, vegetables and sour beer as the cheapest way to give them nutrition.

had watched the storming of the Bastille from the windows of their home in the arsenal. At first, Antoine worked with the new Republic, helping compile systems and data for national bookkeeping (that is, a basic census). He was also put in charge of a programme to build a universal system of metres, litres and grams, and as funding was rather erratic he paid for much of this from his own pocket. But by 1793, the political situation was getting stickier and warrants for arrest were drawn up for the members of Lavoisier's company, the Ferme Générale. With others in the company, Lavoisier was accused of stealing money through corrupt tax collection, but he was also accused of adulterating tobacco (most likely entirely falsely, and possibly due to tobacconists he'd caught smuggling in the past conspiring a way to get back at him). There was a brief window of confusion when they arrested the wrong Monsieur Lavoisier and he hid in the by-then-deserted Académie, but he was soon put in front of a judge. It's often said he pleaded for his life, arguing for all the good he could do for France as a scientist, only to be told 'the Republic has no need of scientists' and sent to the guillotine. There's no evidence that this exchange ever actually happened, but Lavoisier was executed on 8 May 1794.*

Back in Birmingham, Priestley received a letter from his son William, which included two stones from the ruins of the Bastille along with an excited report of the revolution. Revolutionary rumbles were also building in Britain and Priestley was increasingly associated with them in the popular press. Following a banquet organised in sympathy with the French Revolution in July 1791, an anti-revolutionary mob started to grow in the centre of Birmingham. Priestley hadn't even been at the banquet, but was still declared the villain of the day and the clear target. His New Meeting House was first to go, the pulpit torn out, pews taken apart, cushions, books and

---

* Marie-Anne dedicated herself to canonising Antoine's memory. In 1805, she married Count Rumford, possibly hoping to recreate the scientific partnership she'd had with Lavoisier, but the marriage wasn't happy (after one fight she poured boiling water over his cherished rosebush) and ended after just two years. She kept running salons, like the dinners for scientists she'd hosted with Antoine, renowned for her mix of charm and direct rudeness.

wood all piled outside to make a massive bonfire. The Priestleys managed to hide with friends before the angry crowd hit their house, but before long the doors were broken down, the windows shattered, the library pulled to pieces and the precious laboratory smashed, including the lens Priestley had used to isolate oxygen six years before. The rioters were furious not to have found Priestley, some accounts suggest they'd brought a spit, planning to roast him alive. Furniture, books and papers were all piled up in the yard, with trees and shrubs in the gardens pulled out to make the bonfire higher, ready to burn an effigy of Priestley they'd cobbled together when they couldn't find the man himself. It took them a while to find a fire at first, going from house to house offering to buy a candle, but when they did, the whole place went up with an almighty roar.

Eventually the army arrived, but not for several days, by which time four dissenting churches and 27 more houses were attacked, many razed to the ground. Only 17 rioters were charged in the end – out of hundreds – and just four were sentenced. The official response was largely that the dissenters got what they deserved. It didn't help that many local magistrates and clergy were among the rioters. *The Times* published false information (which it later retracted), as did caricaturist James Gillray, suggesting Priestley and others had plotted the beheading of George III. After four days on the run, Priestley and his wife arrived in London, settling in Lower Clapton, Hackney. He taught for a while at the dissenters' academy there, but it wasn't easy. He was accused of poisoning the drinking water and seemed to be the scapegoat for pretty much anything that went wrong. Finally, in early 1794, some compensation for the riots came through. It was much less than he'd hoped for, but enough to buy passage to America. When President Jefferson was inaugurated in 1801, Priestley said that for the first time in his life he lived in a country where he approved of the leadership. He died in Northumberland, Pennsylvania, three years later.

The cosy space for experimenting and sometimes radical debate the Lunar men had enjoyed in Birmingham contracted. Even Boulton and Watt, who largely stayed out of politics, started arming their workers. The attack on Priestley might have been focused on a particularly radical member of the Lunar Society, but it was hard not to see it as an attack on their intellectual pursuits in general. Books had been strewn throughout the streets, with the message 'No philosophers – Church and King forever' chalked on walls. It was more than just elites pushing back against Priestley's call for change. Colonialism and the first waves of the Industrial Revolution had boosted the economy in towns like Birmingham, but they still had very high levels of poverty. Boulton had money, but that wasn't true for everyone in the city. Moreover, as the machines had got larger and harder, so had the work. Factory workers would put in 12-hour days in an environment thick with heat, noise and dust. Science had given a better life to Wedgwood and some of his middle-class customers, but not the majority of the population. Some looked to the revolution happening in France, but others wondered about a past golden age (mythical or otherwise) before the development of these newfangled gases and steam engines. Foreshadowing fights scientists would find themselves tangled up with in the future, Priestley's sometimes quite abstract science meant he was seen as out of touch with the working man and this feeling had been seized by a violent defence of the status quo.

★ ★ ★

In the summer of 1816, to hide from debt and scandal back in London, the poet Lord Byron rented a mansion near Lake Geneva with his friend and physician John Polidori. Another friend, the poet Percy Bysshe Shelley, was staying nearby, along with his future wife Mary Wollstonecraft Godwin and her stepsister Claire Clairmont. The weather was grim and the group had sheltered together indoors to tell ghost stories, from which came the basis of Mary Shelley's *Frankenstein* and Polidori's *The Vampyre* – the first vampire novel. In fact, the weather had been bad for a while, the so-called 'year without summer', and not just around Geneva either. In April 1815, Indonesia's Mount Tambora had exploded in one of the largest volcano eruptions in recorded history. Lava levelled the island, killing pretty much all plant and animal life, including all but 26 of the 12,000 people who lived there. The immense clouds of thick, black dust it belched out choked the area, settling into piles over a metre high. Much of this dust made its way into the upper atmosphere, where it circulated for several years. It's often credited as blocking out sunlight and lowering temperatures around the world, including over Lake Geneva the summer of Byron and the Shelleys' holiday. For that reason, *Frankenstein* is sometimes described as the first climate-change novel, although the context was global cooling, not heating.

Samuel Taylor Coleridge described it as 'end of the World Weather', and the gloominess so many people lived through has been traced not just by scientists spotting chemical markers of the volcanic dust in the geophysical record, but in poems, paintings and music of the time. The famine caused by this lack of sunshine not only contributed to food riots but, as horses died in large numbers, the invention of the bicycle. Another spin-off of this 'end of the world' weather was, arguably, the idea of ice ages. With that, we developed an awareness that the world's climate had been different in the past and might possibly be different in the future too; crucial for our later understanding of what we now know as our modern climate crisis. By 1818, a series of hard winters had caused a dam of ice to form over the Dranse River, a tributary of Lake Geneva, and young engineer Ignace Venetz was sent to investigate. He spent a night in June 1818 lying awake on the glacier, listening to the cracking ice, terrified it was about to crash below him. He survived, but the following afternoon it burst, flushing 18 million tons of water through the valley, shattering houses and killing 50 people. A warning from history, maybe, but more importantly for the development of climatology, it led to some correspondence between Venetz and local chamois hunter Jean-Pierre Perraudin. Perraudin knew the area well and had developed a theory about the giant boulders that could be found scattered in odd places, some teetering precariously on top of others, and occasionally of a very different type of rock from others around them. The standard explanation for these so-called 'erratic rocks' was that they'd been sent there by a giant flood (possibly the flood in the Bible) but Perraudin thought they must have been carried by ice.

As Venetz travelled to other valleys in Switzerland, he started to wonder if Perraudin's theory could be applied elsewhere. However, if glaciers had filled so much of the country, the climate must have been much colder. He told his friend Jean de Charpentier, director of the salt mines at Bex, who developed the idea further and in turn shared it with his friend Louis Agassiz. Agassiz was initially very dismissive of the idea, so Charpentier must have been slightly surprised when in the summer of 1837 Agassiz made a theory of past ice ages the topic of a high-profile lecture to Swiss scientists. Charpentier wasn't the only one raising an eyebrow. Botanist Karl Schimper had also developed research along these lines and, although he'd not got around to publishing it, had discussed the topic with Agassiz. But for most geologists at the time, the whole thing was pretty ridiculous. The Earth, they believed, was much more likely to be cooling over time, an ageing star, perhaps, gradually losing its fire. The audience at the lecture were shocked. They had been expecting a talk on fossilised fish.

Perhaps they shouldn't have been surprised. Agassiz had a tendency to throw around wild speculations on subjects he knew little about. But he was also not one to admit an error, and so doubled down and continued. Luckily this particular flight of fancy worked out. When Agassiz studied the subject in more detail, he found he could make a convincing case, publishing a seminal paper on glaciers in 1840. He travelled to the UK, extending this idea to mountain ranges in Scotland, the north of England and Ireland, and by September 1846 he set off from Liverpool to Canada. As soon as he landed in Halifax, Agassiz started to look for evidence of an ice age in the Americas. Benjamin Silliman Sr invited him to talk at Yale, where he excited students with his lecture on glaciers. In 1847 he was offered a professorship at Harvard. He spent the rest of his life there, founding the Museum of Comparative Zoology and marrying Elizabeth Cabot Cary (a natural history researcher in her own right and co-founder of Radcliffe College). The couple's grave is one of the Alpine 'erratic boulders' that had provoked the idea of ice ages in the first place.

Before we leave Agassiz, we should at least note the racism he folded into his science. In 1850, he commissioned the earliest known photographs of enslaved people, forcing them to strip for the images, to 'prove' the inferiority of Black people. In 2019 the descendants of two of the photographs' subjects sued Harvard for unspecified damages. In 2002, an elementary school just north of Harvard that had been named in Agassiz's honour was renamed following a campaign led by a ninth grader who felt the Agassiz legacy was one of hate and it was time for change. The school is now named after Maria Louise Baldwin, a leading Black educationalist who had worked as principal of the school in the late nineteenth and early twentieth centuries. We have no further use for Agassiz either. He popularised the idea of the ice age and that's all we needed him for. That piece of the puzzle completed; we'll move on to the greenhouse effect.

★ ★ ★

This part of our story starts in Auxerre, a small French town around 100 miles south-west of Paris. Today, we'd put the date as July 1794. But this was the French Revolution and they did things differently then, with a Republican calendar that both attempted to decimalise time and remove religious and royalist influences.* So it was their eleventh month, Thermidor, meaning heat, and a young man named Joseph Fourier was in a very hot position indeed. He was in jail, facing a trip to the authorities in Paris and, ultimately, the guillotine. It's hard to tell exactly how or why Fourier ended up behind bars. We do know he hadn't been that interested in the revolution at first, it seemed like an annoying distraction from the science he yearned to work on. From childhood he'd dreamt of becoming the next big thing in science, the Newton of heat, but as the orphaned son of a small-town tailor, he didn't have many options about how to get there. He tried being a monk, but the revolution dissolved the monasteries before he could take orders. He found a job teaching at his old school and, despite his initial prejudices that politics was all a distraction from science, found himself inspired by the hope of the new politics. As the years rolled on his enthusiasm waned a bit, especially when he heard of the execution of Lavoisier. He also lost some friends when he spoke out in support of

---

* It consisted of 12 months of 30 days each, with the five spare days (six in a leap year) dedicated to festivals. Stripped of religious and monarchist symbols, the new names of months were inspired by the weather, with references to fog, frost, snow, rain, wind, flowering, harvest, heat and fruit. The conservative British press took the piss, dubbing the new months: Wheezy, Sneezy, Freezy, Slippy, Drippy, Nippy, Showery, Flowery, Bowery, Wheaty, Heaty and Sweet.

the radical *sans-culottes*, which was possibly what eventually landed him in jail, although emotions were running extremely high in France by this point – a period known as 'the Terror' – with suspicion everywhere, so it's tricky to tell for sure.

One of the more romantic versions of the story has Fourier running off to Paris after the order was issued for his arrest, pleading eloquently with leader of the Terror Maximilien Robespierre for his release who nonetheless turned him away (this seems too neat an echo of the story of Lavoisier pleading for his life, so worth taking with a pinch of salt). Whether this is true or not, Fourier ended up back in the jail in Auxerre, convinced he was about to be executed. But the guillotine got to Robespierre first, and Fourier was released as part of an amnesty at the end of July. The fall of Robespierre offered Fourier more than just his release. It opened up a fresh chapter of his life. A new university, dubbed the *école normale*, was about to launch in Paris and Fourier was offered a prestigious spot as a student. It had ambitious aims: to repair the breakdown of higher education, bring together students from all over the country and unite all branches of contemporary learning. In practice, it was incredibly chaotic and closed after a few months. Still, Fourier impressed the tutors and was offered a place teaching at a military school. In March 1798, Fourier's hopes for a quiet life, building his reputation as the Newton of heat, were interrupted with a letter from the Minister of the Interior calling him to join a secret mission to Egypt, led by Napoleon Bonaparte. Keeping this secret was no mean feat; as well as 160 members of the scientific and literary commission Fourier was to be part of, it included 30,000 soldiers and sailors. They piled into a fleet of 180 ships, setting sail in the spring of 1798. The British, led by Horatio Nelson, knew something was up, scouring the Mediterranean for the French fleet. This didn't seem to bother Napoleon, however, who was too concerned with bouts of seasickness, invading Malta, and running the occasional intellectual soirée with academics like Fourier while he prepared to land in Egypt. After adventures that included finding the Rosetta Stone, losing it again to the British and setting up an Institut d'Egypte (ostensibly modelled on the Académie des Sciences back in Paris, but really a sort of think tank for Napoleon and his troops), Fourier found himself on the boat back to Paris in November 1801.

Fourier hoped he could return to his work studying heat, but Bonaparte had other plans. Fourier would spend the next decade and a half in Grenoble, running the Alpine region of Isère, near the Italian border. This was an onerous job, leaving very little time for science.[*] It was during his time there that Fourier started to really understand the problem of heat. He felt it physically too, not just as a question of maths and geophysics. His various trips with Napoleon had given him the chance to experience quite different climates. Moreover, somewhere along the way he had picked up an unshakable feeling that he could never keep warm. He'd been prone to sickness since his teens and may also have caught malaria in Egypt. In 1810, he wrote to the government in Paris to say the move from Egypt to the Alps had caused painful rheumatism, something he'd suffer from for the rest of his life. He was famous for wrapping himself in thick woollen clothes, never going anywhere without a large overcoat, even in the hottest weather, and keeping an extra jacket with a servant nearby just in case he wanted another layer.

In the summer of 1815, Napoleon was defeated at Waterloo and Fourier was finally free to move back to Paris. A former student working in local politics slipped him a cushy job as director of statistics, a post offering enough money to live on and, most importantly, enough spare time to work on science at home. Scribbling away in his new apartment in Paris, stoves raging to keep it as warm as possible but still wearing several coats, his arms poking out just enough to write, in 1822 Fourier published a book, *The Analytical Theory of Heat*. This was followed by a slightly less well-known 1827 paper on the temperature of the Earth and interplanetary space, which is where we can see the origin of the idea of the greenhouse effect. Considering the size of the Earth and its distance from the Sun, Fourier deduced it should be a lot colder than it was. It was cold, granted, he was wearing two coats, but surely it should be much, much colder still? Fourier wondered if the

---

[*] He made do, putting on scientific experiments as local entertainment. He also managed to sneak some time away to write a report of his time in Egypt, assuring the Ministry of the Interior that the rumours he had taken an illegal holiday were entirely false and he was doing important work for Napoleon (although he was probably doing some heat maths on the side too).

atmosphere worked a bit like the lining of Saussure's heliothennometer, a layer of insulation between the Earth and interstellar space. He didn't use the words 'greenhouse effect', but when scientists came to develop that concept in later decades, it was Fourier's idea of a solar oven Earth they would credit.

★ ★ ★

The next stage in this story was the realisation that adding particular gases to this atmosphere might change the temperature of the Earth. As we saw in the introduction, it was Eunice Foote who in 1856 first spotted the heating power of carbon dioxide. However, no one really paid much attention to her and when scientists later picked up on carbon dioxide's role in climate change, it would be John Tyndall they'd cite. Working in London around the same time as Foote, Tyndall published a paper with very similar results a few years later in 1861. There's no evidence Tyndall read Foote's work, and Tyndall expert Roland Jackson feels it would have been very out of character for him to deliberately conceal knowledge of someone else's work. Still, Tyndall was an editor of the London-based *Philosophical Magazine*, which republished a paper by Elisha, Eunice Foote's husband, from the *American Journal of Science and Arts*. The husband and wife's papers had been printed next to each other in the American journal, so whoever was on editing duty at the *Philosophical Magazine* couldn't have missed Eunice's while picking out Elisha's. It's quite plausible Tyndall was that editor, and simply spotted the female name and skipped right over to the next page. He wasn't known for his respect for women's intellectual abilities. Or maybe he read it and though he largely forgot it, it planted something. Other scientists in Germany were also playing around with similar experiments, so it's also reasonable to assume Tyndall made these discoveries on his own.[*]

---

[*] Jackson has published a couple of papers on Foote and Tyndall – you can find full references for them in the source section at the end of this book. Jackson has also written a great biography – *The Ascent of John Tyndall* (2018) – which mixes detailed scholarship with clarity and warmth, and has been a key source in writing this book. Read to the end, you don't want to miss the story of Tyndall's death.

Tyndall was born in County Carlow, Ireland, in 1820, his parents both hailing from reasonably prosperous landowning backgrounds, but disinherited by their own parents who disapproved of the marriage on religious grounds. So although Tyndall inherited a bit of cultural capital, he was low on ready cash and had to make a living for himself. His route into science wasn't exactly smooth and came via a job for the Ordnance Survey, evening classes at a mechanics institute in Preston, and then teaching. The school he taught at was Queenwood College in Hampshire, one of the first in England to follow Priestley's lead and offer practical laboratory classes in science. Tyndall quickly forged a close friendship with science teacher Edward Frankland. The two of them would spend long hours in the lab, occasionally getting high on their concoctions, sometimes with the students. But Tyndall was soon frustrated by the school, taking particular offence to the headmaster's wife's involvement in the management of the school. In the summer of 1848, Frankland and Tyndall took a trip to Europe and discussed the possibility of moving to Germany to study. Frankland knew Robert Bunsen (of burner fame) at the University of Marburg. Frankland could study ozone there and Tyndall would be allowed to study for a PhD without having previously attended university. By the summer of 1851 Tyndall was back in London, his PhD in hand, along with a burgeoning scientific reputation. Still, it was hard for a man without independent wealth to build a career in science. He had to return to teaching to clear his debts and thought about moving to Canada or Australia. But after some canny networking and a dollop of luck, in 1854 he landed a post at the Royal Institution in Mayfair, London.

One of the many friends Tyndall had made while trying to land a science job in London was Thomas Henry Huxley. Like Tyndall, Huxley's parents were members of the middle class who had fallen on hard times, and he also needed to earn a living if he was going to work in science. Today, Huxley is best remembered as Charles Darwin's 'bulldog' for his defence of the theory of evolution, but he was a high-profile Victorian scientist in his own right and central to the development of scientific education in nineteenth-century Britain (as well as to views on race and brain size, which has recently led to his name being removed from memorials). Huxley was also the force behind something called the X Club, a group of scientific friends who met monthly for dinner at a hotel just along from the Royal Institution. They gave themselves X-themed nicknames like Xalted Huxley and Xcentric Tyndall, the X apparently chosen because it committed the members to nothing in particular. Still, they scheduled their dinners just before meetings at the Royal Society, with an eye on influencing the scientific establishment, and many went on to play leading roles in Victorian science. In terms of our story, it's a good example of the sorts of networks Foote couldn't join, and it must have felt especially important for Huxley and Tyndall who lacked the sort of powerful networks many of their colleagues had simply inherited. Tyndall loved the club's meetings and rarely missed one.

It was also Huxley who was responsible for Tyndall's love of the Alps, which eventually led to his interest in what we now call greenhouse gases. Tyndall had been puzzling over questions about the formation of slate rock one evening at the Royal Institution in the early summer of 1856 and Huxley suggested further study of glaciers might help him understand the problem. Frankland and Tyndall had been planning a trip to Switzerland anyway, so added a stop in the Alps with Huxley to their itinerary. When they arrived, Huxley got a couple of donkeys to help Tyndall, himself and his then heavily pregnant wife up the Wengen Alp. It was so stunning Tyndall couldn't sleep: 'I never saw anything so beautiful,' he declared. 'The white cone so high up in heaven covered with snow of perfect purity.' At dawn, he scrambled up to the Guggi glacier. Startled by the ghostly rattle of tumbling ice debris, he started to take samples from the fissures and the banded structure of the ice. He was hooked, puzzling over the material structure of it all.

Back in London, Tyndall continued to be fascinated by the glaciers, wondering how they might have been formed; their veins and how flower-shaped structures could be formed with a focused beam of light. He'd give lectures at the Royal Institution on the formation of glaciers, moulding ice under pressure on stage. He battled writing a book on the topic and it was as he was finishing this off in the spring of 1859 that he started to develop his interests in the absorption of heat. He'd been pondering broader questions of heat for years, riffing off the ideas of Priestley, Lavoisier and Count Rumford, and the glacier work rekindled this interest. He developed new equipment that allowed him to measure even tiny differences in the heat of gases that had previously been undetectable. For weeks, he puzzled away in his lab in the basement of the Royal Institution, exploring a range of different gases. When it came to oxygen and nitrogen, the main gases in our atmosphere, he noticed infrared radiation went right through them. But when he tried the coal gas that provided the lighting for his laboratory, the heat got stuck. He presented this work at one of the Royal Institution's Friday evening presentations, chaired by Prince Albert, in June 1859, and continued to study the topic, exploring whether chlorine and bromine might trap heat in the same way. There's even a story about him drinking ale and brandy with his lab assistants and exploring the thermal qualities of their alcohol-tinged breath. In February 1861, he submitted a paper to the Royal Society, putting his study of the heat absorption of gases in the context of ideas of the ice age, suggesting that even slight changes in the chemical composition of the atmosphere might have led to different temperatures in the Earth's past.

He also wrote it up as a bestselling popular science book, *Heat Considered as a Mode of Motion*. The science writer Mary Somerville was delighted, exclaiming: 'Although the quantities of vapour are minute beyond imagination, the experiments give the conclusion that they and their molecules have as true an existence as the objects around us.' The studies of gas and heat by people like Priestley and Lavoisier were starting to come of age and were opening up an entirely new way to see the world in the process. It must have all felt quite wondrous. And yet, in all the public discussion of Tyndall's work, no one seemed to think to talk to him about Foote's very similar conclusions. Even if Tyndall hadn't read the paper himself, surely someone would have sent it to him? There was Joseph Henry for one, the first Secretary of the Smithsonian who read Foote's paper at the 1856 AAAS meeting. He later swapped letters with Tyndall, but never thought to bring up Foote's paper. Plus, as Tyndall expert Jackson points out, Tyndall had made a few scientific enemies over the years;

you might expect them to have pounced on a chance to shout plagiarism, but they didn't.

Today, Tyndall is celebrated as a founding father of climate change science, with a major British research centre named after him. He'd been well known in his day, forgotten for a large part of the twentieth century but remembered in recent decades, as climate change started to bite and his work seemed, in retrospect, so visionary. In contrast it took until 2016 before anyone thought to remember Foote. It is worth remembering, however, that it is only with our modern eyes that either Foote or Tyndall seem important to a story that we now look back on as the discovery of global warming (the same is very much true for Fourier too). Still, Foote and Tyndall spotted something interesting about heat and carbon dioxide, and both connected this to the heat of the Earth, which is pretty remarkable. Neither had any clue that humans were managing to emit enough carbon to create global warming and they certainly wouldn't have thought it was going to be dangerous.

★ ★ ★

Something that might almost be considered an early warning came via Swedish scientist Svante Arrhenius. Arrhenius is most famous for winning the 1903 Nobel Prize for chemistry for work on salts and helping establish the field of physical chemistry. More importantly to this story, in 1894 he married his student Sofia Rudbeck and, when a few years later they went through a stressful divorce, Arrhenius found temperature calculations a soothing distraction. Plus, his friends at the Stockholm Physics Society were really interested in something called 'cosmic physics', which attempted to bring a study of the seas, atmosphere and solid Earth into the domain of the physical sciences (and involved fun new ways of collecting data like balloon trips to the North Pole). They'd been arguing over the causes of ice ages, and Arrhenius picked up Fourier's work on terrestrial heat along with Tyndall's work on carbon dioxide to see if that would give him something new to present to the society. He also had some 1880s research by American astronomer Samuel Langley, who'd developed new apparatus to investigate solar radiation while trying to work out the temperature of the moon and described the atmosphere as acting 'like the glass of a hot-bed' (in other words a sort of small greenhouse).

For months, Arrhenius scribbled away with his pencil calculating the atmospheric moisture and radiation entering and leaving the Earth for each zone of latitude. This was not necessarily the sort of work that, at the time, was seen as a sensible thing for a scientist to be devoting their energies to. The data he based his calculations on wasn't even that rigorous and he simplified the climate system immensely. But it wasn't necessarily meant to be scientifically significant. It was just something to do instead of worry about his divorce and a way to pick up an idea to throw around with friends, which, if it looked promising, could be studied properly later. Like many of his contemporaries, Arrhenius was more worried temperatures would fall rather than rise, so started by working out the impact of halving the amount of carbon dioxide in the air. It would, he found, cool the world by 5°C – enough to bring on another ice age. His colleague Arvid Högbom had been studying the way volcanoes added carbon dioxide to the atmosphere and suggested he might consider it the other way around too: what if carbon dioxide was added to the atmosphere? Arrhenius ran the maths again and calculated that a doubling of atmospheric carbon dioxide would raise the Earth's temperature by 5°C or even 6°C.

He presented a paper at the Royal Swedish Academy of Sciences at the end of 1895 and the following spring it was translated into English in the *Philosophical Magazine*, making its way to the UK and US. Arrhenius didn't use the word 'greenhouse', although he did cite Fourier to suggest the atmosphere functions a little like what he called a *drivbänk* (again, a sort of mini greenhouse). At first, he wasn't talking about anthropogenic global warming – the more modern idea that humans cause climate change with excess carbon emissions. It wasn't until right at the end of the century, in 1899, when another colleague, Nils Ekholm, pointed out that if people kept burning coal at current rates, we might hit that doubling of atmospheric carbon dioxide sometime soon, at least in a few centuries. The idea that humans might change climates wasn't new, Prussian explorer and scientist Alexander von Humboldt had warned back at the start of the nineteenth century of the devastating impacts of deforestation.* Similar ideas could be traced back to antiquity and by the turn of the century were being talked about in a variety of contexts, from the desiccation of land in Turkestan to ideas coming out of America that people might create rain by deliberately setting fire to forests. But such debates were always centred around local climates, changing the atmosphere of the whole of the Earth was something else.

Arrhenius included this observation in his 1908 book *Worlds in the Making*, noting that 'the slight percentage of carbonic acid in the atmosphere may by the advances of industry be changed to a noticeable degree in the course of a few centuries'. Still, he wasn't too worried. As historian of climate science Spencer Weart argues, like most prosperous nineteenth-century Europeans, he tended to see technological change as a good thing. If it caused new problems, the scientists and engineers would create new ways out of them. Moreover, soon enough there was lab work – by another Swedish physicist, Knut Ångström – which seemed to refute the idea, arguing that water vapour trumped carbon dioxide's warming power when you examined it up in the atmosphere. Cylinders of carbon dioxide sitting on Foote's windowsill were one thing, the larger and messier system of the greenhouse effect in action was another. Other scientists argued convincingly that the oceans would soak up the carbon, that volcanic dust was the main problem or that clouds would reflect the sunlight back into space. Leading American geologist Thomas Chrowder Chamberlin had initially been excited when he read Arrhenius's paper in the *Philosophical Magazine*, incorporating it into his work on ice ages, but as criticisms of the carbon dioxide theory stacked up he repeatedly expressed regret that he'd ever been taken in by it. People thought Arrhenius was just a clever guy playing around with numbers to see what would come out. Climate calculations weren't his only side project – he also had ideas about volcanoes, immunity and all sorts. When he got into

---

* After a lot of thought I decided Humboldt's a bit of a side character in the climate crisis, so I've largely left him out of this story. He's a fascinating figure in the broader history of ideas of nature though. For a great read on the topic, see Andrea Wulf's beautiful book *The Invention of Nature* (2015).

electricity, he ran experiments to see if it could help children grow, placing them in classrooms with wires carrying high-frequency alternating current. Arrhenius's book *Worlds in the Making* was mainly famous for his suggestion that life on Earth started with seeds being transported from interstellar space by the pressure of light. In a decade or so most scientists had pretty much thrown out Arrhenius's work on climate change entirely.

CHAPTER THREE

From Whale to Shale

Long before we built offshore rigs for fossil oil and gas, we mined our seas for whales. We grew a massive, multinational industry to slay and strip these incredible sea monsters, using energy stored inside their flesh to light our houses, streets, factories and shops. And then we stopped. This story is sometimes used for inspiration for moving away from fossil fuels; we kicked one bad energy habit, so we can again. As we'll see, the truth's a bit more complex. Still, whales are part of the climate crisis story. They remind us that humans' problematic relationship with energy is bigger than simply fossil fuels. They also set the scene for how the modern oil and gas industry would develop.

Humans have hunted whales for thousands of years, but it was Basque sailors in the seventh century who really organised it into a business. They hunted whales around the Bay of Biscay, finding a steady market in Christians who had concluded whale flesh didn't count as meat (and so could be eaten on one of their many holy days). As stocks dwindled and their sailing skills developed, they followed whales north, up into the Arctic and further west too. Basque whalers possibly landed in the Americas before Columbus – they just kept quiet about it, worried their French and Dutch competitors might also discover all the good whaling in the area. In 1607, London-based traders the Muscovy Company sent an explorer to find a trading route through the Arctic to Asia. The company had been founded a few decades before as the Mystery and Company of Merchant Adventurers for the Discovery of Regions, Dominions, Islands, and Places Unknown. After establishing a trading route to Moscow, it shortened its name but kept a drive for exploration and was keen to find another route up and over to the West. Like many other similar trips to find the fabled 'Northwest Passage', they failed.* However, they did bring back reports of large numbers of whales in the area and were granted a monopoly on whaling by their king, guarding their find from competitors with heavily armed whaling ships. The Dutch responded by claiming a whaling monopoly too, backed up with their own heavily armed ships. Eventually the English and Dutch divided the whaling grounds between them rather than risking war, thinking little about the claims people living in the area might already have.

We don't know exactly when the colonialists started to boil whale blubber for oil, but it was commonplace by the 1630s. Not long after the *Mayflower* pilgrims landed, they saw Native Americans 'busie about a blacke thing' and figured they could use the blubber for lighting, just as they had with the fat of cows and pigs back home. Initially, they stuck to the whales they found beached on the shore and shared the oil among the local community. However, it wasn't long before both the church and government back in London were demanding a cut. The market grew larger than the beached whales alone could provide, so people started taking to boats to hunt. These finds out at sea were squarely for private profit. Private traders also realised they could send oil to be sold in England, using the ships to bring back European goods on the return trip across the Atlantic. As Eric Jay Dolin describes in his excellent (2007) book on the American whaling industry, it was a competitive business, with oil producers trying to get the edge on each other with the highest quality whale oil candles all packaged in coloured tissue paper with carefully engraved labels. Taunton, Massachusetts, was even home to a proto oil cartel at one point, as makers of whale oil candles banded together to fix prices and stop potential rivals building a business. This group was riven with infighting and ended during the American Revolution,

---

* Most famously was a trip led by Sir John Franklin in 1845. After Franklin disappeared, taking more than 120 other men down with him, he was speedily martyred by the British for his contribution to science and exploration, the trip framed as all about knowledge, not trade, glossing over the failure and ignoring the many Inuit witnesses who had seen the survivors resort to cannibalism). As Arctic waters have warmed in recent years, trading routes through the Arctic have opened up and the remains of Franklin's mission were finally found in 2014 (and analysis suggests the Inuit were right about the cannibalism).

but it was one of the earliest American monopolies, foreshadowing actions of the fossil oil industry in the twentieth century.*

Whale oil is not very thick, thinner than olive oil, and it varies in colour from a bright honey-yellow to a dark brown, depending on the species and way it is processed. Initially, whalers brought their catch back to port for rendering – the process for boiling down blubber to collect oil – but around 1750 they realised they could do the business on board, allowing them to travel further and further from land. Building a brick-lined furnace into the centre of the main deck of the ship, they effectively turned their boats into floating factories where they could cut, render and store whale oil all in one place, throwing the waste carcass overboard when they were done to make room for the next catch. Although the key product was oil, whalers would also harvest baleen, the fringed part of the whales' mouths used to filter fish (also known as whalebone, although it's keratin, like our fingernails, rather than bone). Before the development of synthetic plastics, baleen's flexibility and strength made it a popular material for corsets, collar stiffeners and umbrellas. From the early-eighteenth century onwards, whalers off the coast of Nantucket, Massachusetts, also started to hunt sperm whales, harvesting the cloudy substance found in their head known as spermaceti, which was especially prized as a good burner.†

By the time the American Revolution broke out in the mid 1770s, several hundred whale ships were running out of colonial ports, with a cumulative cargo capacity of 33,000 tonnes. The war cut off the supply of whale oil to Britain, so the British government offered £500 a pop for British merchant sailors to shift to whaling. A new generation

---

* As well as Dolin's *Leviathan*, Philip Hoare wrote a (2008) book of the same name, which lyrically tells the story of the whale industry, putting it in a slightly wider context and weaving in some of the attempts to curtail the whaling industry in the 20th century. Both are brilliant reads, and key sources for my research.

† This isn't sperm; someone several hundred years ago just thought it might be, hence the name. No one's quite sure what spermaceti does for the whale exactly; it might help control buoyancy, or maybe they use it to echolocate. It's more of a wax than an oil, and glows especially brightly when burnt. After catching a sperm whale, they would haul it on deck, cut off its head and bail out the spermaceti using a bucket.

of British whalers like William Scoresby (the inventor of the crow's nest) grew a trade working out of the Arctic, in and out of Scottish and northern English ports like Whitby, as well as London. In 1788, Samuel Enderby, acting on information from James Cook, sent out the first British custom-made sperm-whaling ship into the Pacific, opening an office in Sydney a few years later. An established trader, Enderby's ships had previously sent tea to the Americas and brought back whale oil (he's mentioned in *Moby Dick* and sometimes said to have been involved in the boats at the centre of the Boston Tea Party). This time around he'd fixed up a deal with the UK government to take convicts on the outward trip to Australia.

The American whale oil industry continued to thrive and, despite this new competition, the market by this point was large enough to accommodate a range of players. London alone had 5,000 whale oil streetlamps, all hungry for fuel. At the centre of the American whale industry was the port of New Bedford, Massachusetts, known as 'the city that lit the world' for the whale products it exported. In its heyday of the late 1850s, the New Bedford harbour was home to more than 300 whaling vessels, bringing in over $12 million a year, making it the richest city in the US. The complex comings and goings of whale products offered a good hiding place for escaped enslaved people – possibly facilitated by the fact many whalers were Quakers, opposed to the slave trade and happy to turn a blind eye – and New Bedford became a vital stop on the Underground Railroad.

There's a version of this story that goes 'and then we found oil in the ground, and the whales lived happily ever after'. Indeed, there's a superb *Vanity Fair* cartoon from 1861 of a grand ball hosted by the whales in honour of the discovery of rock oil wells in Pennsylvania. Waited on by frog butlers and surrounded by banners proclaiming 'oils well that ends well', the whales dance and hold champagne flutes, dressed up in white tie and ballgowns (presumably without whalebone corsetry). It's a tempting spin on the story, especially for those who want to work a campaign line: 'we dropped one barbaric oil industry, we can do it again'. But the truth of technological transition is rarely so simple. There was the American Civil War, for one thing, which disrupted the whale trade. Plus, after so many decades of aggressive hunting, whales were harder to find around America. As we'll see at the end of this chapter, if anything fossil fuels led to the killing of more whales. Moreover, in the short term, the gas industry would compete for the lighting market, and to trace that we have to go back to the pneumatic scientists we left in the last chapter with their discoveries of carbon dioxide, oxygen, hydrogen and nitrogen.

★ ★ ★

At first, these new gases chemists were pulling out of the air were largely seen as fun and games. There was Joseph Priestley's fizzy water, Tiberius Cavallo's soap bubbles and a new take on fireworks too. Dutch instrument maker Charles Diller first presented what he called 'philosophical fireworks' at the Pantheon in Paris in 1787. He'd send a variety of airs from bladders through copper pipes, which he'd then manipulate to display flames in the shape of suns, stars and geometrical figures. He promised fireworks without the explosions or smoke (although they did stink a bit), along with some cutting-edge scientific commentary too. Diller presented his fireworks to royalty, scientific academies and public theatres across Europe and the US, and his coloured gas flames soon became a theatrical science-themed sensation – not as popular as automata or mesmerism, but a sensation nonetheless.

Perhaps the most exciting application of these new gases, however, was the prospect of flying. In the summer of 1783, Benjamin Franklin had excitedly written to Joseph Banks, president of the Royal Society, with news from Paris: chemist Jacques Charles was filling large silk bags with inflammable air and promising he'd soon be able to lift people up in them too. He launched his first balloon from the Champs de Mars, just south of the spot where you'd now find the Eiffel Tower. It amazed onlookers as it rose so high it was lost in the clouds, before it crossed the Seine, landing a little over 20km away in the village of Gonesse. There, the understandably terrified local people thought it was a demon from the sky, attacking it with pitchforks and knives. The Montgolfier brothers followed, this time simply heating common air. They'd be the first to manage human flight too, with a spectacular launch of a monster 70ft-high blue balloon decorated with pictures of mythical creatures. This carried two men over the Paris skyline for just shy of half an hour, narrowly missing a couple of windmills. Ten days later, Charles made the first assent in a hydrogen balloon, attracting the largest crowd seen in pre-revolutionary Paris. This 30ft pink-and-yellow balloon carried a chaise lounge covered in bunting, upon which travelled two men as well as a host of scientific instruments including a thermometer, barometer and telescope (and a couple of bottles of champagne).[*]

Ballooning, like Diller's gas-fireworks, often used spectacular public displays not just as entertainment in their own right, but as a way to show off science. As with the gas-fireworks, the demonstration of new scientific ideas was seen as part of the fun. But the early plans of scientific ballooning were soon displaced by military ones. It was a way to spy on your enemies – an eye in the sky – and also a psychological weapon, a giant globe ominously looming over the target. Lavoisier at his arsenal labs had found a way to produce hydrogen reasonably cheaply, passing water over red-hot iron, supplying a new French military academy for ballooning. This pioneered the 'mile-high club' (as young recruits to the new balloon regiment would invite local girls for rides in their balloons), but not a lot else. Napoleon took balloons on his 1798 campaign to Egypt, but they were destroyed before they could be deployed. Despite a flurry of plays, poems and cartoons in Britain that imagined an airborne invasion led by Napoleon, it wouldn't be until the First World War that they'd be attacked from the air (by Zeppelins). Still, it was the start of air-based warfare, which a little over a century later would drive aviation innovation and suck up so much oil with it.

The British largely dismissed ballooning as a fanciful activity of the French. Oxford baker James Sadler was keen to explore more though. He'd built a small laboratory out of the back of his shop, and conducted a series of balloon flights not only from Oxford, but later Bristol, Cambridge, Hackney and Dublin too. He failed in various attempts to cross the Irish Sea, but did publish a vivid account of his efforts, calling for more support for scientific ballooning, arguing it

---

[*] Richard Holmes provides a delightful and in places truly thrilling account of these stories and others in the history of ballooning in *Falling Upwards* (2013). Make sure you read to the end – the final chapter is especially good.

could teach us more about electricity, magnetism and chemistry, and also 'throw light on the obscure science of meteorology'. His son William joined him on several trips and even managed the Irish crossing. However, after William died tragically in 1824, his balloon tangled in a chimney in the Pennine hills, and James never set foot in a balloon basket again.

One of Sadler's supporters was Thomas Beddoes, reader in chemistry at Oxford University. A radical in science and politics very much in the vein of Joseph Priestley, Beddoes felt stifled and bored in Oxford, both scientifically and politically. He employed Sadler to improve the university's labs, but still itched for more. When, in 1792, his radical politics earned him a spot on a government list of British Jacobins, the university asked him to leave and he was more than happy to go. But what next? Priestley had started to unravel the role of oxygen and carbon dioxide in our bodies, maybe taking some of these or other 'fractious airs' might be beneficial for health? Beddoes hoped to cure cancer, or failing that tuberculosis, and started to dream of a new pneumatic clinic, one that could provide healthcare for the masses, covering the cost with the sale of portable gas-inhaling equipment to aristocrats. He enrolled Sadler in the scheme, moved to Bristol and started looking around for investment.[*]

Beddoes had links with the influential Lunar Society set. He married one of Richard Edgeworth's 22 children, Anna, and befriended Thomas Wedgwood, youngest son of Josiah. An accomplished scientist in his own right, Thomas had made use of his father's incredible home laboratory and was an early pioneer of photography. He was also in very fragile health, with stomach pains and headaches that would confine him to a darkened room for days. Exhausted after a trip to check out the revolution in France, he took treatment from Beddoes and would become one of the pneumatic clinic's most dependable financial supporters. The clinic needed gas-inhaling equipment though, and Beddoes' ambitions outran his and Sadler's skills. They could administer airs, but it was hard to

---

[*] Mike Jay's *The Atmosphere of Heaven* (2009) provides a very readable history of this pneumatic clinic and Beddoes as a rather unlucky visionary way ahead of his time.

regulate pressure and purity. He appealed to James Watt Sr, who by this point was in his late 50s, semi-retired, tinkering away in a large attic lab he'd built in his home north of Birmingham. Watt's teenage daughter Jessie was sick with tuberculosis and her brother Gregory was showing signs of it too. Watt had started to obsess about the disease and sent Jessie to Beddoes' clinic for treatment with the airs. It didn't do much good and by June she was dead. Looking for distraction from his grief, Watt set his mind to Beddoes' equipment problem, and within a month had designed and built a device for extracting, washing and collecting poisonous and medicinal airs. Unusually for the careful businessman, he let Beddoes know he wouldn't apply for patents or make a claim for copyright.

Beddoes also employed the skills of a young apprentice apothecary from Penzance, Humphry Davy, who had honed his chemistry skills with Tom Wedgwood. With an inhaling device that drew on both Sadler's balloons and Watt's engineering expertise, Davy tried out a variety of airs, first on himself and then on patients. A turn with carbon monoxide nearly killed him, but the most significant discovery was the effects of nitrous oxide. When Davy first inhaled this gas, he started leaping and running around the lab, intoxicated. Taking the gas soon became a regular part of Davy's routine and he'd often be seen carrying a green silk bag of the stuff around with him on walks at night by the Avon, possibly as a distraction from a love affair with Beddoes' wife Anna (or at least his passionate desire for one). At the same time, Davy was increasingly aware that his association with the political radical Beddoes could damage his long-term career plans. And aside from fun with nitrous oxide, the airs didn't seem to have much medical use. Davy caught the eye of a group of scientists in London who were planning a new Institution for Diffusing Knowledge (later known as the Royal Institution). When Count Rumford asked Davy if he wanted to be a chemistry lecturer at this new institution, Davy leapt at the chance. We'll return to Davy in Chapter Five. For now, pneumatic medicine might be a dead end, but the Watt family hadn't finished with systems for holding and moving gases around.

★ ★ ★

In late 1801, James Watt's younger son Gregory visited Paris and wrote home with a warning: 'Tell Murdoch a man here has not merely made a lamp with gaz [sic] procured by heat from wood or coal but that he has lighted up his house and garden with it and has it in contemplation to light up Paris.' The Murdoch he was referring to was an engineer at Boulton & Watt, William Murdoch, who had also been playing with gaslighting systems; first putting one to use in his cottage in Redruth in Cornwall, and later moving to the firm's headquarters in Birmingham for further development. Murdoch used a device similar to the one Watt had built for Beddoes, but it ran on coal gas and you set fire to it rather than breathing it in.

The gas lamp Gregory had spotted was the work of Philippe Lebron. Born in the Champagne region, he had enrolled in the French civil engineering corps and started investigating uses for combustible materials including lighting. In 1798 he wrote a paper on the topic and applied for a patent for a device he called the Thermolamp. He'd initially tried to get the government to buy into the project for city street lighting, before turning to public subscription. He leased a hotel in an upscale neighbourhood and installed two of his lamps – one to light five rooms inside, the other for the gardens – and advertised demonstrations in newspapers at three francs a visit. A large crystal globe sat in the centre of the room where a fire would usually glow, giving off a bright and clear flame as well as some heat. Tubes ran around the room connected to candlesticks fitted with metal candles. Lebron promoted it as a labour-saving system – as you wouldn't have to lug coal or wood around the house – and the advertising material promised 'no more sparks, cinders or soot; no more heavy buckets'. Demonstrations were popular, attracting thousands, but sadly few were inspired to invest. As one German newspaper report noted, it smelt quite bad. Lebron sold at least one, but nowhere near enough to run the project further and instead moved into the tar business.[*]

---

[*] Lebron died in 1803, still in his mid 30s. It's sometimes said he was murdered in a dark street in Paris, used to make a point about how the city should have invested in his innovative street lighting after all. However, as Leslie Tomory notes in his book on the origins of the gaslight industry – *Progressive Enlightenment* (2012) – there is no evidence to support this and he more likely died of some disease.

Inspired by Gregory Watt's reports of gaslighting in Paris, Boulton & Watt added a public display of their gas lamps to a celebration they'd already planned to mark the signing of the Treaty of Amiens in March 1802. Britain's war with Napoleon might be over for the time being, but Boulton & Watt would be damned if they'd be beaten to the gaslight market by a French inventor. Two large gaslights, each in copper vases, one at each end of the factory, sat alongside hundreds of coloured oil lamps, candles, fireworks and three Montgolfier balloons. It wasn't long before they had a customer too; the Phillips and Lee cotton mill in Salford, one of the largest in the country. Like many factory owners, George Lee was keen to be able to keep his workers going long after dark without the cost, cleaning and danger of lamps that ran off animal fat. By 1807, he extended the contract to include not just the whole Phillips and Lee factory, but his personal home and a private road too.

Boulton & Watt had more competition at home in the UK too. In 1805, James Watt Jr wrote to Matthew Boulton Jr with reports of gaslighting spotted in some shops in Glasgow. Murdoch's assistant Samuel Clegg went freelance and built a gaslighting system for a cotton mill in Halifax too. If Boulton didn't at this point wish Watt had been more careful with his gas patents, he would soon. There were also more and more reports of gaslighting in London. The Golden Lane Brewery, a co-op on the edge of the City of London run by publicans, had installed coal-gaslighting inside its buildings and on streets nearby. It used what is thought to have been a Boulton & Watt system, again possibly adapted by one of the firm's staff working freelance.

Then there was German entrepreneur Frederick Albert Winzer. He'd been inspired by Lebron's Paris display and thought he'd try his luck selling the idea in London. Reinventing himself as Fred Winsor, he launched a vigorous PR campaign with displays at The Lyceum (the same theatre where Diller had presented his gas fireworks not long before), as well as pamphlets claiming gas would alleviate pulmonary diseases and the plight of unhappy chimney boys while bringing investors enormous profits. In 1806, he leased a house on Pall Mall, close to gentlemen's clubs and the official residence of the king at the time, St James's Palace. He ordered more than 5,000ft of pipes and started laying them under the streets. At first, the London establishment were disdainful of gaslighting in general and Winsor in particular. Davy impishly suggested they could use the dome of St Paul's Cathedral as a gas holder. The *Anti-Jacobin Review* dismissed it as simple quackery. There were also concerns that replacing whale oil would threaten the security of the nation; the British whaling industry had become a useful source of recruits for the Royal Navy. Still, *The Times* was more supportive and the Prince of Wales agreed to have his new residence built just off Pall Mall fitted out with Winsor's new tech. A group of prospective investors smelt profit, and in July 1807 met in the Crown and Anchor pub on the Strand to found a new company. Maybe to assuage concerns gaslighting might damage the navy, they named it the New Patriotic Imperial and National Light and Heat Company, later simplified to the Gas Light and Coke Company or the GLCC.

Boulton & Watt, worried all their efforts to grow a gaslighting project were about to be gazumped, started to marshal their lobbying power. They put Murdoch up for a high-profile Royal Society medal that had been sponsored by Count Rumford a few years before. This, they figured, would help build Murdoch's reputation as the great inventor and also underline the image of Winsor as a bit of a quack. While Murdoch read an account of his gas experiments to the Royal Society, Winsor's works were dismissed as a mere 'puppet show', more akin to Diller's fireworks than a serious way to light the city. The GLCC had their own lobbying skills, however, promising the Houses of Parliament gaslighting for free. Eventually, to Boulton & Watt's fury, the right to dig up London streets was given to the GLCC, effectively a monopoly on the London gas business. The GLCC's first paying client was St John's Church on Smith Square, which lit its first gas lamps at the end of November 1813, the gas piped from the world's first gasworks on Horseferry Road.

Still, it wasn't exactly smooth running for the GLCC. The hype Winsor had provided got things started, but he was now becoming a bit of a liability, so the GLCC pushed him out. The product wasn't quite up to scratch yet either. Parishioners at St John's Christmas services that year had to cough through several early experiments of different types of gas designed to try to keep the smell to a minimum. There was an explosion at the gasworks too, and Samuel Clegg (Murdoch's assistant who'd gone freelance and was by now working for the GLCC) was badly burnt. The chance of some good PR came with a celebration planned for the summer of 1814 to mark the fall of Napoleon, led by Tory MP, inventor and artillery chief Sir William Congreve. Pall Mall would be lit up with gas, and there'd be a large firework display in St James's Park, complete with a gaslit pagoda. Plans were going well, with pipes laid throughout the park and the lamps up and running two days before the show. However, one of the rockets set off by Congreve for the event hit the pagoda; burning, it collapsed in the lake, killing two people on the way down.

The GLCC was terrified of the PR backlash, and to this day there's a plaque in the park blaming the company, rather than Congreve's rockets, for the disaster. Still, it didn't seem to do much damage. Gaslight was bright and unlike whale oil it didn't stink of fish; it soon caught on. As well as letting factory owners keep workers on after dark, it opened up evening leisure time, at least for those who could afford it. The new tech allowed more sophisticated and less intrusive lighting in theatres, and window-shopping emerged as a pastime too, rather than just a necessity, as shops could light up their window displays and be open after dark. People also felt safer with this brighter light. As industrial cities swelled and the smoke pollution grew thicker, people were more concerned about crime. As Parliament sat through a series of committees about whether or not they should set up a police force for London, gaslighting seemed like a quick way to make the streets safer. Before long, the GLCC was lighting up Westminster Bridge, around the new bridges at Vauxhall and Waterloo, and on the new Regent Street too. Theatres and shops, office buildings like Somerset House, various Inns of Court and East India Company headquarters all signed up for a GLCC contract. Within a year, the company had laid 30 miles of gas mains in London, with cities around the world soon following with gasworks of their own too.

★ ★ ★

Gaslighting ushered in a new era in energy. Connecting to a gas network meant homes were part of a larger infrastructure. Rather than running domestic lighting off packs of store-bought candles or homemade rushlights, homes were on a network, connected to their neighbours. Gas pipes weren't the first infrastructure project. Indeed, in many ways Winsor and the GLCC based their model on the water companies. But they were the start of something new for energy, something the electricity industry would later run with.

With the GLCC and the many other gas companies that followed, gas had also moved on from play with fireworks and balloons, or scientific spats about theories of heat, and started making money. This was not in the utopian model Beddoes and Winsor had imagined, but one that would start with customers like the Prince of Wales and if not 'trickle down' then at least gradually expand (the pollution and sometime fatal explosions arriving for the working classes long before the benefits). It was tamed too, made less smelly, more controllable, and with that it also got quite boring. If anything, its boring-ness was to be celebrated, aspirational, a sense that it was serious and safe – having put aside the fun and games of fireworks or talk of revolution – and less likely to explode in your face. It would become invisible again, sometimes deliberately buried underground in pipes, but also hidden simply by its normalcy. What had once been so wondrous became something no one bothered to wonder about, the everyday matter of turning a switch.

One nuisance that was hard to hide was the coal tar pollution. They tried dumping this oily, sticky and smelly by-product in rivers, but it soon piled up and killed the fish. It was exciting chemically though, and scientists started wondering what they could draw out of it. The largest by-product was creosote, used to preserve timber sleepers in the ever-expanding railways. In the 1820s Scottish chemist Charles Macintosh also realised he could use it to waterproof fabric. Then in 1856, teenage chemistry student William Henry Perkin produced a purple dye from coal tar. He'd been looking to produce a synthetic alternative to quinine, used to treat malaria, and stumbled across the dyestuff almost by accident. The *New York Herald* dubbed him a 'coal tar wizard', as a host of new coal-based colours travelled the world in brightly coloured postage stamps, hair dyes, clothing, sweets and more. It provoked a brief fashion for mauve dresses, put the yellow in Alfred Bird's new eggless custard and made frankfurters ooze pink.

Coal-tar derivatives would go on to support medical and forensic innovation too, and the British admiralty would soon be securing stocks for the production of explosives, as Perkin's discovery kick-started a whole organic chemical industry.* Between them, Boulton & Watt, the GLCC and the other gas companies might have tamed the gas industry from its days with political radicals and theatre shows, but the chemistry of the stuff was harder to control. It was bursting with colour and possibility.

★ ★ ★

As we'll see in Chapter Five, electricity would eventually take the lighting market and gas would shift to heating (and electricity generation), but in the short term, there was another competitor: kerosene.

In south-west Trinidad there's a lake that's literally black, filled with an estimated 10 million tons of sticky, black pitch. A semi-solid form of petroleum, it's the same substance found in oil sands, also known as tar, bitumen or asphalt. Walter Raleigh used Trinidadian pitch to caulk his boats in 1595 and roads around the world have since been asphalted with it. You've probably walked down one. In 1816, a young Canadian named Abraham Gesner visited the lake. Only 18 at the time, it was just a stop off on an otherwise ill-fated trip to sell horses in exchange for rum (which also involved him being shipwrecked, twice). But Gesner took a sample of the stuff as a souvenir, along with rocks, shells and other minerals for his geology collection. Back home in Nova Scotia, he tried to make a living in farming, failed and was sent to London by his father-in-law to train to be a doctor. This provided him with some scientific education, but his heart was always with his collection of rocks and, upon returning to Canada, he cobbled together a mix of work in geology, with a small bit of farming and medicine on the side.

Gesner's interest in lighting started in the summer of 1847, when he delivered some public lectures on the idea of caloric. For one of

---

* For an engaging telling of the story of Perkin and his dyestuffs (with a beautiful use of colour in the prose), see Simon Garfield's *Mauve* (2000).

his demonstrations, he distilled gas from that sample of Trinidadian pitch he still had from his teenage rum-trading exploits. Newspaper coverage of the lectures caught the eye of Admiral Thomas Cochrane* who, in his 70s by this point, had become obsessed with the business potential of the Trinidad pitch lake. With Cochrane's support, Gesner continued his experiments with the pitch, and by June 1849 had patented an illuminating oil and called it 'kerosene'. The name was a spin on the Greek for wax, *kenos*, designed to also sound like camphene, a plant-based illuminant customers would already be familiar with. Gesner was successful selling the idea in New York and Philadelphia, and the Nova Scotia government seemed interested in buying some kerosene for their lighthouses. But he was about to hit a stumbling block or two. He lost his patron when Cochrane decided to go home to England. More seriously, there was the question of asphalt supply. There was plenty to be had – as well as the Trinidadian pitch lake, it could be found across Mexico, Cuba and Texas – but it was hard to transport up north as most sea captains refused to carry pitch for fear it would burst the ship's bottom. There was an asphalt-like material to be found in New Brunswick, known as 'the Albert mineral', but Gesner ended up being drawn into a series of lawsuits, partly relating to British land law but also over whether it was coal or not. Finally, he gave up and packed his family off to New York.

He was to find fresh legal battles waiting for him there too. Scottish chemist James Young had already patented something he called 'paraffine' and argued this covered the manufacture of Gesner's kerosene. Young had trained in chemistry in Glasgow, at a college set up in the memory of John Anderson (the professor who had given James Watt a model Newcomen engine to fix back in 1764) and after a stint at the newly established University College

---

* Cochrane was quite the character. After a career in the navy during the Napoleonic Wars and a stint as an MP, he left the UK in disgrace having been accused of fraud, going on to support independence battles in Chile, Brazil and Greece. He's a possible model for Lord Byron's *Don Juan*, as well as influencing C. S. Forester's *Horatio Hornblower* and Patrick O'Brian's *Jack Aubrey*, and appearing as himself in one of Bernard Cornwell's *Sharpe* books.

London ended up in Lancaster to work in the chemical industry. In 1847 one of his old college friends, Lyon Playfair, invited him to examine a petroleum seepage in a coal mine at Riddings in Derbyshire. Playfair thought this seepage could be distilled to use as a lubricant or in lamps, but was called back to London to work on the Great Exhibition, and left Young to build a pilot oil plant. When the oil ran out, he experimented with different types of coal, discovering that any coal could, more or less, be distilled to produce a crude oil under a slow, low heat. He initially marketed it as a lubricant, as Playfair had planned, but found he could sell it as lamp oil too. Young won the paraffine versus kerosene patent fight in New York, so Gesner had to pay him royalty fees. Still, it's the brand name kerosene that's made its way into everyday language today, and Gesner continued to perfect his invention, so it burned brighter and smelt less, building a factory in Brooklyn. His company brochure claimed Gesner's kerosene burnt six times as bright as sperm whale oil and 'four times as bright as even that paragon of the age, the gaslight'. It was cheap too, a dollar a gallon, six times cheaper than sperm whale oil and half the price of gaslight. Imitators sprang up, and by 1860 there were around 70 coal oil plants operating up and down the Eastern Seaboard, producing several million gallons a year.

★ ★ ★

It wasn't long before Gesner and the other kerosene entrepreneurs started looking to the 'rock oil' that could be found underground. Although generally just seen as a nuisance, it was used as a medical treatment in several parts of the world; topically for pains or burns, or as a treatment for stomach upsets and coughs. People were known to set fire to it, as part of religious ceremonies, or shoot it as flaming missiles during war. But to use oil or gas at scale, you need technologies to drill, store and transport it. And even if you managed that, it needed refining before it made an effective illuminant. For the short term at least, producing oil from coal seemed more straightforward.

The oil industry as we know it today started with a bottle of the black stuff sitting on a shelf in an office at Dartmouth College, New

Hampshire. It had been left there by Francis Beattie Brewer, a physician who'd recently moved to Titusville, a small town in northeast Pennsylvania, to join his father's lumbering firm. The area was known as 'oil creek' due to the seepages of oil regularly found there. Native American Seneca people would purify the oil by burning it and use it as an ointment for various aches and pains, and Brewer noticed local people using it as a remedy for everything from toothache to worms. He gave a sample of it to his uncle, a professor at Dartmouth, and there it was spotted just a few weeks later by a former student visiting his alma mater: George Bissell. Bissell wasn't, initially, all that impressed, but it planted a seed.

Self-supporting from the age of 12, Bissell had worked his way through Dartmouth College by teaching and writing, before moving to Washington DC to work as a journalist. He'd then ended up in New Orleans where he became principal of a high school and later superintendent of public schools. In his spare time, he taught himself law and several languages. But ill health had forced him to move back north in 1853 and he dropped by Dartmouth while visiting his mother, which is where he spotted Brewer's small bottle of rock oil. Bissell figured that he could try selling this oil found in the ground, not as a medicine but for lighting, piggybacking the kerosene market. He convinced a few others to form the Pennsylvania Rock Oil Company and commissioned a report from Yale chemistry professor Benjamin Silliman Jr. His father, Benjamin Silliman Sr, was the chemistry professor who'd brought Joseph Priestley's soda water to the US, and both father and son had long subsidised what they could get from Yale for academic work consulting for mining companies. Silliman Jr had been pulled in as an expert witness in some of Gesner's legal battles and knew the business well. If anyone could establish there was a decent lamp oil to be found in Pennsylvania rocks it was a Silliman. What's more, Bissell and his associates knew the Silliman name would help them sell the idea and leverage more finance. It would be as much marketing material as scientific research.

Using fractional distillation techniques developed by his father, Silliman Jr got to work on the samples of Pennsylvanian rock oil. The process was quite slow and tedious, but he gradually realised that it contained several different components, all of varying colours, viscosity and smell. It was through this distillation that he 'struck oil' in among the chemistry equipment; that is, he identified a substance within the crude that'd work well for the lighting market. Silliman would usually spend just a few days on this sort of project, and charged for his time accordingly, but this project for Bissell took months, at huge cost to the rock oil entrepreneurs. Silliman let Bissell know he'd found something good, but that it had taken a long time and he wouldn't release the report until he'd been paid in full for all those hours. Bissell eventually found the money and it was worth it, even at that high price. The publication of Silliman Jr's report would become one of the key moments in the history of oil, helping the Pennsylvania Rock Oil Company raise finance for the next stage; getting enough of the oil out of the ground to warrant refining it.

Getting oil out of the ground was easier said than done though. Small seepages were fine if all you wanted was a small bottle to sell as a remedy for worms, but it wasn't enough to build a kerosene market. In parts of Eastern Europe, people had started digging oil pits by hand, but that was incredibly laborious. Bissell's key innovation was to use technologies for salt drilling, which had come to the US from China a few decades before. The idea apparently came to Bissell when he saw an advert for medical rock oil that featured some salt-drilling equipment in the background. Rock oil was often a by-product of salt drilling, so it was no surprise it featured it on such ads, but it sparked something in Bissell's mind. Why not make the extraction of oil the primary venture? Instead of digging for oil, drill.

The Pennsylvania Rock Oil Company employed Edwin Drake as president, commissioning him to hire a salt driller and find oil in Titusville. Drake had no expertise in either salt drilling or oil, or even local knowledge of Pennsylvania. He'd simply met James Townsend, a partner of Bissell's, when they were both living in a hotel in New Haven, Connecticut. Townsend wanted Drake to be taken seriously in Titusville, so sent letters ahead for him addressed to the 'colonel'. Drake had no claim to such a title, but it was a name that would stick with him and an early example of the oil industry's canny use of PR. In Titusville, Drake hired salt driller William 'Uncle Billy' Smith and together they went oil prospecting. At first, they had little success – the investors had all but given up. Eventually, in August 1859,

Townsend wrote to Drake to close the whole project down, enclosing $500 to cover any final bills. Fortunately, the oil arrived before the post did. On a Saturday evening at the end of August, Smith broke through a layer of rock. It was promising, but they'd been disappointed before and it was the end of the day, so he pulled the drill string and called it a day. He must have been curious, however, because even though he usually reserved Sundays for church and rest, he went to check it the following day and found the hole filled with oil. The news spread like wildfire, with farmers running into Titusville shouting 'the Yankee has struck oil'.

The population of Titusville doubled and land prices shot up. But the price of oil itself soon plummeted again, not least because there was more oil than anyone knew what to do with. For a while, the most prized objects in the area were whisky barrels, just because everyone was so desperate to find a place to store all this crude coming out of the ground. At one point, barrels cost twice as much as the oil stored inside them. This too was short-lived though, as infrastructure for processing, moving and selling oil caught up, including railways and pipelines as well as refineries. In 1861 they struck the first real gusher, flowing out at a rate of 3,000 barrels a day. As it rumbled up something in the escaping gases ignited, setting off a massive explosion that killed 19 people and blazed for three full days. The outbreak of the American Civil War eclipsed news of this explosion, but it didn't dampen the appetite for rock oil. If anything, quite the opposite: as the war cut off supplies of camphor grown in the South and the Confederate States Navy hunted down and destroyed Yankee whaling ships, the market was left relatively open.

Some of the work needed to build a market had already been done; you could use rock oil in lamps you might have bought already for Gesner's kerosene. As the market grew, practical advice to the US market on how to use this new fuel came in 1869 from Harriet Beecher Stowe (better known as the author of *Uncle Tom's Cabin*) who assisted her sister in writing a book entitled *American Woman's Home or Principles of Domestic Science*, which gave advice on which kerosene lamp to buy and to avoid poor quality fuels. However, there were still thousands of deaths in the mid 1870s due to accidents involving kerosene, with regulation of both the fuel and lamps patchy. And yet it was popular. Kerosene was much cheaper than whale oil and reached more parts of the country than gas networks, meaning more people could work, read or play after dark more cheaply. Even in the gaslit towns, kerosene started to build a strong market. In 1885, New Yorkers were told they could light their home with oil for $10 a year, whereas they might pay that much a month for gas.

* * *

As people drilled for oil, they'd often find flammable gases with it. A few areas managed to replicate Carlisle Spedding's experiments with 'firedamp' in Whitehaven in the mid-eighteenth century, lighting buildings with gas they found in the ground rather than the stuff that was produced in gasworks from coal. But on the whole, gases that came straight from the ground were too hard to transport to be of much widespread use. When oil drillers hit gas, they'd burn it off. Improvements in pipelines made it easier to exploit after the First World War and a few decades on from that technologies for liquifying the gas would mean it could be transported even further. Burning naturally occurring methane deposits was less obviously polluting than gas manufactured from coal and it soon became popular with politicians as a 'cleaner' fossil fuel. As we'll come back to in Chapter Eleven, during the energy crisis of the early 1970s, Nixon would dub gas 'America's premium fuel'.

As the oil and gas market grew, drillers started looking in what gets dubbed 'unconventional' spots, because they are slightly harder to extract from, like shale (a type of rock made of fine grains of compressed mud, where gas doesn't flow so easily, but can be very plentiful). New extraction techniques were developed too. Fracking, for example – injecting rock with high pressure fluids and sand – has been used commercially since the late 1940s. At one point in the 1960s, the American gas industry even experimented with nuclear explosions to open gas deposits. 'Project Gasbuggy', as it was called, detonated a 19-kiloton blast two-thirds of a mile below the ground near Farmington, New Mexico. However, after protests from local labour groups, Native Americans and business magnate Howard

Hughes, the project was closed (it didn't help that the process was very expensive and the gas extracted remained radioactive).

There's a Monty Python sketch about people having to change their cookers to accommodate the new type of gas, but on the whole most people have forgotten this transition, if they ever knew it happened at all. Still, the coal-gaslighting industry has left its mark – in the networked way we run our energy systems, in language like the expression 'to gaslight' and, perhaps more pertinently, in pollution that's still with us. Bits of some of the old infrastructure for producing gas from coal still pepper many cities; a set of luxury flats in King's Cross built around an old circular gasometer, for example, or in Seattle's Gas Works Park, which was an active coal gasification plant for the first half of the twentieth century. There's even a bit of East London named after one of the GLCC's former chairmen, Simon Adams Beck: Beckton. The gasworks there was demolished in the 1980s, briefly playing the role of Vietnam in *Full Metal Jacket* before it went down. As one of the film-making team later told the *Guardian*, once they discovered the gasworks had been constructed by the same company of architects that built in the city of Hué, Vietnam 'all we had to do was dress it up, put signs on it and blow it up'. The pile of toxic ash the gasworks produced remains still, known locally as the Beckton Alps due to its use as a dry ski slope in the 1990s (frequented by, among others, Princess Di).

★ ★ ★

And wither the whale? One thing is clear, the Pennsylvania Rock Oil Company didn't save the whale. Whale oil was already on the way out by the time rock oil came around. It was being pushed out by coal gas and kerosene, and the demand for whalebone had dropped too, partly because people were inventing plastics, but also because corsets had fallen out of fashion. It was getting harder to find whales around the coasts of America, they'd been hunted so extensively, but there were plenty of whalers working in other parts of the world and they found other ways to sell their wares too. Around the turn of the twentieth century, processes for treating oils with hydrogen removed the fishy smell and helped transform the relatively thin whale oil into solid or semi-solid states, opening up new markets. In the First World War whale oil was still used to treat trench foot and, along with by-products from coal gas and fossil oil industries, to produce nitroglycerin for explosives. If anything, fossil fuels helped power all this, as whalers took first to coal-powered steamships and then diesel ones. By the outbreak of the Second World War, huge ships were harvesting 500,000 tonnes of whale a year, all running on fossil fuels. After the war, even though by that point there was an increasing call from scientists to halt whaling, the business didn't stop. Indeed, in 1951 alone, more whales were killed wordwide than New Bedford's ships took in a century and a half of whaling.

People weren't burning bits of whale to light their homes anymore, they were spreading it on toast and painting their faces with it, in margarine and lipstick. With the smell taken out, and a large, complex chemical industry between the harpoon and the high street, consumers were less aware of where whale oil-based products came from. Today most of these products largely run off rock oil, but bits of dead whale continued to sit in our fridges, on the make-up counter and in the DIY store, in crayons, ice cream, detergents, photography equipment, fertiliser, medicines and more, right until the middle of the twentieth century. It's sometimes said the Hubble telescope orbits the Earth lubricated with spermaceti, staring deep into the Universe with the aid of a bit of a dead whale. When NASA historians conducted an internal enquiry in the late 1990s, they concluded whale products had been replaced by other oils by the time Hubble was launched, but it is plausible whale oil was used for earlier space projects. More down to Earth, Unilever – formed after a 1929 merger of Dutch margarine company Unie and the British soap maker Lever Brothers – relied on whale oil for a lot of its early business and later in the twentieth century would be pulled into the politics of a global moratorium on commercial whaling.

Among all the stories of whales and rocks, we need to remember a third type of oil, not animal or mineral, but vegetable. The first winter the *Mayflower* settlers spent in Cape Cod, they used an oil from pine trees for lighting, not the whale they'd found on the beach. As America grew, the camphor industry continued to compete with whale oil, coal gas and rock oil. It's one of the reasons kerosene is called kerosene, as Gesner wanted his product to sound like the already popular camphene. Plant-based oils weren't left in the nineteenth century either. In 1919, as people started to worry all this oil we were sucking from the ground might run dry, the Anglo-Persian Oil Company (today's BP) invested in several acres of Jerusalem artichoke fields in Dorset, hoping it could use plant-based alcohol as an alternative motor fuel. Today, the idea of 'plant-based' oils might seem relatively innocuous, but they are still a concern for environmentalists. And the palm oil tree, like rock oils, has its own story of colonialism. Native to the area between Angola and the Gambia, they were first taken to South East Asia by the Dutch and British in the nineteenth century. Used by Victorians as a machine lubricant, as well as in soap, it was one of the many colonial products on display at the 1851 Great Exhibition as Britain sought to show off the riches of its Empire. In 1911, one of the Lever brothers, William, signed a treaty with the Belgian regime in the Congo, buying nearly 2 million acres of forest for palm oil production. For all that Lever was a proponent of 'moral capitalism', investing heavily in a suburb of the Wirral, Port Sunlight, for his British workers, the Congo plantations used Belgian systems of forced labour* (strongly criticised at the time as well as today). By the mid-twentieth century, palm oil helped the margarine industry when it was eventually forced to give up whaling and became a key part of several post-colonial economies. But the size of today's palm oil industry comes at a huge biodiversity cost, threatening a host of species from chimpanzees in Nigeria to orangutans in Malaysia.

So just as Unilever was once pushed on its use of endangered whale oil, now it's criticised for the environmental problems of palm oil. Arguably, in saving the whale we simply shifted our destruction elsewhere. Whales are still hunted, and they're also disturbed by the noise of modern shipping. We need them alive though, not least because they sequester carbon. Science journalist Maddie Stone puts it well when she describes whales as 'giant swimming trees'. They munch up carbon via their food, storing it in that fat we used to light our homes with, before dragging it deep down to the bottom of the

---

* For more on this, see Jules Marchal's *Lord Leverhulme's Ghosts* (2008).

ocean when they die – a literal carbon sink. Their large plumes of excrement also bring nutrients like nitrogen and iron into the water, helping stimulate the growth of phytoplankton, which also suck carbon out of the air via photosynthesis (a process 'geoengineers' try to replicate, but we could achieve more simply if we only let whales quietly take a shit).

CHAPTER FOUR

The Weather Watchers

The weather of 1911 was weird – or so reported the March 1912 edition of *Popular Mechanics*. Almost buried under reports of new submarines, zeppelins, an electronic hearing aid from France and the innovative use of canaries in coal mines in Tennessee, there's a four-page illustrated feature on climate change.

The author Francis Molena describes a heavy heat having begun to dig in around June: 'The cities baked and gasped for breath, while the burning Sun and hot winds withered the corn and cost the farmers a million dollars a day.' What had started in the US soon made its way to Europe. Whalers brought back reports of once icy Arctic regions full of water. Then, around the middle of the summer, 'the flood-gates of the heavens opened'. Kentucky was deluged while a cyclone devastated Costa Rica and the Philippines were 'more thoroughly drowned than they had been before since the time of Noah'. By this point, temperature records had been kept in the US going back four decades and a graph illustrated how temperatures in 1911 had been beaten in each month but November. It's the sort of reporting we're all too used to today – but this was 1912.

He notes 'a general impression among older men' that the 'good old-fashioned winters' they knew in their youth – snow 15ft deep, lasting six months – had gone. The weather just wasn't what it used to be. Molena reminds readers that once upon a time parts of the Earth had very different climates, referring to fossilised plants found in Greenland or oddly placed boulders Agassiz referred to in his work on ice ages, before taking them through a basic explanation of the greenhouse effect and the warming role of carbon dioxide. With a tip of the hat to 'the great Swedish scientist Arrhenius', he asks whether, as we know burning coal produces carbon dioxide, might we be producing sufficient quantities to alter the climate?

'It is perhaps somewhat hazardous to make conjectures for centuries yet to come,' Molena concludes, but 'it is reasonable to conclude that not only has the brain of man contrived machines by means of which he can travel faster than the wind, navigate the ocean depths, fly above the clouds, and do the work of a hundred, but also that indirectly by these very things, which change the constitution of the atmosphere, have his activities reached beyond the near at hand and the immediate present and modified the cosmic processes themselves.' Still, with *Popular Mechanics* never keen to pour pessimism on mainstream ideas of progress, he takes all this extra warmth as largely for the good: 'It is largely the courageous, enterprising, and ingenious American whose brains are changing the world. Yet even the dull foreigner, who burrows in the earth by the faint gleam of his miner's lamp, not only supports his family and helps to feed the consuming furnaces of modern industry, but by his toil in the dirt and darkness adds to the carbon dioxide in the Earth's atmosphere so that men in generations to come shall enjoy milder breezes and live under sunnier skies.'

It wasn't the only place the topic popped up that year. In August, a newspaper in New Zealand picked up similar estimates for carbon emissions as Molena had used, summarising that the carbon emissions from burning coal creates a 'blanket for the Earth and to raise its temperature', the effects of which might 'be considerable' in a few centuries. There was a similar piece in an Australian paper a few weeks before, and back in the spring there'd been a short exchange of letters in *Scientific American*. Are we using our sky as a giant slag heap, the magazine's readers wondered, the air filled with carbon dioxide just as coal tar had clogged up the rivers? Can we rely on the oceans and plants to breathe it all in, clearing up the mess? Was it time to start a movement for the conservation of the atmosphere, similar to calls to save the forests?

Later that summer, *Scientific American* dug out an old 1904 book by Wisconsin geologist Charles Van Hise, *A Treatise on Metamorphism*, dubbing it 'a prophecy' of the recent hot summers. It's a long book, but *Scientific American* had found a few pages in the middle that contained a reference to Arrhenius and the idea that by filling the atmosphere with carbon dioxide humans could bring about a marked increase in the temperature of the globe. Van Hise was still talking about effects we might notice dramatically in a thousand or so years, but *Scientific American* warned that since his book had been published consumption of coal had doubled and so 'there is little wonder that Prof Van Hise's prognostications in regard to the increased temperature should have been so swiftly verified'. They avoided *Popular Mechanics*' celebratory tone, but still there wasn't any particular sense of terror or urgency. It'd probably be good for plants.

★ ★ ★

Before we could see any change in our climate, we needed decent weather data. Today, we take a deep understanding of the weather for granted. Charismatic weather people on the news or an app on our phone provide us with remarkably accurate forecasts, backed up with a dazzling network of human expertise, powerful supercomputers, and observation stations on Earth and dotted around it in outer space. Meteorological data, equipment, societies and expertise would all be vital to the growth of our knowledge of climate change in the twentieth century. It'd also be meteorologists who were some of the first to laugh at the idea of anthropogenic global warming too. But all that needed to be built.

The first documented project to systematically collect weather data was a scientific institution founded by Grand Duke Ferdinand II of Tuscany in 1654, the Accademia del Cimento. This established a series of 12 recording hubs spanning northern Italy and into central Europe and would send out sets of calibrated instruments along with specially printed forms for taking standardised records of air pressure, wind direction, temperature and more. It collected a fair amount of data but did little with it and fizzled out in 1667. Around the same time, Robert Hooke, the curator of experiments at the then newly instituted Royal Society in London also briefly became excited by the idea of keeping weather records. In 1665 he proposed 'A method for making a history of the weather' with guidelines for the collection of quantifiable data on wind force, temperature, barometric pressure and humidity, as well as space to note in words the appearance of the sky and any other phenomena like thunder or dew. But he never managed to get enough people to fill in their observations. The volunteers, mainly his friends, got bored with what felt like pointless administration. Even Hooke himself gave up after a few months. All this collecting of data was good empirical work in the British Baconian model the Royal Society was set up within, but the project arguably needed a bit more theory, poetry or sense of mission to motivate people to take it further.

It would be a much less official society that would kick-start meteorological thinking – not the Royal Society, but the Askesian, founded in London in March 1796. The Askesian Society's origins could be found in a group of dissenter scientists that had first convened around lectures given by Irish chemist Bryan Higgins. Higgins had got around the tight barriers of British science of the day, graduating from the Dutch University of Leiden in 1765 and then setting up an independent School of Practical Chemistry in Soho, London. His audience was largely dissenters of the merchant class – shopkeepers, artisans or mechanical workers. Joseph Priestley had been a regular for a while (before Higgins had started claiming he, not Priestly or Lavoisier, had discovered Oxygen, and they'd fallen out). Higgins sold his lectures – and beyond that, science itself – as something for the young, the way to take hold of the future, and became a bit of a cult figure with a new generation of young dissenters who were eager for change. When Higgins packed up shop in 1797 to work on rum production in Jamaica some of his fans decided to keep meeting, naming themselves the Askesian Society after the Greek *Askesis*, meaning 'training' or 'application'.

They met out of office hours, in labs owned by one of the group, William Allen, who was building a career for himself as a pharmacist. The Askesians proudly engaged in loud, diverse and sometimes slightly messy conversations, occasionally passing around bladders of 'the new gas, from nitrate of ammonia' (Davy and Beddoes' nitrous oxide). As Richard Hamblyn described in his beautiful (2001) book *The Invention of Clouds*, the Askesian paper that really caused a stir was Luke Howard's, on the naming of clouds. Howard, a Quaker from London, had befriended Allen through the Higgins set and ended up working for him, running a factory in Plaistow, to the east of London. Howard had adored watching clouds since childhood and so when, in December 1802, it was his turn to speak to the Askesians, he picked it as a topic. Riding in from Plaistow on a borrowed horse, Howard was exceptionally nervous. He hated public speaking at the best of times, and the room was full of strong personalities more successful in business and science than himself. Holding a bundle of handwritten papers and watercolour drawings, he apologised that the topic might seem silly, useless even, but argued that if we were to understand the weather we must be able to differentiate types of cloud – and that's what, over the course of about an hour, he'd do. Hooke had talked about the need to classify 'the faces of the sky' as part of his brief attempt at weather recording, but it was Howard who really started to name them. There was cirrus (Latin for 'fibre' or 'hair'), stratus (Latin for 'layer'), cumulus (Latin for 'heap' or 'pile') and nimbus (slightly less poetically, Latin for 'cloud'). The Askesians loved it. Alexander Tilloch, founder of the *Philosophical Magazine*, grasped Howard by the hand and insisted he visited his office the next day to discuss a version of the paper in print. This would become the 15,000-word essay 'On the modification of clouds', published the following year and later republished as a popular pamphlet. You can trace Howard's new language of clouds through the development of meteorological science and also the writings of Shelley, Coleridge and Goethe, and the paintings of Turner and Constable.

After Howard's clouds, came Francis Beaufort's wind scale. Slightly less poetic, perhaps, but just as important. Beaufort was born in 1774 in Navan, about 30 miles north of Dublin. His father was a clergyman with a tendency to get into debt, and young Beaufort's formal education suffered somewhat from the family continually having to pack up home and move town. He got a job at sea in his teens, and there he might have stayed if it hadn't been for serious injuries sustained attacking a Spanish brig near Fuengirola in 1800. He was shot at point-blank range and then set upon with a sabre. He lost all feeling in three fingers and was left with shrapnel in his lung and several holes torn in his arm (one of which he'd joke he could stick an ink bottle into). Aged 30, Beaufort ended up pensioned back home and soon slipped into a deep boredom. However, his sister's husband, Richard Edgeworth (one of the Lunar set) had started building his new idea for an 'optical telegraph', a warning based on a chain of towers that could pass semaphore messages to each other. Edgeworth had first played with the idea back in 1760 – as a way to send messages from horse races at Newmarket to men placing bets at a coffee house in Chelsea, 25 miles away – and picked it up again as a way to raise the alarm if Napoleon tried to invade Ireland. Beaufort worked on this telegraph system for two years, but ultimately the politicians decided the idea was too expensive to properly invest in. This was a blow to Beaufort, who wrote to his brother angrily saying: 'I am nothing, I have nothing, I expect nothing.' Still, his scientific work with Edgeworth provided training he'd later put to use.

Two years later, and Beaufort was working at the Royal Dockyard in Deptford, commanding the *Woolwich*, a gunner that had been turned into a storeship. Again, he felt rather bitter about this, writing to friends: 'I have spilled my blood, sacrificed the prime of my life … For a storeship.' Feeling acutely left out of more exciting battles happening at sea at the time, he started keeping ever more detailed weather diaries, developing a systematic way to record the wind, a form of which is still used today. Throughout, he maintained a correspondence with Edgeworth, at one point complaining that the Admiralty frittered away all the amazing data they had in ships' logs. In 1810 he was finally promoted to captain and sent to complete a hydrographic survey on the coastline of what is today Syria and Turkey. Again, this ended abruptly when the ship was attacked and Beaufort was badly injured, and he was pensioned off home once more. Back in London, he was able to settle into the sort of prestigious scientific life he'd never had access to in his youth. Edgeworth wrangled Beaufort's election to the Royal Society and he was appointed hydrographer to the Royal Navy, with a suite of offices in the centre of town. In his lyrical history of early weather research, Peter Moore describes this as the early-nineteenth-century equivalent of NASA: 'On behalf of the wealthiest and most powerful nation on Earth he was conducting explorations at the very edge of human knowledge.' And it was at his hydrography desk at the Royal Navy that, in 1830, he met Robert FitzRoy, a promising young captain with aristocratic lineage. Beaufort commissioned FitzRoy to take his ship HMS *Beagle* to survey the coasts of the southern part of South America, returning via Tahiti and Australia on the way back. Conscious that such a long journey could be lonely, and that the previous captain of the *Beagle* had shot himself, FitzRoy asked Beaufort to find him a scientific gentleman companion. The Lunar networks kicked in once again and Beaufort suggested the grandson Erasmus Darwin shared with Josiah Wedgwood – Charles.

★ ★ ★

Back on land, for all Sadler's hopes that scientific ballooning could transform our understanding of the atmosphere, for the first half of the nineteenth century, balloons were largely left to the showmen – something for the Vauxhall Pleasure Gardens, but not serious science. In the summer of 1804, chemist Joseph Gay-Lussac managed to get hold of an old balloon that had survived Napoleon's trip to Egypt and filled its basket with scientific instruments. He managed a height of nearly 23,000ft before realising he was finding it hard to swallow, figured he must be close to the limit of a breathable atmosphere and descended. Still, he was the exception to the rule that serious scientists didn't go up in balloons. However, ballooning historian Richard Holmes argues that the British Ordnance Survey – the first state mapping programme in the world – was partly inspired by the new vision of Earth the early balloon trips opened up, so it did have some impact. For Holmes, this new tech of the air challenged humans to see the Earth in a new way – from up high, looking down at a distance. That 'Earth from above' imagery we're now reasonably familiar with – from satellites to when we look out of the window on a plane trip – was new and let people see, for the first time, quite what an impact industrialisation was having on the landscape, as forests shifted to an orderly patchwork of fields and then smoky, densely populated cities. Holmes compares this to the way the 'Earthrise' photo, taken by Apollo astronauts in the 1960s inspired a wave of environmentalism in the late-twentieth century, suggesting that the impact of man on nature was, for the first time, laid out for these early aeronauts.

Luke Howard went to see one of Sadler's balloon trips from the Mermaid Tavern in Hackney in August 1811 and must have wondered what it would be like to travel up into the clouds himself, but clearly didn't. He did, however, use his new cloud fame to try to bring together people interested in meteorology. The September 1823 issue of the *Philosophical Magazine* advertised an inaugural meeting of a Meteorological Society of London, to be held at the London Coffee Shop on Ludgate Hill. They continued to meet regularly over the next few months, corresponding with other scientists interested in meteorology in Europe – including Gay-Lussac in Paris – but after the initial excitement, it fell by the wayside. Howard moved to Yorkshire, and of the other founder members, George Birkbeck became preoccupied setting up a London Mechanics' Institute (now Birkbeck College). It was revived again in the mid 1830s, with a young John Ruskin an enthusiastic member. In 1839 he ambitiously laid out the society's aims, declaring it had been reformed 'not for a city, not for a kingdom, but for the world. It wishes to be the central point, the motivating power, of a vast machine, and it feels that unless it can be this, it must be powerless; if it cannot do all, it can do nothing.' But that kind of synoptic meteorological research power was still a long way off. After collecting odd bits of data from under trees and inside hen houses, they produced a UK map of rainfall before going the way of their predecessors and disbanding.

When it came to researching the weather, the western European states were a bit limited by their size. The US on the other hand was huge, with a steady supply of exciting, sometimes deadly, weather events to investigate. Pretty much as soon as European settlers had arrived in the Americas they complained about the weather. The rains, the winds, the heat and cold were all, well, just a bit much. Early theories included the idea that this weird weather was because it was a 'New World', not just new to their discovery, they imagined, but perhaps it only recently emerged from the sea, or possibly from under a giant ice cap? Others wondered if they could quieten the weather slightly by 'taming' the environment, draining marshes and felling forests. Edmund Halley wondered if maybe Hudson's Bay was so cold because it had once been further north, only recently having shifted due to the impact of a comet. As American science grew, their ideas settled slightly, but the weather didn't. So perhaps it's no surprise that American science kicked the world's meteorology into the modern era with a fight about storms.

On one side of the 'storm wars' was engineer William C. Redfield. He'd made his name running steamboats between New York and Albany, but had an interest in science on the side. He first developed a rotational theory of storms, where wind blows around a centre of low pressure, after a hurricane had flooded much of the New Jersey coastline and several streets in Manhattan in September 1821. Out walking with his son not long after the storm had passed, he noticed that trees in one side of the storm had fallen to the north-west, whereas in other spots they'd fallen south-east. Science not being his full-time occupation, he sat on it. Then, about a decade later, he'd recognised a Yale professor of physics on a steamship trip, struck up a conversation on atmospheric physics and ended up convinced he should publish his theory. Redfield's opponent in the storm wars was ambitious Franklin Institute scientist James Pollard Espy. A keen member of the Kite Club of Pennsylvania, he'd often noticed how his kite would be tugged up and started to build a theory based on columns of rising air. The fight between the two was acrimonious at times and not especially edifying. However, it prompted Espy to embark on a series of lectures for the general public, gaining himself the title 'the Storm King' and in 1841 he published a non-technical intro to the topic, *Philosophy of Storms*, which among other things outlined a method by which he argued he could make it rain by setting fire to forests. This caught the attention of newspaper editors, who supported his campaign for federally funded meteorological studies. In 1842, Congress decided to invest $3,000 for meteorological observations at military posts, and Espy was chosen to oversee the project. He'd be the first official meteorologist of the US.

Redfield had his own fans though, not least at the British Colonial Office. Hurricanes weren't something the British had to worry about at home, but they'd had to get used to a regular hurricane season in the Caribbean, and saw knowledge of storms as part and parcel of colonial management (it was bad for trade, if nothing else). Not long after Redfield's paper on the 1821 storm was published, a large hurricane hit Barbados. British Army officer William Reid, who had been tasked with rebuilding government offices, read Redfield's paper and, hoping to understand this force that had hit the island so hard, he started to collect data from local ships' logs. After several years of study, in 1838 he presented scientific papers and published a practical work with tips for mariners, *An Attempt to Develop the Law of Storms*. This was a hit with the Colonial Office, which promptly gave Reid the governorship of Bermuda (where he'd be very well placed to continue his studies on storms). They also started to make diplomatic approaches to the American government, for some sort of US–UK collaboration for sharing any useful data that might be found in ships' logs.

This ended up on the desk of Matthew Fontaine Maury. Maury was a former sailor who – after a stagecoach injury had put an end to his career at sea – ended up stuck in Washington DC with the job of sorting through US Navy logbooks. There were military logs, but records from whalers too, giving Maury fascinating snippets of information not just on weather conditions, but the movement of these huge, waterborne mammals. He pored over the data, pioneering a new brand of practical meteorology, gaining the nickname 'Pathfinder of the Seas'. In 1854 he was appointed the first superintendent of a new United States Naval Observatory and Hydrographical Office; if Beaufort had set up a 'Victorian NASA' in London, there was a new player in DC. When Reid's friends at the British Colonial Office got in touch, Maury proposed not just a system for Anglo-American data sharing, but a whole new international, universal system of meteorology. That synoptic vision of weather research 'not for a city, not for a kingdom, but for the world' the wide-eyed young Ruskin had waxed lyrical about in 1839 might finally be taking shape. Maury convened a conference in Brussels in August 1853, inviting France, Russia, the Netherlands, Norway, Portugal, Denmark and Sweden, as well as the Brits. These were countries much more used to going to war with one another than co-operating, but Maury was persuasive. The wind charts he'd developed had reduced the time it took to get between the east and west coasts of the US by a third. The European powers, all too aware that time was money, wanted these efficiency gains for themselves. They agreed to cooperate with logs of basic observations of wind force and direction, air pressure and temperature at set times every day. The British delegates were wary of Maury's expanded idea of international cooperation but took the idea home to the Royal Society which agreed that a more uniform system of observations was at least a good idea. It was a long way from today's World Meteorological Organization, but it was a start.

★ ★ ★

Maury wasn't the only meteorological innovator in DC. On the National Mall, Joseph Henry was setting up shop at the Smithsonian Institution, finding space to house a load of animal and plant specimens brought back from an American expedition that circumnavigated the globe in 1842. More pertinently, perhaps, people kept talking about the new electric telegraph, which had been so dramatically launched by Samuel Morse from the Capitol building in May 1844. As the telegraph network grew, several American storm experts, including both Redfield and Espy, started wondering if it could be put to meteorological use. Mathematician Elias Loomis (who was also interested in storms and, sidestepping the Redfield vs Espy storm wars, pioneered the idea of plotting weather data on a map) wrote to Henry suggesting he run such a set up out of the Smithsonian. Henry had been involved in a medium-sized weather network run out of the University of the State of New York back in 1825. He liked Loomis's idea and in late 1847 agreed a plan with the Smithsonian board for a large-scale network utilising electric telegraph technology.

Henry's network recruited several hundred observers. They'd record their weather data in specially printed forms sent out by the Smithsonian before telegraphing them back. An 1857 report of the project notes they'd grown to 15 clerical staff, many of them women, who were inundated with upward of half a million separate observations, each requiring mathematical analysis. As well as the telegraph helping them collect data, they had one big advantage over other previous projects: they were a museum, right at the centre of DC, giving them a clear route to communicate with the public and politicians. They set up a large map in the museum hall, where clerks would hang small cards on iron pins about current weather conditions: black for rain, grey for cloudy. Henry noted in his 1858 report that visitors seemed curious to see what the weather was like in other parts of the country and how it varied. Weather warnings would be displayed on the Smithsonian castle tower and politicians would privately consult Henry for weather predictions. But the project was interrupted by the American Civil War and as they were starting to rebuild the programme a fire destroyed much of the Smithsonian building. After the war, they also found the telegraph companies were less willing to donate their services for free. Henry was more than happy to leave the work to the army, but in 1870 Congress passed a joint resolution utilising the Signal Service, a unit of telegraphic engineers, to issue storm warnings (which, in 1890, would become the US Weather Bureau, today's National Weather Service).

Over in London, James Glaisher was running something similar out of the Royal Observatory in Greenwich. Born in Rotherhithe, south-east London, Glaisher's father was a watchmaker, and he'd grown up with an interest in precision instruments and a passion for measurement. After a stint working for the Ordnance Survey in Ireland, he got a job as a mathematical computer at the Cambridge University Observatory (this was back when computers were people). When his boss was appointed Astronomer Royal in 1835, he brought Glaisher to Greenwich with him. A few years later he was promoted to head the newly opened Department of Magnetism and Meteorology at the Royal Observatory. Excited by this new post – effectively Britain's first government meteorologist – Glaisher got busy collecting weather data, establishing a network of 60 volunteers across England. His weather readers were mainly doctors and clergymen, people connected to a particular local area who the Observatory felt they could rely on to take regular readings. Kitted out with calibrated instruments to ensure standardised measurements, the network would take its readings at 9am every day, telegraphing them over to Glaisher at the Royal Observatory for noon.

Glaisher was in many ways the first weatherman. In March 1844 he was asked to write a feature on his department at the Observatory for the *Illustrated London News*, and he soon became a semi-official meteorological correspondent for the paper, providing expert comments when needed. When, in the summer of 1848, the editor moved to Charles Dickens's new *Daily News*, he asked if Glaisher could offer a regular weather update there too. Glaisher replied he could do better, with his network of regional correspondents all working on the new electric telegraph, he could now provide a full weather report of the previous day's weather across the country. This was a long way from today's forecasts but, like the big map in the Smithsonian, it proved popular. When the Great Exhibition opened in 1851, Glaisher teamed up with the Electric Telegraph Company to present a same-day weather map, charging a penny a piece for prints. Glaisher's boss disliked this foray into PR though, appalled that he'd signed his name, alongside that of the Observatory, in an early piece for the *Illustrated London News*. Other scientists dismissed Glaisher as too interested in self-promotion and somehow unserious about science by association. It didn't hurt in the long-term though; his boss stopped complaining and Glaisher was elected to the Royal Society in 1849.

Glaisher also brought back scientific ballooning, with a series of ascents in the 1860s with aeronaut Henry Tracey Coxwell. These usually launched from Wolverhampton, partly because it was near the centre of England, but also because it contained a very supportive gasworks owner who'd fill the balloon. Their most celebrated ascent took place in September 1862. As they approached the 23,000ft point Gay-Lussac had declared the limit of a breathable atmosphere, they found it noticeably harder. But they decided to keep going. Around 29,000ft the basket became tangled in a valve. By this point, the men were both suffering from oxygen deprivation and couldn't read the scientific equipment, even with a magnifying glass, and were losing the power of their arms and feet too. Coxwell managed to climb into the balloon to disentangle the system. Hands frozen – so much so they had gone black – he managed to get back into the basket, where he found Glaisher unconscious. Pulling the rope in his teeth, he released the valve to descend. They lost five of their six carrier pigeons, but otherwise landed safely and walked seven miles to the nearest village inn where they could have a pint of beer and find a way home on land.*

* * *

---

* The 2019 film *The Aeronauts*, inspired by Holmes's *Falling Upwards*, plays with the history of this trip somewhat, with Glaisher a youthful nerd accompanied not by Coxwell but a character based on another aeronaut, Sophie Blanchard. It's a science history romance that floats some way from the truth, but that's not necessarily a bad thing.

Meanwhile, a joint British–French expedition at Balaklava, in the Black Sea, had been devastated by a sudden storm in November 1854. Trivia fans might already know this gave us the balaclava facemask, but more pertinent to our story, it alarmed European governments and it was generally agreed that better investment in meteorological research might have saved lives. The British government, for their part, decided to set up a Meteorological Department. This would manage the record keeping they'd agreed with Maury in Brussels and, they hoped, might help save them from another Balaklava too. It would be the start of the UK's national weather service. More importantly, it would be the start of weather forecasting.

When we left Captain FitzRoy he was setting off on the *Beagle* with Charles Darwin. The trip took five years, but they returned heroes. FitzRoy was highly praised for his impeccable logs and perfectly kept instruments. The Geological Society awarded him a Founder's Medal, their highest honour. He was appointed an Elder Brother of Trinity House, the body responsible for overseeing lighthouses. He got married, was elected an MP and as a fellow of the Royal Society and, perhaps even more exclusive, The Athenaeum Club, allowed to skip their usual 16-year waiting list. Then in 1843 he was convinced to pack up and take the post of governor of New Zealand. This didn't go well and he was recalled a few years later. He got home – after a hair-raising voyage where he nearly died in a storm alongside his wife and children – broke. He needed work. Then his wife died, leaving him with four young children. When he heard a new Meteorological Department was being set up and the Board of Trade was looking for someone to run it, he pounced at the opportunity. A naval man who knew his way around both science and the government, he was perfect for the post. FitzRoy set up offices on Regent Street in August 1856 and was given a larger permanent home in Westminster the following spring.

From FitzRoy's offices, captains were kitted out with high-quality instruments and given Maury's wind charts in exchange for collecting data. There were further prizes like telescopes for the most diligently kept records. They recorded air and sea temperature, pressure, humidity, wind, cloud, soundings, currents, magnetic variation and unusual conditions like auroras, ice or shooting stars. Back in London,

FitzRoy's team would compile reports at 11am every day, sending them out to the military, insurers, lifesaving bodies and several newspapers. In October 1859, a devastating storm hit parts of England and Scotland, wrecking nearly 350 ships, including iron-clad steamer the *Royal Charter*. Some of the passengers were returning from the Australian gold rush and there were stories of desperate men trying to swim to shore weighed down by the gold. It had caught the public attention, and there was a sense, a bit like after the Balaklava storm in 1854, that lives could have been saved if only these clever weathermen with all their data could issue a warning. FitzRoy knew what it was like to be caught in a storm all too well and started to campaign for his department to run storm warnings.

The first of these warnings was issued in February 1861. Within weeks, they had issued eight more and by August were providing daily forecasts for five regions of the British Isles. It was expensive to run. There was the cost of telegraphy, plus they needed physical space to store all the data they were amassing. The validity of the exercise was also controversial. Scientists at the time were wary of forecasting the weather, unsure that they really knew enough to make predictions. It seemed a little too magical, not scientific enough, like government-funded fortune-telling. The public wanted to know what the weather was going to be like, sure, but they also wanted to know if they'd meet their true love or if their father would recover from his illness. Science wasn't there to feign certainty it didn't have. It didn't help that there was already an established market for less-than-scientific weather predictions, which ran alongside horoscopes in popular almanacs. Perhaps the most interesting publisher of these was Richard James Morrison, a retired naval lieutenant who had studied astrology, joined an occult association and started to go by the name Zadkiel. He launched *Zadkiel's Almanac* in the late 1820s, which sold well throughout the nineteenth century. Zadkiel seems to have taken the emerging science of meteorology as a way to gain some scientific credibility for astrology, developing what he called 'astro-meteorology', weaving in concepts of electricity, electromagnetism and astrophysics. Zadkiel wanted to look as scientific as possible and arguably genuinely believed he was, if only science would reform to fit him. When FitzRoy started issuing forecasts, Zadkiel painted himself as a sceptical judge, an outsider who was unusually able to represent the people's voice and challenge the exclusive concerns of the scientific establishment. It's a pattern modern watchers of climate sceptics know only too well; outsiders of science are sometimes the keenest to wear science's clothes while also attacking the mainstream science they desperately want to join in with.[*]

FitzRoy was keen to draw a distinction between what he did and the likes of Zadkiel, writing in *The Times* that it was a 'sham and calculated to mislead the public'. In response, the Astro–Meteorological Society challenged 'the gallant admiral to mortal meteorological combat' in an advertisement placed in the *Daily Telegraph* (their second choice, *The Times* refused to take it). But it wasn't just the astrological end of meteorology that was challenging FitzRoy's foray into forecasting. Maury expressed scepticism over FitzRoy's work, as did Glaisher. The latter was possibly annoyed the Meteorological Department job hadn't gone to him, and aware FitzRoy's new forecasts were competition for a daily weather map business he was trying to set up. But Glaisher was far from the only one raising concern over FitzRoy's work, so it might simply have been concern of the misuse of science too. Ship owners also started to complain when they were told to keep their ships in port during a storm warning, losing them money. Questions were raised in parliament and FitzRoy started to feel under attack, writing letter after letter to *The Times* trying to defend himself.

In the autumn of 1864, the Board of Trade took an axe to his budget, stripping funding for eight vital telegraph stations. Feeling starved of data, FitzRoy was also increasingly in bad health. He'd had

---

[*] In 1844, he set up BAAAS, the British Association for the Advancement of Astral Science and Protection of Astrologers, presenting himself as a guardian of an ancient science reformulated for a new age. He even briefly ended up as president of the Meteorological Society of London too, when it reformed for a while in the late 1840s. Indeed, one of the reasons this society kept reforming and collapsing was fights with the more astrological end of things. For more details on the cultures of almanacs, FitzRoy, Zadkiel and characters around them see Katherine Anderson's brilliant book on Victorian meteorology, *Predicting the Weather* (2005).

a breakdown in 1850 and by April 1865 seemed to be in another, deep depression. He moved his family to Upper Norwood for some peace and quiet (not far from the spot on which the Crystal Palace was being rebuilt) and for a while brightened up, only to be plunged back into depression when he went back to work. He was also heavily in debt. He'd never been reimbursed for the cost of his trips on the *Beagle*, which had put a dent in his private fortune, and that brief post in New Zealand had meant dropping several lucrative posts that would have filled the gap. He couldn't sleep. He tried an opium pill from the doctor, but it only made him sicker. He had been perturbed for some time by the route Charles Darwin had taken since the *Beagle*, feeling great discomfort at the idea of natural selection. He now also started to obsess about the assassination of Abraham Lincoln. One evening at the end of April 1865, he came home in a state of high agitation after a meeting with Matthew Fontaine Maury, who was in England to explore electrical torpedoes. The following morning FitzRoy cut his throat with a razor, bleeding to death on his dressing room floor.

Speculation about the future of the Meteorological Department began circulating before FitzRoy's body was cold. Surely Glaisher was heir apparent? *The Times* seemed to think so. But Whitehall had a larger review in mind and appointed Francis Galton to run a formal investigation. Galton came of fine scientific pedigree, his grandfathers were Lunar men Erasmus Darwin and the Quaker gunmaker Samuel Galton (he'd even grown up in the home Joseph Priestley had vacated when he left Birmingham, although they'd changed the name to avoid any association with radical politics). Galton would later go on to write a book on hereditary genius, so he might like us to draw a comparison between his diverse research interests and Erasmus's, but we'll just stick to the fact that the family gun business left Francis Galton a very rich man, so he was free to study science rather as he pleased.[*] The more playful ends of this still get trotted out by scientific press officers: a formula for the best cup of tea or how to use maths to

---

[*] For more on the story of the Galton gun fortune, see Priya Satia's *Empire of Guns* (2018).

cut a cake. On the other side, he coined the term 'eugenics' and once plotted a 'beauty map' of Britain, standing on street corners with a special glove he'd made so he could count and rate women without being noticed. He was, to put it mildly, a bit of a creep.

Galton had been involved in weather work for a while, working with an observatory at Kew Gardens in west London, publishing a series of weather maps and identifying anticyclones. FitzRoy had been one of the people to sign Galton's nomination for a fellowship of the Royal Society, but this generosity wasn't returned. Galton had long had it in for FitzRoy, believing the Meteorological Department was simply stockpiling data and not making good use of it. Appointing Galton to study FitzRoy's work was never going to end with a glowing review and the report, published a year after FitzRoy's death, was as damning as expected. It argued that what had started on a strong statistical basis in the spirit of Maury's Brussels conference had fallen into chaos, diverting from the advice of the Royal Society and lured to 'prognostications of weather' without distilling any clear methodology. The word 'prognostications', not 'forecasts', was deliberate; Galton's report wanted to paint FitzRoy as more of a fortune-teller than a scientist. It underlined the popular image of FitzRoy as a bit of a gentleman fraud, out of his depth; all bluff, no better than the astrologers trying to look scientific. The forecasts ended; the storm warnings put on hold.

Zadkiel was appalled. No fan of FitzRoy, he nonetheless strongly believed that meteorologists had a responsibility to pursue forecasting rather than simply accumulating statistics. On this occasion, the political establishment agreed with him, with several MPs pushing for the return of storm warnings. The compromise was 'storm intelligence'; not a warning per se, but if an active gale was spotted then word would be spread by telegraph across the coast. Over in the US, the Signal Service had grown Henry's project at the Smithsonian to 290 stations, and in 1871 started issuing 'probabilities' of coming weather, claiming 90 per cent accuracy in their storm warnings (FitzRoy's got it right a little more than half the time). By 1875, Galton was helping *The Times* with daily weather maps and by 1879 the Royal Society felt comfortable enough to allow the Meteorological Department to resume its forecasts. FitzRoy had perhaps launched weather forecasting a little early and not in a manner the scientific establishment felt comfortable with, but it was here to stay.

★ ★ ★

A lot of early weather research started within state projects like FitzRoy's, but there was a side to weather work that could never stay bound by national projects, that yearned to stretch its arms further outside. As John Ruskin had declared breathlessly in 1839, the study of weather must grow to understand and connect the entire planet: 'Not for a city, not for a kingdom, but for the world.' Maury's 1853 conference in Brussels had only really encouraged individual countries to collect better records, rather than share them. But by the 1870s, staff managing these collections started to itch for more. In 1873 an International Meteorological Congress was held in Vienna and, from that, an International Meteorological Organization was established (which would become part of the UN in 1950, renamed the World Meteorological Organization). They published and met as an international group and started planning for a coordinated international effort to investigate the Arctic and Antarctic, the 'International Polar Year' of 1882. A new era of international meteorological cooperation had started.

There was plenty a researcher could do within a state meteorological department as it was though. You could go up, for a start. The Americans started measuring the upper atmosphere using kites in 1893 when Charles Marvin developed a large box kite that could carry a small technical payload to record temperature, pressure, humidity and wind speed. They could use balloons – smaller ones, simply carrying equipment, as well as the larger investigations Glaisher nearly died in – and, when the tech came, they'd use planes and rockets too. Researchers could go deep, intellectually speaking, questioning the nature of fluids and friction, and all the multiple other phenomena underpinning weather. And they could go wide, depending on your country, or rather how much of the rest of the world you'd invaded. It's that old line that has been said about empires all the way back to Ancient Greece: 'The Sun never sets on the [insert your favoured dominating power here] Empire', they straddle so many lines of longitude.

Studying the environment in the eighteenth and nineteenth centuries was always about empire, and meteorological work was no exception. Some studied the new climates they encountered to better exploit resources. Others used studies of climate to help define a sense of national unity; this is our weather over here, it's different over there. Climate was also used as an argument for white supremacy, arguing genius needed the right type of climate (and of course the 'right' types were those found in European latitudes). As Deborah Coen argues in her 2018 book *Climate in Motion*, the sheer scale of Empire by the nineteenth century offered imperialists a chance for learning too. It was, all too often, a process of unlearning, as colonisers suppressed or appropriated existing local environmental knowledge. But imperial networks for exchange of information laid the basis for the more globalised approach to climate science of the late-twentieth century, just as they did in studies of plant or animal life. Indeed, we could argue it offered the emergence of a form of whole planet thinking. It's a planetary consciousness based on colonial values of control and exploitation, not cooperation and equality, but one that nonetheless helped scientists think about the global climate at large, not just the bits they lived in.

In her 2019 book *Waters of the World*, Sarah Dry offers up the example of Gilbert Walker. He'd grown up in Croydon, where his father was the borough engineer (known for his pioneering use of concrete). In 1903, after scholarships to St Paul's School and Trinity College Cambridge, he'd ended up at the Meteorological Observatories in India in his mid-30s. He knew nothing much about meteorology. His most famous paper had been on the maths of boomerangs, and he'd been working on applying it to golf, cycling and billiards. Indeed, it's possible his appointment reflected a lack of commitment to meteorology on the part of the Colonial Office. There had been a series of droughts in India the decade before. Although it's hard to tell precisely how many millions had died, *The Lancet* estimated 19 million. Disasters are never natural, and in this case the British had dismantled traditional systems of grain storage and mutual relief that otherwise would have supported people through the hazard of bad weather, to disastrous consequences. It wasn't the first time either. After the Bengal famine in 1870 the viceroy Lord Lytton had taken the view that events like these happened every now and again to balance the Indian population and keep grain prices high. It was the poor who died and they were better off that way. If you sent relief, he argued, you'd only make it worse, contributing to the problem of over-population.

After another famine in the late 1870s, a commission had been set up to explore what might be done to avoid such a disaster in the future. It decided India needed more trains and telegraphs. But trains just moved grain to where it could be sold at the highest price, aided by telegraphic lines that helped investors work the grain market. So, finally, at the start of the twentieth century, it was time to see if better monsoon forecasting might at least help. Walker arrived at his base in Shimla – the Himalayan town the British used as their 'summer capital' – to find an office that took a fair number of observations but had little scientific basis for the production of their forecasts. They received data from a network of several thousand rain gauges across India, along with information on pressure, temperature and wind speed from a few dozen observatories. But Walker knew he needed the oceans to really understand monsoons, so sent clerks to harbours at Mumbai and Kolkata for ships' logbooks. Keen to learn more about the upper atmosphere too, he launched kites and balloons, attaching cards promising rewards for their return. Moreover, he started to pull in data from observatories around the world. Monsoons couldn't be explained from the boundaries of India, large as the country was; they were part of something bigger, so he must be too. He built up regular correspondence with observatories in Mauritius, Zanzibar, the Seychelles, Dar es Salaam, Cairo, Zomba, Entebbe, Durban, Sydney, Buenos Aires and Santiago.

And all this left him with another problem, one many other climatologists suffered from – much more data than actual analysis. Dry suggests this might have been where Walker's ignorance of weather research was a boon. He wasn't bogged down by old theories of monsoons and so could bring in something new. All he really had was his maths, so he applied a statistics device developed by Karl Pearson, a correlation coefficient to help him find patterns. It was exactly the sort of use of statistics Francis Galton had wanted back when he was complaining about FitzRoy.* The approach Walker developed off the back of Pearson's tool was much more efficient than anything his predecessors had tried to wade their way through all the data. As Napier Shaw (the first professor of meteorology at Imperial College, a post Walker would later inherit) noted in his 1926 *Manual of Meteorology*, it was a 'kind of searchlight for sweeping the meteorological horizon'.

Still, if anything, Walker's story is an example of how little the British Empire made of the potential to study weather. When Shaw retired as director of the Met Office in 1908 he made the same complaint Francis Beaufort had made almost exactly a century before: they wasted so much data. Coen points us to the large, Eurasian land empires of Russia and Austria as areas where climatology really thrived in the late-nineteenth century, in particular studies of the Eurasian steppe – grassland plains running from Hungary to northern China. Scientific study of this started in the eighteenth century, with Alexander von Humboldt exploring differences between Asian climates at one end and European ones at the other, what he called *vergleichende klimatologie* (comparative climatology) to show climates weren't necessarily uniform along lines of latitude, but might shift with altitude, landmass, proximity to oceans, glaciers or forests.

Concerns had developed in the 1890s about long periods of drought over in the eastern end of the steppe, in Turkestan. The area relied on runoff from melting glaciers to irrigate the land, but as the glaciers shrank back, their water supply did too. This coincided with discovery of glacier retreat in the West too, with Viennese geographer Albrecht Penck making detailed measurements of glaciers close to the high-altitude Sonnblick Observatory and concluding the Alps had experienced a warming trend over the previous quarter century. In the early twentieth century, this had been picked up by Munich-born Heinrich von Ficker. His work was interrupted slightly by the First World War when he was taken prisoner by the Russians early in 1915,

---

* Indeed, Pearson was a protégé of Galton's. When Galton died, he left much of his estate to UCL to set up a Galton Professorship of Eugenics with Pearson appointed the first 'Galton Professor'. Pearson was even more racist than Galton, and in 2020, after a long campaign from staff and students, UCL announced they'd be renaming lecture theatres bearing Pearson and Galton's names.

after a failed escape by balloon from a besieged Galician fortress, and interned in the Russian city of Kazan. Undeterred, Ficker studied Russian and was permitted to consult Russian meteorological publications, and in 1923 published a paper warning Turkestan was not only 'dying' but this was part of a larger problem of environmental destruction that stretched all the way along the steppe to central Europe. This was a crucial development from earlier studies of local climates as distinct from one another. As Coen puts it: 'Instead of a contrast between enlightened West and uncivilised East, he saw two regions suffering a related fate, linked by a global climate system.'

With more evidence of glacier retreats and droughts, the idea that the Earth might be warming was starting to establish itself, but people were less clear as to why. Something to do with the Sun, or dust, or possibly the saltiness of oceans. For all the spurt of press on the carbon dioxide theory around 1912, scientists had pretty much thrown the idea out by that point. By the end of the 1920s, people tended to agree with George Simpson (a colleague of Walker's at Shimla before becoming director of the Met Office) that even when variations in atmospheric carbon dioxide did occur, they had no appreciable effect on the climate. The fact that weather offices around the world were registering increasing temperatures while richer countries were also burning through more and more oil and coal was mere coincidence.

★ ★ ★

In 1938, Guy Stewart Callendar decided it was time to look again at the carbon dioxide theory. He was an amateur when it came to climatology. His day job was studying steam for the British electrical industry and this was possibly one of the reasons he was ignored, at least at first. As James Rodger Fleming describes in his biography of Callendar,[*] Guy's

---

[*] Pretty much everything by Fleming is worth reading if you're interested in the history of climate science and you'll find several references to him in the sources section of this book. His short biography of Callendar is by far the most accessible (not least because you can download it from Colby College's website) and as well as telling Callendar's story it does a great job of explaining the context of climate change research up to that point.

father Hugh was by far the more famous scientist during their lifetimes. Hugh Callendar had worked at the Cavendish lab in the 1880s, patenting a new high-precision thermometer, and in 1893 took a job as professor of physics at McGill in Montreal. This is where Guy and his two older siblings were born, although the family returned to England after his father landed a plumb job at the University of London. They moved to a large home in Ealing, with space for a tennis lawn along with a large greenhouse that was soon transformed into a home lab.

The children were all encouraged to play in this greenhouse lab, sometimes to disastrous consequences. Aged five, Guy was partially blinded by his older brother Leslie, who stuck a pin in his left eye, and Leslie later blew up the greenhouse trying to make TNT. This eye injury kept Guy out of active service during the First World War though. Instead he worked first in his father's lab at Imperial College, X-raying aircraft engines to look for hairline cracks. After the war, he trained at Imperial College before getting a job back in his father's lab, working on steam engineering. He met his wife at the Ealing tennis club, had twin daughters and moved to Worthing. Amongst all this, studying weather records then became a hobby for Callendar. In 1938, he presented his now classic paper 'The artificial production of carbon dioxide and its influence on temperature' to the Royal Meteorological Society. Callendar estimated that since the end of the nineteenth century (roughly his own lifetime, also roughly the time since Arrhenius had written about the issue) humans had added 150,000 million tonnes of carbon dioxide to the atmosphere by burning fossil fuels. This, he argued, had exceeded the limits of natural carbon sinks like the oceans and three-quarters of it remained in the atmosphere, causing about a third of a degree of warming. Still, he wasn't necessarily trying to ring any alarm bells. The combustion of fossil fuels, he concluded, 'whether it be peat from the surface or oil from 10,000ft below' is beneficial to humankind. It gives us heat and power, and this warming could be useful too. It might be good for agriculture and 'in any case the return of the deadly glaciers should be delayed indefinitely.' Global cooling was still more of a worry for Callendar, and he imagined all these carbon emissions would save us from another ice age.

The paper was published along with the questions that were asked afterwards, giving us an insight into contemporary reviews. They largely agreed warming was happening, but were less convinced it was a matter in the rise of carbon dioxide. Indeed, they could well be read as rather patronising. One, for example, 'congratulated Mr Callendar on his courage and perseverance' ('courage' being British scientist code for 'absolutely ludicrous behaviour'). Reading through them, you get the sense that the meteorologists, having finally managed to professionalise their field, were still annoyed by all these 'helpful' enthusiasts from outside. Callendar wasn't like the astro-meteorologist Zadkiel, he wasn't even another FitzRoy – they could see he'd put the hours in and was a skilled scientist, but he wasn't their kind of skilled scientist. Callendar gave as good as he got though. In response to the suggestion he'd ignored the natural movements of carbon dioxide, he replied he had 'actually written an account of these, but it was just eight times as long as the present paper'. As Fleming puts it, Callendar was clearly 'accustomed to rubbing elbows with scientific elites and was already a seasoned veteran of scientific debate'. He'd grown up with it.

Callendar went to war the following year. He ended up working for the Petroleum Warfare Department, work that brought together a range of different experts: businesses like the Anglo-Persian Oil Company, the Gas Light and Coke Company, General Electric, the railways and the water companies, as well as military and university-based scientists. He ended up sharing a patent for something called Fido (Fog Investigation and Dispersal Operation), which helped clear fog so aircraft could land in bad weather conditions. An initial test for Fido was set up at the well-named Moody Down in Hampshire. A local fireman climbed to the top of a ladder, disappearing into the fog as he went; then, as two burners were lit either side, the fog began to clear and the fireman came into view. Callendar went back to London to test his improvements, using a large skating rink; the ice-making machinery was repurposed to produce synthetic fogs and a large wind tunnel used to recreate the conditions of an airfield. When it was finally applied in the field, Fido burnt through 6,000 gallons of petrol in the four minutes required to land an aircraft, totting up to a total of 30 million gallons over the two and a half years it was in operation. There's some pathos to this: the academy threw out Callendar's work on anthropogenic climate change, so he went to work burning as much fossil fuel as possible in an attempt to deliberately control the weather. But we'll stick a pin in the story of climate science for a few chapters, because there was a lot more burning to be done before anyone would really start to take Callendar seriously.

CHAPTER FIVE

Electric Avenues

If the 1851 Great Exhibition is remembered for its splendid Crystal Palace, a vision in industrial plate glass and cast iron, then the 1893 Chicago World's Columbian Exposition left a glow of intensely dazzling white light. Known as the White City, a possible inspiration for L. Frank Baum's Emerald City in the Land of Oz, the 14 neoclassical-style buildings constructed for the Fair had all been thoroughly sprayed a specific shade of white. Inside, the 27 million visitors were introduced to a cornucopia of inventions, products, ideas, techniques, tastes and smells from around the world. Most striking was the 264ft Ferris wheel that could hoist 2,000 people up in the air at a time – Chicago's attempt to better the tower Gustave Eiffel had built for Paris's 1889 Exposition Universelle. You could travel along a moving sidewalk, try Juicy Fruit chewing gum or a chocolate brownie for the first time, have lunch made to the highest scientific principles in a model 'Rumford' kitchen built by Ellen Swallow Richards (the first female graduate of MIT), catch a Houdini show or watch the very first game of night football. You could visit a Moorish mosque or an English pub, watch Egyptian belly dancing or South Sea Islanders canoeing in the lake, eat a chapati or pay 25c to have your photo taken with an Inuit baby (before the baby died after the 'Eskimo village' was hit by an outbreak of measles and a load of them packed off back home in disgust at the conditions they'd been kept in, that is).

And shining over this vision of oh-so-modern America was an immense quantity of electrical light. Never had so much artificial light been produced in one place. You could see the glow a hundred miles away, thanks to a 6ft searchlight on the roof of the Manufacturers Building, its 57,000-candlepower lamp (the biggest in the world) increased to nearly 200 million with a parabolic mirror. Inside the Exposition's park itself, the rim of the Ferris wheel was studded with 3,000 lightbulbs, the reflections bouncing off the lake and nearby fountains – which themselves were powered by electricity – each shooting 22 thousand gallons of water a minute into the air and surrounded by 250,000 candlepower lights fitted with rotating filters so the water could be bathed, successively, in green, yellow, blue, red or white light. It totted up more than 200,000 lightbulbs in the outdoors alone, a long way from the 1851 Great Exhibition which had to close at dusk.

Lighting wasn't the only electrical wonder on offer. You could move around the park using the electric railway, take a ride in an electric boat, have a go on an electric crane or ride one of the electric elevators instead of climbing the stairs. General Electric had mangled its chances of the contract by pitching rates three times higher than its local branch had set the year before, and so its key competitor the Westinghouse Electric Corporation pounced at the opportunity. It wasn't just the prospect of the large contract (the expo used three times as much electricity as the city), but the PR that would come with it. Electricity was such a central part of the vision of a bright future on display in the Fair there was a dedicated Electricity Building, complete with a statue of Benjamin Franklin at the entrance, kite in hand. Inside, General Electric had erected an 80ft model of its lightbulb, lit by 5,000 laboriously installed smaller bulbs that pulsed in rhythm to music and flashed different colours. You could also listen to opera transmitted from New York in the Bell Telephone Company exhibit, play with electric dolls, test an electric sewing machine and enjoy the cooling effects of an electric fan. For some reason, the Western Electric Company had chosen an Ancient Egyptian theme, so in their section you could watch women dressed as if they lived at the time of Ramesses II operating the telephone board.

By the 1890s, the people of Chicago had become quite used to choking their way through the coal smoke of industrialisation. Founded on flat marshland, Chicago couldn't rely on waterpower like other parts of the US and coal had been key to the city's growth. Steam-powered transport by boat and train brought goods and people in and out. Steam transport also worked within the city, pulling cable cars and the elevated Loop, allowing it to grow in size beyond a 'walking city' (i.e. one where your commute was limited to how far you could walk to work), and steam-powered pumping stations gave them drinking water. But coal smoke wasn't to be allowed to smudge the carefully whitewashed White City. So, just as Beijing shut down factories for the 2008 Olympics, the 1893 Exposition ensured its electricity supply came from burning oil, delivered by pipeline from Ohio.

Some 'smudges' were harder to hide though. The future on display at this expo might gleam, but the country was on the brink of a financial panic. By the end of the year, 500 banks across Gilded Age America had failed, as had 150 railroads and 1,600 businesses. And then there was the racial politics of the event, which did not go without comment at the time. A Dahomey village (the Fon people of modern-day Benin), for example, was described by Frederick Douglass as being 'as if to shame the Negro ... to exhibit the Negro as a repulsive savage'. Years before the event had even opened, Native American leaders sent in a petition at the news that it was to honour Columbus. They followed this up with a letter in February 1893 addressed to Frederic Putnam, the Harvard anthropologist tasked with overseeing the Fair's ethnography section, underlining that they'd be watching out for any Wild West show performed at the expense of the dignity of Indigenous Nations. Putnam gave assurances that there wouldn't be any such shows in the Fair. But it wasn't a promise he could make. 'Buffalo Bill' Cody rocked up and, although he wasn't allowed to be an official part of the expo, camped right alongside and made himself a tidy million-dollar profit.

Concerned that the whitewashing of the expo wasn't just a matter of spraying the walls, pioneering investigative journalist Ida B. Wells pulled together a collection of essays: *The Reason Why the Colored American is Not in the World's Columbian Exposition*. Distributed through the Haitian Pavilion, the essays pushed against the conceit of great American progress on display at the Exposition. The challenges the collection presented are some that today's science promotion could ask itself too. How could America pat itself so smugly on the back for 'progress' when people were being lynched? Where were the Black Americans in this vision of modern America? As the preface puts it: 'The labor of one-half of this country has always been, and is still being done by them ... The wealth created by their industry has afforded to the white people of this country the leisure essential to

their great progress in education, art, science, industry and invention.'
As one chapter outlined in detail, it's not as if there was a lack of Black
Americans excelling in education, invention, business, literature,
journalism, art and music to showcase, if only the organisers had
thought to look.

★ ★ ★

One of the many Black workers involved in the production of this
new electrical future was Samuel Parker. When, in 1832, Joseph
Henry moved to a post at the college that would later be better known
as Princeton, he was disappointed by the quality of the laboratory
equipment available and set about improving the research facilities. In
1841 he wrote to Elias Loomis that the trustees had finally 'furnished
me with an article which I now find indispensable', by which he
meant Parker, 'whom I have taught to manage my batteries and who
now relieves me from all the dirty work of the laboratory'.

Parker was paid $48 a year, working in Henry's lab and home until
Henry moved on to the Smithsonian in 1846. Julia Grummitt, writing
as part of the Princeton and Slavery Project in 2017, notes that from
Henry's letters it seems Parker would secure materials for experiments
and solve technical issues in Henry's work. Still, Henry seems to have
continually referred to Parker as a servant, suggesting that he saw him
in a lower status than a scientific lab assistant because he was Black.
Henry also wrote to his wife when he was travelling for work to say
she shouldn't allow Parker out at night. One student of Henry's
remembered students 'snapping sparks from his [Parker's] nose and
chin' and that Parker 'stood for it like a hero, in the interests of science'.
It's worth remembering that Henry used other 'human galvanometers'
(devices for detecting electrical charge), including himself. As we'll
see later, it was hardly an unusual scientific practice at the time. But
it's unlikely that Parker had much choice in the matter, and a group of
rich, white students surrounding you to draw electrical shocks from
your face can't have been exactly pleasant, even if it was 'in the
interests of science'.

Grummitt argues that although Henry was an anti-slavery
sympathiser, he clearly believed in a hierarchy of races and notes that

in 1862 he barred Frederick Douglass from speaking at a series of lectures from prominent abolitionists. We don't know what happened to Parker, we do know he was vital to Henry's work though. When Parker became ill briefly in June 1842, Henry had to halt experiments entirely. We also know that Parker was well known around campus for his sense of style, rumoured to own more than a hundred suits. He'd wait on college students, bringing them turkey dinners at night in exchange for clothes, and would change his outfit several times a day, enjoying the attention. So perhaps he was more a frustrated tailor than an electrician and we could hope that's where he went on to find more enjoyable work. History hasn't recorded his life in the detail we have for Henry, so we don't know.

Electricity had been buzzing long before either Henry or Parker got their fingers on it though. The Ancient Greeks had played with it, rubbing amber and seeing how small, lightweight objects like feathers might float towards it, or treating gout by standing on electric eels. The remains of a sort of battery have been found near Baghdad, dating from the Sasanian period (circa AD 225–640), although no one is sure what it was used for – medicine perhaps, or to provide small shocks from religious idols. Electrical research rumbled on without much development for several hundred years. Then, in 1600, Queen Elizabeth I's chief physician William Gilbert coined the term 'electric' in the course of his study on magnetism, *De Magnete* (a crucial topic for a country so obsessed with its navy). He'd taken the word from the Greek for amber, *elektron*, but also identified that many other hard materials could create a charge, offering new avenues for research.

Like gas and heat and so much else, it was in the eighteenth century that studies of electricity really opened up. In 1706 Francis Hauksbee, a Colchester draper turned air-pump manufacturer, placed his hand on a spinning glass globe to draw light, and then to attract and repel nearby pieces of metal. He showed a professor at Oxford and the experiments started to circulate around the scientifically interested clientèle of London's coffee shops. It caught the attention of Stephen Gray, a Kentish dyer who'd managed to build a junior career in science as an assistant at the Cambridge Observatory before ending up at Charterhouse – a school and almshouse in London built for

poor boys and 'impecunious gentlemen'. Charterhouse gave Gray
some freedom to pick up his electrical research and in its grounds he
also found an assistant in the poet Anna Williams, who'd been
assisting her father, another Charterhouse resident, with experiments
on magnetism. Gray started experimenting with how far he could
send electricity, playing with a range of materials including feathers,
hair, silk, wool, soap bubbles, beeswax and cubes of oak, soon
spotting that metal wire was excellent for carrying 'electrical vertue',
though he could manage a good effect with packing thread too,
eventually sending it 650ft. Historian Simon Schaffer suggests that
Gray possibly had an especially good 'knack' for pulling electricity
because of his training in dyeing, and notes that Hauksbee also started
in the cloth trade. Gray is perhaps best remembered for his tricks,
notably 'the electric boy': suspend a small boy from the ceiling using
silk strings, touch his toes with an electrified tube and you could
make his hair stand on end. When the experiment was replicated by
French chemist Charles du Fay he discovered that if you touched the
electric boy's nose not only might you see a spark, but you'd both
most likely feel a slight shock. Williams would later claim she'd
discovered the electric shock years before while assisting on one of
Gray's experiments. However, she made this claim after Gray had
died and no one else could verify it.

Both philosophising about electricity and the more spectacular
demonstrations were given a boost with the development of the
Leiden jar, a way of storing charge. This had been discovered by
scientists in the Dutch city of Leiden when someone accidentally gave
themselves a severe shock from a wire protruding from a jar of
electrified water. On further study, experimenters realised it could be
improved by lining the jar with metal foil. If you lined up several jars
together, you'd have enough charge to kill a small bird. Before long,
electro-magicians were touring the parlours of Europe with Leiden
jars to show the scientifically engaged bourgeois of the eighteenth
century everything from feathers flying up, to an electric boy, to
sparks flying around a gilt picture frame, or rum being set on fire from
a Leiden jar. With an increased charge, the electric boy was also
developed into the 'electric kiss' – a young woman would be put in
place of the boy and men would be invited from the audience to kiss

her on the cheek. Sometimes the charge would be strong enough to crack teeth (eighteenth-century teeth, but teeth nonetheless).[*]

In the 1740s, Benjamin Franklin saw the electric boy demonstrated. His interest sparked and, a shipment of books and electrical equipment later, he was engrossed. By April 1749, he was excitedly writing to a friend in London that with balmy weather coming in the summer he was planning an 'electric picnic' by the banks of the Schuylkill River. The party, he imagined, would ignite the alcohol in their drinks by sending a spark from one side of the river to the other (using the river as a conductor). They would also kill a turkey using a shock, roasting it on an electrostatic motor Franklin had devised just for the occasion, over a fire kindled with discharge from a Leiden jar. To round things off, they'd drink to famous electrical researchers in Europe from 'electrified bumpers' – glasses tingling with static to shock the drinker's lips – all while firing guns again ignited by electricity. He coined the term 'battery' to describe a row of Leiden jars lined up to give him enough charge for these sorts of tricks, like a battery of cannons would be lined up side by side to batter a target. It was still a way away from modern batteries, but as the idea developed Franklin's term for it remained. He also coined the term 'electrician', although this meant 'experimenter' at the time, not 'fixer'. The idea of electricians as fixers would need consumer electrical devices and infrastructure first.

Sometime in the autumn of 1752, Franklin embarked on his infamous experiment with a kite in a thunderstorm. It had been planned for a while and he'd tried similar experiments before. It was more than just a physical challenge. At the time, many people believed lightning and electricity to be entirely different phenomena, the former being a manifestation of God and quite distinct from the terrestrial goings-on of scientists rubbing glass with paper. This was why, people surmised, so many churches were hit by lightning. They also found Franklin's idea of lightning rods as a way to protect high

---

[*] Possibly the most impressive trick was performed by French clergyman and physicist Jean-Antoine Nollet in the 1740s when he lined up 180 of Louis XV's soldiers at Versailles, all holding hands, and watched them jump as the charge passed through them (he later also tried it with 750 Carthusian monks).

buildings as tantamount to blasphemy. Undeterred, Franklin carried his silk kite, hemp rope, key and Leiden jar out for an experiment that showed you could collect 'electric fire' from clouds just as you might usually do by rubbing a glass globe or tube. For his efforts, Franklin was given honorary degrees from Harvard and Yale and elected a Fellow of the Royal Society. Moreover, electricity as a topic was considered to have risen from fun and games to something more fundamental, with practical applications.

Electrical charge was, like pretty much any new discovery, soon hailed as a medical cure by respectable doctors and quacks alike. Possibly due to vestiges of ideas of lightning coming from God in the sky, many believed in a link between electricity and life. So, there was little surprise when in January 1791 Luigi Galvani – a physician and natural philosopher living in Bologna, Italy – announced he'd found electricity in animal tissue,[*] and developed a theory of the brain as the source of 'animal electricity' that would then be conducted around the body via nerves and could be stored in muscles a little like a Leiden jar. Alessandro Volta, who by then had moved on from experiments with marsh gas, was sceptical. He repeated the experiment and decided that all the flesh did was conduct electricity, it wasn't coming from the animal itself. The debate raged back and forth for the next decade or so, but when Napoleon invaded northern Italy Galvani refused to swear loyalty to the new government, leaving him stripped of his professional positions and with that his finances, dying depressed and in poverty in 1798. Volta, on the other hand, was happier to cosy up to the new regime and used his research into electrical conductivity to build an improvement on the Leiden jar, a sort of battery made from a pile of disks of zinc and copper, interspersed with cloth soaked in salt solution. Reports of this new piece of

---

[*] There are various legends about how this discovery came about. In some versions, his wife Lucia was preparing a dinner of frogs' legs during a thunderstorm and Luigi noticed the legs seemed to twitch with each flash of lightning. In others, one of them was dissecting frogs on a table where an experiment into electricity had also been set up, and Luigi spotted that when the electrostatic machine was turned the frogs' legs jerked.

electrical kit spread; Volta was lauded by scientists across Europe and made a count in the Napoleonic government.

The idea of animal electricity wasn't entirely dead yet though. Galvani's nephew, Giovanni Aldini, was determined to defend his uncle's reputation. Weaving his way through various wars in Europe to tour scientific institutions, he used the heads of decapitated frogs and dogs to argue that brains and muscles worked in the same way Volta's batteries did. In January 1803, Aldini even managed to run this experiment on the body of George Forster who'd been recently hanged for murder in London. He connected Forster's remains to 120 plates of zinc and copper while a public audience watched with amazement as the body's jaw seemed to start to shake, before muscles contorted and at one point, an eye opened. The Royal Humane Society wondered if it could be used to bring people back from drowning and it wasn't long before you could pay an electro-physician £50 a night to stay on a 'celestial bed', where electrical powers were said to cure impotence.

* * *

When we left young Humphry Davy, he was setting up shop at Count Rumford's new Royal Institution (RI) in Mayfair. Back at the controversial Bristol clinic for taking the new airs, Davy had become enamoured by both Thomas Beddoes' wife and inhaling nitrous oxide. The new job offered him a fresh start and he grabbed the opportunity to build a new image of himself as an establishment chemist. Although the RI had initially been seen as a way to educate the masses, it had ended up more like a posh version of a mechanics' institute. A way for the upper classes to enjoy science and be seen by each other to do so. The RI's amazing funnel lecture theatre was built so the audience could look down on the demonstrations and get a good view of what they were presenting. But it also has the advantage of allowing you to see other members of the audience: the scientific equivalent of a box at the opera.

Davy's first lecture at the RI was on galvanism. Taking to the stage with the skill of a conjuror, his lyrical approach to scientific performance was popular and he soon built up a large following. that the road outside the RI became London's first one-way street as a way to manage the congestion caused by all the carriages arriving to watch his lectures. Davy also built up rather a strong image of himself, buoyed by friends like the poet Coleridge who flattered him as not just a great scientist but a great writer too. Despite his relatively humble background, Davy soon started to treat anyone he saw as his social or intellectual inferior with contempt. Careful not to lose his social position and wanting to avoid anything that might be seen as too hungry for spectacle, he avoided playing with dead bodies, human or otherwise, and rooted his electrical research in chemistry rather than biology. By 1808 Davy had built a gigantic battery of two thousand plates in the RI's basement lab, using it to decompose chemical compounds, extracting first potassium and sodium, and later magnesium, calcium, barium and strontium using electricity. Not entirely against spectacle, as long as it had a sufficient scent of respectability about it, within a few years Davy was also astonishing audiences with brilliant arcs of light made from two charcoal sticks connected to a voltaic pile; a dazzling sight in its own right and a serious step towards electric lighting.

Davy is just an *amuse-bouche* for the real electrical hero of the RI – Michael Faraday. Faraday's story is sometimes overused as an example of the meritocracy of science, but it's still quite the scientific fairytale.[*] He was born in 1791 in a place called Newington Butts, a bit of south London that is today better known as Elephant and Castle, and back then would have had farm animals grazing on it. While Faraday was still a child, his father got a job with an ironmonger near Marylebone and the family moved north of the river. He had the chance to attend a school, albeit nothing particularly special, and would explore the local area, watching excitedly as the canal from Birmingham was built at Paddington, pondering how an iron barge filled with coal could still stay afloat. Not long after his 13th birthday, Faraday started work running errands and deliveries for a local bookshop. He'd often read the books he was sorting or binding and developed a strong

---

[*] For more on Faraday (and, to some extent, also Davy), see Frank James's (2010) *Faraday: A Very Short Introduction*. An entertaining, detailed and clear biography, it packs in a lot in a few pages.

interest in science. He'd also go to evening lectures and would play around with experiments at home. He was encouraged by his employer, who'd get the young Faraday to show off his notes to customers. One customer was so impressed he gave Faraday tickets for some of Davy's lectures at the RI.

By this point, it was the spring of 1812. Davy had been knighted and married an extremely wealthy widow who'd fallen for him after seeing one of his lectures. Still only 34, he was now rich enough to take semi-retirement and these lectures Faraday had tickets for were to be his final performances. The young Faraday sat in the gallery of the RI's lecture theatre, by the clock, looking down on the performance from one of the highest points. He took meticulous notes and, using his bookbinding skills, presented them to Davy in a beautifully bound 300-page book the following Christmas. Not long after, when one of the RI's chemistry assistants was dismissed for fighting, Davy remembered this keen young bookbinder and offered him a job. Faraday was paid less than he got at the bookshop, but it came with a room at the RI – and he got to work on science. The following autumn, Davy announced he was going on a tour of Europe and Faraday was invited along as his 'philosophical assistant'. Davy had quite the scientific grand tour planned, managing to use an appeal to the importance of scientific debate to get through the politics of the day (much of the continent, including Britain, was at war with Napoleon at the time). In Paris, Davy demonstrated that iodine was a chemical element, in Montpellier he experimented on electric fish and at Vesuvius they partied in celebration of the allied victories over Napoleon. Davy's trip was cut short when Napoleon escaped from Elba the following year, but Faraday stayed on, meeting with Volta, who gave him a battery that you can still see on display at the RI's Faraday Museum today.

Back in London, Faraday started to build a reputation as a skilled chemist. He assisted Davy on some of his work, including experiments on a sample of 'firedamp' (methane) he'd picked up in Newcastle to develop a safety lamp for miners. In September 1821, Faraday was commissioned to look into the phenomena of electromagnetism, which had been discovered by Danish chemist and physicist Hans Ørsted the year before, when he noticed that a compass needle was moved by an electric current. Faraday found he could make a wire carrying an electric current move around a magnet. In his excitement, Faraday published results without acknowledging that he'd worked alongside Davy. This strained their relationship and it's possible that Faraday avoided working on electromagnetism for a while because of this, embarrassed by the experience. Faraday was busy with other work too though. There was an unsuccessful project with father and son Marc and Isambard Kingdom Brunel to use gas liquefaction as a source of power, and Davy pulled him into helping work out the ventilation and lighting system for the new Athenaeum Club. In 1825 he also set up a series of Christmas lectures for children at the RI that still run today.

Whether it was because Davy died in 1829, or just that Faraday found a bit of spare time, he returned to electromagnetism in August 1831. Continuing experimenting over the autumn, by October he found he could generate an electric current by moving a magnet in and out of a coil. In 1821, he'd made things move using electricity, now he could make electricity with movement. It was really inefficient, not practical as a power source itself. Still, it was the start of something. Virtually all the electricity we use today is generated using the principles developed by Faraday. In 1833 Faraday was appointed the first 'Fullerian professor' of chemistry at the Royal Institution. At this point, it's worth noting that even Faraday's rags-to-riches tale is a good example of one of Ida B. Wells's criticisms of the Chicago Exposition – that when white people made great progress, they often did it off the back of the slave trade. Just as James Watt got his first break at the University of Glasgow when a consignment of salty astronomical equipment arrived from a Jamaican plantation, this professorship came via gifts to the RI from John Fuller, whose inherited fortune was built on a mix of slave ownership and gunmaking.

★ ★ ★

By the 1840s, batteries had developed well enough to support the development of the electric telegraph. As telegraph lines proliferated, so did jobs in electricity, with training systems, societies and magazines soon following to build this new profession. By the 1860s telegraphy was a major industry on both sides of the Atlantic, with companies like the Berlin-based Siemens & Halske establishing hubs in several other countries as well as a host of international contracts. It was an exciting new industry to be part of too, full of wonder about what more might be discovered and created. There was something mysterious, mystical even, about electricity, possibly because it never shook off the association with God. The University of Cambridge set up a centre for electrical research at the new Cavendish Lab (named after Henry Cavendish, following a grant from one of his rich relatives), with Scottish mathematical physicist James Clerk Maxwell appointed the first Cavendish professor. In Germany, Werner von Siemens and Hermann von Helmholtz convinced German Chancellor Bismark to fund a new imperial technical institute so new opportunities in electrical research wouldn't simply be captured by the British.

There were experiments with electric transport as early as the 1830s. Thomas Davenport, a blacksmith in Vermont, made a workable model of an electric car based on battery-stored power, and German-born Moritz Hermann von Jacob also sailed an electric boat (and 14 passengers) along the River Neva in St Petersburg, powered by an engine working off a battery developed by Welsh electrician William Robert Grove. In November 1842, Aberdonian chemist Robert Davidson managed to run an electric locomotive, the *Galvani*, on the Edinburgh to Glasgow line (although it could only manage 4mph and no passengers or goods). In DC, chemist Charles Grafton Page managed to convince Congress to invest $40,000 in his project to build an electric locomotive. It took him a decade and when he did finally set off to Baltimore in late April 1851, the battery cells smashed on the bumpy ride, only making it a few miles into Maryland. It'd be a few more decades before electric transport would go mainstream. Light was the big prize for the time being, although it would still need work too. Davy's invention of an intensely bright arc light had found a home in some lighthouses and the odd avant-garde country pile, but it was generally seen as too bright for domestic use. An experimental arc light was installed on the Place de la Concorde in Paris in 1844, but it was so strong people had to put up their umbrellas for shade.

The science of electricity retained its sense of spectacle and, just as gaslighting had started with a scientific twist on fireworks, electrical lights were often first seen on stage. In mainstream theatre, the Paris opera used an arc light in 1849 to create the effect of a sunrise in a specially commissioned ballet *Electra*, designed to showcase their new powerful lamp. They later employed their own electrical expert L. J. Duboscq to design spotlights, rainbows and 'luminous fountains', and in 1877 he published a catalogue of these to sell to other theatres. There were also 'Geissler tubes', a sort of early neon sign developed in the 1850s by German scientific instrument maker Heinrich Geissler; fantastical glass shapes filled with a variety of gases to glow in different colours. In London, Professor John Henry Pepper of the Royal Polytechnic Institution used light to impressive effect with his famous 'Pepper's Ghost' trick, first presented on Christmas Eve, 1862. The Polytechnic was similar to the Royal Institution, but catered for less-well-off Londoners and was less reticent of putting on a spectacle. As well as Pepper's shows, it had a diving bell you could ride and became well known for its magic lantern pantomimes. Pepper, always with an eye for how to turn science into a show, built a 'monster' great electric induction coil in 1869 with which he could produce a flash of indoor lightning that was 29in long.

People found the flashes of electricity exciting to watch, but there was also a sense that this mysterious force might help unlock the secrets of the Universe. At the very least, it might usurp steam power. Edward Bulwer-Lytton's 1871 novel *The Coming Race* offers a sense of some of the utopian ways electricity was imagined in popular culture. Here, an American engineer finds a subterranean race of superior beings, the Vril-ya, whose culture revolves around a mystical power they call Vril (but readers would recognise as electricity), which provided them with telepathic powers, powerful weaponry and even the ability to control the weather. It soon became popular with Madame Blavatsky and others in her theosophy movement (which Thomas Edison would also briefly flirt with). Much later, in the 1930s, some Nazi fans founded a Vril Society, or 'the Luminous Lodge' in Berlin, which fused eugenics with allusions to the Vril. Although it's likely a fiction that Hitler was actively involved in this, stories like this are a possible inspiration for plots in the early Indiana Jones movies. The book is also where we get the 'vril' in the British 'beef tea' Bovril, the idea being that if you drank Bovril you might be imbued with some of the Vril's modern electric power.

In 1876, America held its first World's Fair expo, the Centennial Exposition in Philadelphia. As well as the right arm of the Statue of Liberty (complete with torch), a cannon from the Franco-Prussian War, a cable to be used in the Brooklyn Bridge, Heinz Ketchup, the first weather map in a US newspaper and a steam-powered elevated monorail train, visitors were amazed by Alexander Graham Bell's new telephone, which could transmit not just dots and dashes like the telegraph but the human voice. After judges William Thompson and Joseph Henry both pronounced it 'the greatest marvel ever achieved in electrical science' it won the expo prize medal, and the idea of telepathic electrical communications must have seemed ever so slightly less a matter of science fiction. Another of the inventions showcased at the 1876 Philadelphia expo was a new arc light from Moses G. Farmer, powered by a dynamo designed by Farmer along with William Wallace, who owned the country's foremost brass and copper foundry. They were, however, beaten to market by young Ohio electrical engineer Charles F. Brush who, by the autumn of 1878 was installing arc lights inside a Boston department store. His lights hissed, but they gave off a brilliant light. It wasn't long before New York's commissioner of public works had ordered 23 lights from the Brush Arc Lighting Company. It was very much still an experiment – in the same year New York bought a thousand more gaslights too – but it was a successful one, followed up by more orders. Still, even the new-generation arc lights weren't going to compete with gas or whale oil in the home or office, they were just a bit too dazzling. Novelist Robert Louis Stevenson (famous for *Treasure Island* and *Strange Case of Dr Jekyll and Mr Hyde*) described them as: 'A lamp for a nightmare! Such a light as this should shine only on murders and public crime, or along the corridors of lunatic asylums, a horror to heighten horror. To look at it only once is to fall in love with gas, which gives a warm domestic radiance fit to eat by.'

There were anxieties people might run out of coal, and as electricity developed so did experimenting with alternative energy sources. At the next World's Fair, the 1878 Paris Exposition Universelle, maths teacher Augustin Mouchot presented his development in steam solar. This included a printing press that, even when cloudy, could produce 500 copies an hour of a special solar-themed publication, the *Soleil-Journal*. Mouchot had started off playing with Saussure's solar ovens as a way to create steam without coal, adding a massive funnel of mirrors to concentrate the sunlight, an approach that is still used in today's concentrated solar plants. The French government gave Mouchot funding to take a scientific mission to Algeria to experiment with the abundant sunshine there (it's not just fossil fuels that have histories wrapped up in colonialism). Still, coal was cheap. Mouchot's research funding was cut and he went back to teaching.

The movement of electrons we'd now generally think of as solar power – those shimmery, blue solar photovoltaic panels your neighbour maybe has on their roof – took a bit longer to get going. But the basic principle that you can make electricity when sunshine falls on particular substances also dates back to the mid-nineteenth century. In 1839, young French physicist Edmond Becquerel (at just 19, working in his father's lab) first observed what we'd now call the photovoltaic effect, noticing he could create a small electric current when plates of some metals were immersed in an acid solution and exposed to sunlight. A few decades later, British chemist Willoughby Smith was working on insulation for submarine telegraph cables. He'd tried using coal tar between the gutta percha (a sort of rubber from South East Asia) coverings of the wires, and while experimenting with other options started playing with selenium rods. He found them entirely useless for his experiment but noticed their conductivity increased significantly under sunlight, writing it up in a February 1873 issue of *Nature*. A decade on, American inventor Charles Fritts built on this observation with what could be described as the world's first solar cell, using selenium and a very thin layer of gold. It was only 1 per cent efficient, so wasn't exactly going to compete with coal just yet, but it was a start, and there followed a flurry of patents and proto-solar entrepreneurs around the turn of the century. As we'll see in Chapter Seven, solar wouldn't reach maturation until the Cold War, but it was born in the nineteenth century.

There were plenty of other renewable sources for electricity though and it was hydro that got going first. William Armstrong had made a name and fortune for himself in hydraulics, guns and warships, and in the 1870s built a large country house for himself, Cragside, in Northumberland in the north of England. A grand palace set in 1,700 acres, it was partly a project for semi-retirement, but also crucial for impressing visiting dignitaries he wanted to sell arms to. Alongside the house, Armstrong built five artificial lakes, constructed elaborate rock gardens and planted 7 million new trees and shrubs. The interior was decorated with the finest marble and carved woods, with stained glass and intricate wallpaper on display alongside Armstrong's impressive art collection. It also featured the latest in domestic technologies; a complex system of plumbing providing hot and cold baths and a shower, as well as heating for the house. Hydraulic machines in the kitchen transported food and provided a rudimentary dishwasher. Electricity provided gongs to summon guests to meals, and there was even an internal telephone system. Initially, there were arc lights too, but they were too harsh, so Armstrong's friend Joseph Swan persuaded him to try his new incandescent lightbulbs (i.e making light from heat). And providing the electricity? Water, as Armstrong had placed a Siemens' dynamo under a nearby waterfall.

Brush was another engineer to dabble in renewable energy in semi-retirement. Having made a tidy sum from selling his arc lights to cities around the world, he built a mansion for himself on Cleveland's high-end Euclid Avenue and, in 1888, built a huge wind turbine that gave electrical lighting to his home, without failure, for 20 years. A cover story from a December 1890 edition of *Scientific American* paints an almost fantastical image of a 60ft wind turbine built from 144 blades, twisted like screw propellers. This fed a basement full of hundreds of jars acting as a battery, powering 350 incandescent lightbulbs, two arc lights and three electric motors. The cover illustration shows two men standing on top of an almost skyscraper-like turbine, having climbed up several flights of internal staircases. Among stories of a pineapple grove in Florida, a trip to the North Pole in a balloon and government debates on the artificial production of rain, *Scientific American*'s wind power feature warns that the wind might be free, but Brush's elaborate set up for turning it into electricity must have been very expensive indeed. Still, they muse: 'There is a great satisfaction in making use of one of nature's most unruly motive agents.'

Brush wasn't the first to try such an experiment. Scottish engineer James Blyth beat Brush by a few months. Blyth was born in 1839 in the village of Marykirk, a little south of Aberdeen. Blyth's parents weren't exactly rich – his father was an innkeeper and farmer – but he was educated at the local school, worked as a schoolteacher and, in 1880, got a job as a professor at Anderson's College in Glasgow. He kept a holiday cottage back home in Marykirk and it was there, in 1887, that he built a cloth-sailed, 33ft wind turbine and used it to charge a sort of battery, which in turn powered the lights. It was the first house in the world to be powered by wind-generated electricity. Blyth tried to sell the idea to the local villagers to light the main street, but they branded these newfangled sparks 'the work of the devil' and turned him away. Still, he got a patent for his invention and managed to build a second, improved turbine for a nearby lunatic asylum where it ran for the next 30 years. One can only imagine what Blyth would have made of the giant offshore turbines competing with oil and gas rigs off the Aberdeenshire coast today.

★ ★ ★

Whether you produced your electricity from water, wind or coal, homes lit by it were still a novelty in the 1880s. Having successfully provided Armstrong's mansion, Swann sold his lightbulbs to a few others happy to experiment – the Savoy Theatre and warship HMS *Inflexible*, for example. But the man who'd really bring electricity home was Thomas Alva Edison. Born in 1847, Edison grew up in Port Huron, Ohio, starting work aged 13 on the railways as a newsboy and then, a few years later, as a junior telegraph operator. His career in the telegraph industry wasn't exactly glowing. While working in Kentucky he'd been fired after he spilled sulphuric acid in the bureau's battery room and it'd eaten through the floor into his manager's office below. In Cincinnati, his colleagues noticed he'd often excuse himself from work, pleading illness but making a beeline for the local library. One of the many pieces of electrical research Edison absorbed was Faraday's. The Londoner became an instant hero to Edison; the poor boy who'd become an international scientific star. In 1869 Edison resigned to make more time for his inventions, first renting space in a corner of a telegraphic instrument manufacturer's in Boston and eventually, in 1876, building an industrial research laboratory in Menlo Park.[*]

Edison started off tweaking already established inventions like the telegraph or telephone, but things changed in 1877 when he invented the phonograph, a machine that promised to record and then replay the human voice. He presented it to the world with a degree of showmanship he'd later be famous for, taking it straight to the editor of *Scientific American*. There, he famously recited 'Mary had a little lamb' and replayed the recording to a crowd that grew and grew, demanding he replayed and replayed it until one of the magazine's owners intervened, concerned the floor would collapse with the weight of the audience. The following day's newspapers were full of reports of this exciting new invention, and also its inventor. It was a friend of Edison's – George F. Barker, a professor of physics at the University of Pennsylvania – who got him into lighting. Convinced Edison would find this area of innovation ripe for development, Barker sent him a steady supply of reports of new projects in artificial lighting. When this didn't work, Barker eventually escorted the inventor to William Wallace's foundry in September 1878 to see arc lights in action. According to a *New York Sun* journalist who came along for the ride, Edison was 'enraptured', running between the instruments like a child, sprawling calculations over a table, working out quite how much coal one might save with the right type of electrical lighting set-up, over a day, a week, a month, a year. On his return, Edison threw himself and the other resources of Menlo Park into lighting and within a week informed his contact at the *New York Sun* that he had invented the first practical incandescent lightbulb. Edison was vague when it came to the details (he hadn't got a patent

---

[*] There are plenty of biographies of Edison if you want to read more about his life. I enjoyed Paul Israel's (1998) and Neil Baldwin's (2001). Alternatively, Jill Jonnes' *Empires of Light* (2003) puts Edison's story in context of other key characters like Westinghouse and Tesla. Jane Brox's *Brilliant* (2010) provides an engrossing tour through the history of artificial light, and was a useful source both for the history of electric light here, and of whale oil, coal gas and rock oil covered in Chapter Three.

yet), but promised it was entirely different from anything that had come before. He'd light the southern half of Manhattan with a 500-horsepower engine that would not only give you light, but heat and power too. He'd build a whole electrical system. Within a month, Edison was ready to show his much-talked-about lightbulb to the world, promising a light without flicker, flame, fumes or smoke, a 'healthy' light. The *New York Sun* was appropriately glowing: 'There was the light, clear, cold and beautiful. The intense brightness was gone. There was nothing irritating to the eye.' Many experts were dismissive of Edison's hype, but gas stocks still dropped at the prospect.

Edison needed more than a lightbulb to show off to journalists, he needed one that wouldn't burn out too fast. The bulb he'd shown to the *New York Sun* gave off a lovely light, but only for an hour or so. He also needed a more powerful generator if he was really going to light up half a city. Stories of 'a lightbulb moment' in innovation are usually make believe, and this is especially true when it comes to the actual lightbulb. While Edison met with financiers in the Menlo Park library, his colleagues in the workshops busied themselves with improving the generator, and tried multiple different approaches to the bulb itself, from ways to blow the glass or how to vacuum air from it. They worked through a host of filaments; they had some made from horsehair, teak, spruce, silk from China, silk from Italy, cork, carbonised cotton-thread, a new material called celluloid, parchment and flax from New Zealand. They even used team members' beard hair, placing bets on whose would last the longest.[*] Finally, the new Edison system – lightbulb, generator, wires, insulators, marketing and more – took shape and, by the start of the new decade in January 1880, was ready for public display. They had to open a few days early, due to the *New York Herald* breaking the embargo on their exclusive before Christmas, but a steady stream of 'electrified' sightseers speedily made their way to Menlo Park to see this great new invention, and investment followed.

---

[*] The Smithsonian has a wonderful display of these bulbs in the National Museum of American History. Row after row of lightbulbs, each ever so slightly different, each representing the work of many hands, many moments; a physical representation of the fallacy of the myth of individual, quick invention.

The first home Edison lit was the Madison Avenue brownstone owned by financier J. P. Morgan, in June 1882. Wires were threaded through the building, hidden between wood panels and plaster, often in the spots where gas piping might have been, all run by a coal-powered steam boiler in the basement, attached to two electric generators. The boiler had to be run by an expert engineer who worked 3–11pm. If you wanted light to fix yourself a midnight snack, you'd still need a candle. It was also very noisy and smelly. The neighbours complained, and when Edison's team went in to try some improvements it caused a fire. But Morgan was eventually pleased, showing it off to his rich friends who soon put in their own orders too. Electric lights modelled on chandeliers, 'electroliers', were developed to let this fine new electric light shine through cut crystal and Tiffany's brought out a range of specially produced table lamps with stained glass shades. Mrs Vanderbilt was known to greet her guests wearing an electric dress covered in tiny bulbs, which would light up when she stood on a special conducting plate placed by the front door.

But all this was just a glamour side project next to Edison's much larger programme of building central electrical power plants that could power homes, offices and streets across a city. Morgan's brownstone was a neat bit of PR, but not a model to grow. Frederick Winsor and the gas companies had built a large central gas plant connected to a network of customers, in contrast to Boulton & Watt's smaller bespoke gas systems for factories. Similarly, Edison didn't want to build small, separate generators in millionaires' basements; he wanted infrastructure to feed the whole city. Indeed, the beauty of electricity was that you could distance yourself from the dirt of the generation of power. Rich men like Morgan didn't want to be bothered by coal smoke and the need to hire their own boiler engineer; that could be pushed outside, the power being wired in. Edison opened his first public electricity station in London on Holborn Viaduct in January 1882. Its coal-powered steam turbine drove a 125-horsepower generator named 'Jumbo' – after P. T. Barnum's famously large elephant – which lit a half-mile strip from the General Post Office building to the jewellers' quarter in Hatton Garden. It ran at a loss, however, and was closed in September 1886. But London was just an early experiment; it was Manhattan Edison had his eye on. At 3pm precisely, on 4 September 1882 – the day before the first Labor Day parade – the switch was pulled at the new Edison plant at 255–257 Pearl Street. Here another Jumbo generator churned out power via several miles of cables laid under a square mile south of Fulton Street.

Laying cables under streets, alongside water and gas pipes, was key to Edison's plans. As arc lights, telephones and electric burglar alarms had become more popular, a spiderweb of cables had grown overhead, an ever-present risk of electrification literally looming over New Yorkers' heads. Edison wanted his system to be as safe and unobtrusive as possible. The cables were wrapped in hemp from Manilla and inserted into 20ft lengths of cast-iron pipe, which themselves were coated in asphalt from that pitch lake in Trinidad, and boiled in linseed oil, paraffin and a pinch of beeswax. Edison's marketing played up an image of ancient 'evil' gaslight in contrast to a new, 'good' electric light, coming in as the knight in oh-so-gleaming armour to lift Manhattan from the dark ages. Gas, Edison's adverts argued, takes our oxygen to burn while also excreting its noxious impurities into our homes, it flickers and it is so much more heat than light. Electricity, in contrast, was painted as the clean and healthy way forward. Customers around Pearl Street were given the special introductory offer of a box of lightbulbs for free, and block by block, the lightly twinkling electric-lit cityscape we are now so familiar with emerged from the shadows.

When the Statue of Liberty was installed in 1886, she was lit up with footlights inside and out, and lamps studding the crown to give the effect of jewels. Fashionable ladies of New York started to adorn themselves with 'luminous electric jewels' too. Like the Statue of Liberty, these were also imported from Paris, the work of Gustave Trouvé who utilised an innovative pocket-sized battery that could be discreetly hidden beneath clothing, powering tiny coloured electric lights attached to brooches, necklaces, hairpins and tiaras, shaped like butterflies, stars or birds. He even had a luminous electric walking stick for sale. And then there were the electrical adverts. The first sign spelling out a word was displayed at an electrical exhibition held in 1882 at Crystal Palace. Designed by William Hammer, an Edison engineer working in London, the bulbs rather unimaginatively spelt out 'EDISON', as did the improved flashing sign Hammer produced for a German expo the following year. By the time a 50x80ft illuminated sign arrived on Broadway, it had more to advertise than simply electricity. A worker on a nearby roof had the job of turning a switch to illuminate the 1,500 bulbs that cycled between ads for fireworks, hotels and other products that wanted their name in lights. In 1892 Heinz added a 45ft illuminated pickle to the mix. Worried they might be left in the shade, local churches invested in electric crosses.

With all the pools of light pouring out from Broadway, it quickly picked up the title of 'the Great White Way'. As David Nye describes in his 1990 book *Electrifying America*, other cities soon sought to emulate the 'White Way' look, seeing an electrified main street as a way to deter crime, bring more people to shops and boost civic pride. Sometimes paid for by shopkeepers, sometimes at public expense, the inauguration of a new White Way could be a community event with speeches, press coverage and a parade. General Electric – the company that had grown out of Edison's – created a specific product for this growing White Way market, the 'Path of Gold' which used their incandescent lightbulbs rather than arc lights, selling first to San Francisco before many others. As Nye describes, the building of a White Way allowed cities to literally highlight their chosen spots. Tall buildings shone against the night sky and bridges were outlined by strings of bulbs. But at the same time, those parts of cities not illuminated were even harder to spot. Businessmen and politicians could shine a light on their favoured parts of a town, while the poor were kept hidden in the dark.

★ ★ ★

One of the many engineers Edison employed as the electric lighting market grew was a bright, if rather eccentric, spark from Serbia named Nikola Tesla. He'd arrived in New York at the start of June 1884, via studies in Graz and a stint at Edison's branch in Paris. An electricity utopian who was convinced it could free people from the pain and worry of hard work and scarcity, Tesla found Edison thrilling, although it doesn't seem to have been reciprocal. By this point, the Pearl Street station was doing well, but the setup – which could only deliver power to a half-mile radius – was proving a harder sell when it came to spots less densely populated than Manhattan. For Tesla, the solution was a generator built on alternating current, rather than a direct one, which could travel further. He pitched the idea to his boss, but Edison was dismissive – he'd tried AC and it didn't work, it wasn't safe. DC was better. Frustrated his ideas weren't taken seriously, Tesla quit after a fight over pay soon after, less than a year after he'd started.[*]

Meanwhile, railway entrepreneur George Westinghouse had watched Manhattan light up and wondered if it was time he got in on the business too. He'd read about AC generators in a British engineering journal in the middle of 1885 and figured, as Tesla had argued, that this could be a way to transport electricity over longer distances. This meant you could sell a central power plant, like the ones Edison was building, to less densely populated areas. It'd also be useful if you wanted to use hydropower in any big way. You could dig up coal and transport it by train to a power station in the middle of a city, but a waterfall was harder to move. One of his team was visiting family in Turin, Italy, so he asked him to investigate a 50-mile hydroelectricity project that was reportedly using AC to light buildings nearby, including the main station. Views on AC were still mixed – Werner von Siemens told Westinghouse's rep it was 'humbug', but Westinghouse figured it was worth a try. He bought the American patents from a group in Budapest, ordered a generator and a transformer, and hired a young physicist named William Stanley to build something sellable.

Stanley got to work improving both the transformer and generator, escaping Westinghouse's base in heavily polluted Pittsburgh to head for the Berkshire hills of Massachusetts. By March 1886, he was ready to give it a go, firing up a 22-horsepower steam engine in an old barn turned electrical lab off Main Street, Great Barrington. Five-hundred volts of experimental alternating current ran through copper wires

---

[*] For more on Tesla, I recommend Iwan Rhys Morus's (2019) *Nikola Tesla and the Electrical Future*, not least because it's as much about the electrical cultures around Tesla as the man himself.

into the basement of a store owned by his cousin, working its way through Stanley's new transformer to power three lights in the storefront above. A local newspaper gushed that this new electrical light was 'so perfectly white, that green and blue can be readily distinguished' (something that was less clear under gaslight). Within a week Stanley had wired up several other businesses too. A year later, Westinghouse had nearly 70 AC stations either built or in planning across the country. Tesla resurfaced too. After a period of poverty and hard labour following his work with Edison, he'd managed to find some financial support for his AC ideas and was filing a steady stream of patents from a new lab on Liberty Street, only a few blocks from Edison's Pearl Street station. Westinghouse, spotting some advantages in Tesla's designs over Stanley's and not wanting the AC competition, bought Tesla out.

Edison looked on at the rise of AC with increasing disquiet. He'd been annoyed by the Westinghouse installation in New Orleans, where he'd had a plant planned too, and was especially twitchy about the many companies (AC or otherwise) infringing his lightbulb patents. Arc light companies like Brush didn't worry him much, he reckoned this was a flawed technology on its way out, but this new AC competitor in the incandescent bulb business could be a problem, especially as they were profiting from all his work on perfecting the lightbulb. All this was happening in the background of rising public concern over electric safety. For all that electric light had been marketed as the great white knight in contrast to grubby gas, 'death by wire' became a repeated concern of the New York papers. In November 1887 Edison received a letter from Dr Alfred Southwick, a member of the New York State Death Commission, wondering if electrocution could be used as a slightly less barbaric alternative to hanging. Initially Edison refused – a lifelong opponent of the death penalty, he didn't want anything to do with such a project. But a month later, when Southwick tried again, Edison replied that, after careful consideration, if the state was going to go about such things, the most effective device would be the 'alternating machines' that were 'manufactured principally in this country by George Westinghouse'. Edition knew exactly what he was doing with this sort of statement.

Early the next year, Edison followed this up with an 84-page diatribe against AC jacketed in red with the title in capital letters: WARNING. When it came to safety, Westinghouse could fight allusions to fire with fire. There hadn't been any fires from its systems, but Edison had suffered several, including destruction of its power plant in Boston and a theatre in Philadelphia. However, the idea that AC was dangerous was helped along by electrical consultant Harold P. Brown, who in May 1888 wrote a long and angry letter to the *New York Post* condemning the 'damnable' AC. This wasn't instigated by the DC lobby; Brown seems to have come to hate AC of his own accord. Still, it suited Edison nicely. He offered Brown use of his new laboratory at Glenmont, West Orange, New Jersey, along with the assistance of his chief electrician Arthur E. Kennelly. A few months later, they put on a gruesome electric dog show in a room at the School of Mines at Columbia University in New York. Here, Brown took a large black retriever, strapped him in a cage and, after measuring his 'resistance' at a little over 15,000 ohms, shocked him first with DC and then AC, showing how much worse the latter was. Brown had been preparing this for weeks, paying kids near the lab 25c to go to find dogs for him to practise on. Many of the spectators left the room, shocked. When the dog died at the administration of 330 volts of AC a reporter from the *New York World* stood up to protest. He was followed by a representative from the American Society for the Prevention of Cruelty to Animals (ASPCA), who put a stop to the proceedings. Brown repeated it a few days later though, killing three more dogs in front of public health officials and journalists. When it was repeated yet again in December in the Edison labs (this time with a calf and then a large, heathy horse) the *New York Times* reported that 'experiments proved the alternating current to be the most deadly force known to science'.

Westinghouse took out adverts in the New York press, arguing AC was safer than DC, that the animal experiments were more about scaring the public than actual science, and that Brown was being paid by Edison. Brown doubled down, challenging Westinghouse to an electric duel: each would pick their current and, starting with an initial jot of 100, gradually increase by 50-volt increments until one or the other 'publicly admits his error'. At which point Westinghouse seems to have decided the best idea was to simply ignore the man. But this bad PR is one of the reasons why the 1893 Chicago Exposition contract was such a coup, offering them a chance to show AC not only as safe but all the fun of the fair. Indeed, one of the electrical highlights of the expo was a trick from Tesla where he passed 250,000 volts through his body; his white tie and tails were engulfed by sparks of dazzling lights, and yet he was entirely unhurt.

★ ★ ★

Niagara Falls was the big prize though. The development of the railroads had made it a popular tourist spot, and as tightrope walkers excited visitors with performances like headstands or cooking omelettes balanced over the great falls, businessmen could see the power of all this tumbling water and lamented that it wasn't put to work. In the early 1880s a local entrepreneur diverted some water down a canal to turn a wheel. Worried this might turn the area into an industrial park, this prompted New York state to create the Niagara Reservation, forbidding the construction of buildings within 500 acres. Regulation so often being the inspiration for invention, this restriction inspired Thomas Evershed, an engineer from the New York state canal system, to develop a much more ambitious plan (ridiculously ambitious some might say). This worked around the rules with a large tunnel under the village of Niagara. The prize, if they pulled it off, would be huge, but it'd take a lot of work and a lot of money. Just to build the tunnel, they'd need over a thousand men to pull out 600,000 tonnes of rock, lining it with a layer of cement and 20 million bricks. Evershed's original plan to raise $1.4 million floundered, but after bringing in New York banker and fixer Edward Dean Adams, he started to attract interest from some of the richest investors of the gilded age, including J. P. Morgan, William Vanderbilt and John Astor. They still had a problem though. They could make a lot of electricity, but where could they sell it? There was Buffalo, but it was 26 miles away and no one had sent electricity that far before.

To solve the problem, Adams brought together an International Niagara Commission. Led by British scientist William Thomson (later Lord Kelvin), this would bring together electrical expertise from around the world with an innovation competition. Westinghouse was scathing: 'These people are trying to secure $100,000 worth of information by offering prizes, the largest of which is $3,000. When they are ready to do business, we will show them how to do it.' His dismissive attitude won out, although only after several more years of fights between AC and DC. Finally, on 6 May 1893, days after the switch was flicked for Westinghouse's glamour project at the Chicago Exposition, the Niagara Falls Power Company declared that AC was, without a doubt, their choice. It was still a leap of faith – with William Thomson still lobbying for DC until the very end. Finally, after more fighting over AC and DC, expensive testing, calibrations and improvements, on 16 November 1896, at the stroke of midnight, the switches for three, 86-tonne 5,000-horsepower Tesla motors (five times the size of the generator at the Chicago World's Fair), were pulled, sending electricity 26 miles down the line to Buffalo. This mysterious force – which only a hundred or so years before people were rubbing glass to play with – had come a long way.

CHAPTER SIX

Tree Huggers

Ever been called a tree hugger? Ever called someone else one? Ever hugged a tree? The term started to pop up around the 1970s, largely as an epithet. It's a way to laugh at someone seemingly more interested in nature than other people. It's sometimes directed at the green movement at large. Sometimes it's more specifically the dreamier, whiter, more privileged ends of environmentalism who can afford to go around cuddling plants as opposed to challenging the political, economic, cultural and social situation around them. But tree hugging has quite radical roots.

The first recorded tree huggers date back to 1730, in Khejarli, Rajasthan, in the north-east of India. So the story goes, khejri trees were due to be felled to build a new palace for Maharaja Abhai Singh, the ruler of Jodhpur. Khejri are a reasonably delicate flowering member of the pea family that can grow in desert regions and held strong cultural significance for the local Bishnois people. They were materially important too. People relied on the trees for firewood or their shade in the dry heat. The fruit was the basis of a local dish, and the leaves, sap and flowers were all used in medicine too. One villager, Amrita Devi, said she'd rather die than see the tree in front of her home felled, and offered her life in exchange for the life of the tree. The axemen took her at her word, chopping off her head and sparing the tree. Her three daughters followed with three more trees, paying the same price. Soon, villagers were wrapping their arms around the trees to put their bodies in the way of the axes, 'hugging' them. This spread to other Bishnois in the region, with hundreds of tree huggers killed before Abhai Singh called an end to the bloodbath, recalling his men and promising the village would never again be compelled to give up their wood to the kingdom. You can still find monuments in Khejarli to Devi and the others who died for the trees, and Bishnois gather every year to pay homage to the people who lost their lives in the massacre.

In the nineteenth century, the British took their axes to Indian forests – for sleepers to supply the railways they were building or for ships, or to clear land to grow crops – and the destruction soon made an obvious mark on the land. Forest acts in 1865, 1878 and then 1927 consolidated British rule over Indian trees, but where ideas of 'conservation' were applied, it was to preserve timber resources for colonial uses, not for the trees themselves or other local economic or cultural ties people had with them. A heavy monsoon in 1920 had led to crop failure, unemployment, hunger and disease in the central district of Madhya Pradesh. Coupled with a growing sense of revolution, this had led to a wave of protests, with people breaking the British forest laws as a statement of non-cooperation with British rule. In 1930, Mahatma Gandhi led a large *satyagraha* – which loosely means a 'force of truth', a form of non-violent direct action – against the British salt monopoly. Madhya Pradesh, not having a coast from which to obtain salt, joined in with a new, larger wave of forest-based *satyagraha*, causing the British government significant financial losses (as well as concern about the mounting rebellion).

Indian independence was far from the end of these protests. In the 1960s, a spate of road building and badly managed logging in the rural Himalayas of Uttar Pradesh had led to a sense that the government was colluding with large industries to take control (and strip) local forests, ignoring the needs and wishes of local people. After the felling of trees was blamed for exacerbating flooding in 1970, a new 'Chipko' movement grew, drawing on the story of Amrita Devi and her daughters two and a half centuries before (*chipko* means to 'hug' or to 'cling to') as well as the more recent Gandhian principles of *satyagraha*. Photos of defiant women creating a circle around thick tree trunks spread around the world. The movement grew across the Himalayas and in 1980 Prime Minister Indira Gandhi agreed a 15-year ban on commercial logging in the area. It's possibly this well-publicised twentieth-century version of tree hugging that inspired the tree hugger epithet. Or perhaps it just fuelled use of the term after it had been coined.

Wherever the term actually came from, tree hugging is far from fluffy. Even if the eco-socialist and eco-feminist aspects of Chipko are sometimes romanticised, their stories – from 1730 through to the 1970s – are as much about women finding power against men, working-class people standing up to a coalition of state and industry, and people protecting the natural resources their lives depend on as they were anything to do with defending the tree. That isn't to say the rights of the tree aren't important, but they're only part of the picture. The same is true of climate change focused campaigns today. This chapter introduces some of the early movements against fossil fuels and for conservation, which would become the basis of the modern movement for action on climate change. It includes a reasonably diverse mix of characters and is rarely a simple story of nature versus humanity.

* * *

In recent years it's become common for climate activists to talk about the fossil fuel industry's 'social licence to operate' – the social status of, for example, a multinational oil company as a normal, even necessary, part of the economy despite outcry over oil spills, reports of human rights violations or concerns about climate change. This status has to be carefully cultivated, through careful relationship-building with press and politicians, for example, or sponsorship of sports teams, arts exhibitions and educational programmes, and it's nothing new. Back in 1914, the Rockefeller family famously brought in the skills of PR man Ivy Lee after the company was implicated in a very bloody conclusion to a coal strike in Colorado.[*] Fossil fuels didn't come with such a social licence already built in. Coal in particular was generally disliked at first, the burning of it banned even by Edward I. The British only turned to the sooty stuff once they had ripped through their forests and needed to preserve wood for ship-building. In his 2016 book *Fossil Capital*, Andreas Malm describes a form of 'Wattolatry' emerging around the middle of the nineteenth century, with people worshipping not just James Watt himself but, more broadly, the steam

---

[*] Amy Westervelt's audio series *The Mad Men of Climate Denial* (season three of her *Drilled* podcast) offers an account of this and other similar stories from the early days of oil PR. It's a beautiful bit of storytelling and very much worth downloading.

power he signified. But this veneration had to be nurtured to grow. It, like the steam mills, had to be constructed.

Some felt it prudent to throw cold water on the coals, so to speak, concerned by the new patterns of political and economic power that seemed to follow these engines, as well as their environmental impact. The Duke of Wellington, the prime minister who George Stephenson had taken on the inaugural ride of the Liverpool to Manchester railway line, remained a steam train sceptic, convinced they'd only encourage the lower classes to move about more. Similarly, the head of Eton College objected to railways linking London and Windsor, lest students might be tempted to take corrupting and distracting trips to London – or worse, out of town riff-raff would visit. Others were worried by the quantity of coal steamships got through, figuring surely at some point soon we'd run out. In 1865 economist William Jevons's book *The Coal Question* warned that increasing the efficiency of engines, as Watt did, only increased their use, meaning we'd end up burning more coal overall, not less. Policy enthusiasts today still talk of the 'Jevons paradox' of energy efficiency. Jevons thought there were still a few decades left of coal-based economic growth and so it was important to invest in educating society to be better prepared for the challenges to come. The book's preface offers a warning many generations since might have taken notice of: 'We are now in the full morning of our national prosperity and we are approaching noon. Yet we have hardly begun to pay the moral and social debts to millions of our countrymen which we must pay before evening.' For those who worked closely with the daily chug and belch of steam engines, there was the more immediate problem of the heat they produced, and suffocating levels of carbon dioxide inside factories, not to mention the risk of exploding boilers (just ask George Stephenson's blinded father). It's hard to get a precise figure for the number of boiler explosions; official government statistics had it as under 500 in the whole of the 1850s, but the *Engineer and Mechanist* magazine estimated it was nearer one a day.

Some people became very rich off the back of steam engines, others less so. In 1842 German steam magnate Friedrich Engels Sr sent his son to the family mills in Salford, hoping this might distract him from the radical politics he'd picked up in Berlin. There, Friedrich Engels Jr met

Irish woman Mary Burns, who took him on a tour of some of the poorer districts in the area. He went on to write *The Condition of the Working Class in England*, the basis for his relationship with Karl Marx. Industrialisation didn't just produce new hazards at work either; poorer people were more likely to live in the more polluted parts of town. The rich could move, or simply found it was never built near them in the first place. If they craved some sunshine because the smog had blocked the Sun or the river got especially stinky in the summer heat, they could get out of town for a few days. The poor had fewer options. After a fatal gas works explosion in 1865 at Nine Elms in south London (an area better known today for luxury housing developments and the new American Embassy), *The Times* compared gasometers to gunpowder stores. The Royal Society suggested gas works should be moved away from residential areas. But the industry dismissed such fears, complaining it was simply an attempt to harass them. Samuel Clegg, the former Boulton & Watt man who'd ended up working for the Gas Light and Coke Company, said he'd be happy to put his bed on top of a gasometer, they were so safe. But of course, he didn't have to.

As hype around the Great Exhibition had grown in the run up to 1851, some religious commentators saw it as a harbinger of the apocalypse. An anonymous tract from the evangelical Plymouth Brethren warned that material objects were being worshipped as God was ignored. 'The gods of gold and of silver, of brass, of wood and of iron, are praised' it stated, quoting the book of Daniel 5:4 and urging committed Christians to avoid the exhibition. There's a story of designer and proto eco-socialist William Morris taking one step inside the Crystal Palace and promptly throwing up at its vision of industrial promise. This is probably apocryphal, but the Arts and Crafts movement Morris helped build offered a popular critique of the way the Industrial Revolution created greater distance between manufactured goods, nature and craftspeople. Strongly influenced by John Ruskin, Morris was worried about the ways in which the machines and division of labour in factories lost not only a connection with nature but also a sense of the product as a whole. For Morris, this not only made for unpleasant work, but also led to less creative, less beautiful products. Guilty, like many of the time, of romanticising the past, especially the medieval period as a model for good living, Morris was, however, also a utopian thinker with a vision of a future under socialism where workers would control their own crafts. However, if Morris did have problems with the Great Exhibition, he got over it enough to agree to decorate the tea rooms of the museum built in South Kensington with its profits. Now the cafe of the V&A Museum, you can still buy a sticky bun in it today, bathing in the beautiful forest-green space Morris produced.

In the US, transcendentalist Henry David Thoreau warned in his 1854 book *Walden*: 'We do not ride on the railroad; it rides upon us.' The transcendentalist movement had grown up around New England in the 1820s and 30s, inspired by a mix of British and German romanticism as well as Hindu philosophy, all put through an American Puritan treatment; it emphasised the role of the individual, the possible goodness of people and gratitude for nature. *Walden* was Thoreau's account of two years spent living ostensibly alone in a cabin in woodland owned by his friend, the fellow transcendentalist Ralph Waldo Emerson near Concord, Massachusetts, immersing himself in nature and reflecting on the ways he felt the world had been degraded by the Industrial Revolution. In a wonderfully scathing *New Yorker* essay in 2015, Kathryn Schulz dubs it 'the original cabin porn: a fantasy about rustic life divorced from the reality of living in the woods, and, especially, a fantasy about escaping the entanglements and responsibilities of living among other people'. As Schulz points out, Thoreau was hardly off grid, dropping in on his family and friends 20 minutes' walk away (his mother and sister also brought him food). Still, it inspired generations of American environmentalists.

★ ★ ★

Some of the earliest and most high-profile reactions against the Industrial Revolution came from the poets and artists of the Romantic movement. William Blake's 'dark satanic mills' side-eye at Boulton & Watt's steam-powered flour factory, for example. However, for all that there were points of tension between romanticism's focus on subjective, individual experience and the more objective thinking of the Enlightenment, scientists found inspiration in a romantic telling of nature (and vice versa). Arguably it's no surprise the Romantics found common cause with scientists and industrialists though – they tended to be of the same class, knowing the same people, living similar lives and benefiting from the same structures in society. Humphry Davy was friends with Samuel Taylor Coleridge and William 'wandered lonely as a cloud' Wordsworth, eagerly inviting them to inhale nitrous oxide and describe the experience back to him, hoping their way with words would help him to better scientifically understand the gas's effects.[*] Through the same social circles, Coleridge and Wordsworth would also forge a strong friendship with Josiah Wedgwood's son Tom. Together, they'd enjoy a lot more than just the new chemicals Davy could cook up; botanist and president of the Royal Society Joseph Banks sent them a bag of hemp for 'experiment' and they had a regular supply of opium via Wordsworth's sailor brother. Tom Wedgwood and his brother Jos had the idea of building a sort of educational centre with the money their father left them, a 'nursery to genius', and employed both poets. This never got off the ground, but they acted as patrons to Coleridge, giving him £150 a year for life.

Much has been made of the romantic poetry of scientists like John Tyndall, Humphry Davy or Erasmus Darwin ('They were *scientists*, writing *poetry*!' as if this should surprise us), but arguably this only ignores the much more interesting work produced by industrial workers of the time. Both the 'mill girls' of Lowell, Massachusetts, and members of the Chartist movement in the UK produced large amounts of poetry, offering a first-hand record of changing relationships with nature and industry that characters like Blake or Thoreau could only imagine. The town of Lowell had been founded in the 1820s as a sort of experiment to take the new British approaches to textile manufacturing to America. The town drew much of its labour from young women, usually in their teens, who had grown up in the New England countryside and left home with the promise of relatively good pay and lodgings in supervised boarding houses. They'd moved from tasks such as berry picking or butter making at home to noisy, hot, crowded and repetitive work in

---

[*] For similar reasons Davy also urged a Dr Peter Mark Roget to take nitrous oxide. He was apparently at a loss for words but did go on to invent the thesaurus. For a beautiful tour through the links between scientists and Romantics of the time, see Richard Holmes's *The Age of Wonder* (2008).

the factories. Some of them would have worked back home spinning, weaving or dyeing cloth, just as they did in the factory, but in an entirely different setting. And they'd write about it. As Chad Montrie describes in his (2001) *A People's History of Environmentalism in the United States*, they'd write poetry and prose, at first in letters and diaries, but soon published in local magazines too, with subscribers across New England. Clearly influenced by the Romantics, they'd play on a sometimes-sentimentalised idea of a past where people had lived well in an idea of nature which was often also associated with the divine, contrasting this with the brutal new machines. The mill girls would often describe themselves as cooped up, wanting to escape the noise of machines and smell of smoke. As one described herself, she was 'like a prisoned bird, with painful longings for an unchecked flight amidst the beautiful creation around me'. Charles Dickens visited them in 1842 and read their literary journals, celebrating the 'strong feeling for the beauties of nature' in the mill girls' work. It's sometimes argued that he took a few ideas for *A Christmas Carol* from them.

The poems weren't shared simply for entertainment or literary value but also as a way to bring the new industrial workers in the area together, raising political points and organising strikes. They were far from the only labour activists at the time to use poetry in this way. Poetry was a big part of Chartist culture and, like the mill girls, was full of imagery of new industrialised work in contrast with a romanticised idea of the pre-industrialised countryside as somehow pure, divine and unsullied by the polluting hand of man. The Chartist name came from a short list – their 'charter' – of political demands, including the right to vote for all men over 21. However, broader issues of workers' rights sat around these calls, not least because workers hoped if they had some political representation, they might get better pay and conditions. As Andreas Malm vividly describes in his (2016) book *Fossil Capital*, steam was a recurring character in Chartist rhetoric, sometimes depicted as ruthless king belching smoke and consuming fire as nature receded in the unbearable heat. One of the Chartist leaders, Fergus O'Connor, fresh out of jail in 1842, lectured crowds on the dangers of the 'smokeocracy' – a pernicious combination of steam power, its owners, the police and national politicians. When he imagined this smokeocracy, O'Connor might well have been thinking of Robert Peel, heir to a mill fortune and great supporter of James Watt, who by this point had founded the Metropolitan Police and was in his second stint as prime minister.

It was 1842 that brought what was arguably the biggest Chartist response to the Industrial Revolution, with the so-called 'Plug Plot Riots'. The name came from calls for strikers to literally pull the plug on their factories' boilers, spilling water on the floor, letting steam fly free up into the air and bringing engines to a halt. Chartism had been growing for a while, especially in manufacturing towns, and in May 1842 they submitted several miles of petition to Parliament, signed by more than 3 million people (a little over a fifth of the population at the time). But the MPs didn't like the idea of canvassing for votes among miners, farm hands and factory workers or, perish the thought, sitting next to one of them in chambers of the House of Commons, and voted against it by 287 votes to only 49. The anger at this rejection provoked a strike that spread to half a million workers in more than 30 counties, the biggest revolt of the British working class in the nineteenth century. *The Times* described it as 'an outbreak like the firing of a train of gunpowder – one spark ignited, and then the entire North in a blaze'.

Arguably, calling the protest the 'Plug Plot Riots' was simply a way to imply a sinister conspiracy and dismiss the larger political arguments of the charter. Still, plugs were key to the whole affair. A speaker whipped up a crowd in Todmorden, West Yorkshire, with cries of, 'And now I ax ye, will ye pull the plugs out?' A rioter in Leeds boasted of having knocked out 13 plugs in a single morning. This was brave. Following the Luddites[*] protests earlier in the century, the Frame Breaking Act of 1812 had made industrial sabotage a crime punishable by death. The 1842 protests were about the charter, but they were also about steam engines: what it was like to live with them and the coal that British society was now so dependent on. The pulling of the plugs

---

[*] Today, the word 'Luddite' is usually applied pejoratively to people who don't like technology. More specifically, the Luddites were a movement of textile workers in the early-nineteenth century who saw the introduction of automated textile equipment as threatening the jobs and livelihoods of skilled workers, and took radical, sometimes violent action, smashing mechanised looms.

was more than just a clever way to clear a factory to publicise the right to vote, it was an expression of the workers' complaints and of their power. As the poetry made clear, as well as a host of dangers at work, they found the environmental destruction caused by the Industrial Revolution emotionally upsetting, yearning for cleaner air and worried about what they saw as a retreat of nature. They also knew how much power coal had come to hold in the economy and that this gave them power too.

In fact, the strikes had started with coal – in the mines of North Staffordshire in early July. A strike had been called demanding pay rises and shorter working days along with a commitment to the charter, and strike leaders set about putting out fires under the boilers that pumped water from the mines and cutting the ropes that led miners down to pits. This quickly starved the local pottery industry of fuel. News spread to mining districts in Lanarkshire, Wales and the Black Country (which supplied the Birmingham manufactories with coal). When the textile workers around Manchester vowed to shut factories there too, crowds of strikers swelled to tens of thousands. A placard at one of the strikes addressed the coal miners directly: 'Without coal the lordly aristocrat cannot cook his luxurious meal,' moreover, 'Without coal this giant monster, the Steam Engine, cannot work. Your labour, my honest friends, supplies it with strength, for without Coal it is powerless. Stop getting Coal, for Coal supports the money-mongering Capitalists.' As mines lay quiet, the price of coal went up and the middle classes started to feel the pinch. The *Manchester Guardian* complained the 'working classes have had everything according to their wish', decrying this 'fiendish scheme developed by Mr Feargus O'Connor', which they believed he'd been plotting for years. For the *Illustrated London News*, it was 'a giant torrent, which, tearing and bursting its furious and misguided way amongst our northern depots of commerce, has swept down the barriers of citizenship and order, converted the labourer into an anarchist, and assumed all the alarming features of systematic insurrection.' Similarly, a *Times* correspondent in Hull complained: 'It is not now a question of easier toil and better wages, to which many were and are sincere well-wishers, but a question of the English constitution against French Jacobinism and American democracy, and men must take sides accordingly.'

Peel sent in the troops on 14 August. Thousands of soldiers arrived by train, making a show of their power and numbers as they marched through the streets. *The Times* devoted several pages to reports from its correspondents across the northern manufacturing towns, outlining the drama as it unfolded. In Wigan 'most of the respectable shopkeepers and tradesmen were sworn in as special constables', but it didn't seem to do much good as the streets were full of protestors. When the army arrived, they were met by a loud group of 600–700 people shouting at them. At midday, 8,000 people marched through the streets, many women and boys, they noted, as well as working men, some armed: 'all the factories were visited by the mob, and the plugs taken out of the boilers'. Manchester and Salford looked quiet, but only after thousands of special constables had marched through the town earlier that morning. When strikers marched to cries of 'stop the smoke' in Preston, several protesters were shot dead.* By October 1842, 1,500 arrests had been made, with 10 times as many by the following year, as local magistrates were on a mission to root out and punish the plug-pullers. The reports of rioting were used as a way to dismiss the political arguments of the Chartists and after another failed petition in 1848 the movement dissipated. Still, it was a context for the Great Exhibition in 1851; one of many reasons why parts of British society were so keen to make a big show and dance of how much of their national power and identity was wrapped up in coal and steam.

In his 2011 book *Carbon Democracy*, Timothy Mitchell goes as far as to argue coal could be credited with helping build the thing we understand today as democracy. The Plug Plot Riots hadn't won the right for the working classes to vote, but later protests would be more successful. British miners were far from the only people rebelling, with labour movements growing around industrial hubs in Europe and America. In April 1882, coal heavers in Port Said in the north-east of Egypt, then the world's largest coaling station, went on strike

---

* A young John Tyndall witnessed this, having just arrived in Preston to work for the Ordnance Survey. Inspired by the strikers' calls for better pay and conditions, he flirted with political radicalism, taking his concerns about the ways Irish labour was being exploited by the Ordnance Survey to the press (and was soon sacked as a troublemaker as a result).

and a wave of coal strikes in Germany at the end of that decade shocked new Kaiser Wilhelm II into supporting a programme of labour reforms. He convened an international conference in March 1890 calling for international standards on coal mining. The same day, more than 250,000 men, women and children participated in mining strikes in England and Wales. Dockers were also organised and powerful at striking, as were railway workers. By the start of the twentieth century, the vulnerability of new industrial systems to worker action was clear. The new networks for extracting, moving and using fossil fuels that had built up in the twentieth century had made a few industrialists very rich, often off the back of incredibly hard working employees suffering poor living conditions, but they'd also offered new contexts for the workers to organise within.

* * *

And yet, if anything, the most remarkable aspect of the nineteenth-century reaction to fossil fuels was that people tolerated them. This is especially striking when it came to the most obvious downside of burning all that coal – the smoke. The Victorians abhorred dirt and yet they seemed to have a blind spot for coal soot, portraying it as 'good, honest' dirt. Coal smoke was a sign of prosperity, a barometer of success and progress that should demand no apology. So, they found ways to ignore or excuse the problem. It's sometimes said this is the origin of the Londoner's characteristic black umbrella, to avoid their clothes getting stained by splodges of soot-filled rain. The English tradition of afternoon tea possibly owes something to coal smoke too. As Victorians found their outdoor air becoming more and more noxious, they looked around for recreational activities they could do indoors, hence the increasing interest in chamber music, galleries and afternoon tea (a 'let them eat cake' approach to pollution if ever there was one). Whether stories like these are urban myths or not, the Victorians were remarkably good at pretending the coal smoke wasn't really a problem. It wasn't just this period either, people would keep on tolerating it for decades into the twentieth century. Londoners would be walking around in air so thick with soot they couldn't see past the end of their nose; they'd be walking into walls, walking into the river, into traffic, falling over the dead that wouldn't be found for ages, and yet no one did anything. And this went on for generations.

The smoke didn't go entirely without complaint. There were government inquiries, gloomy poems and even a few 'smoke abatement' societies, especially in the cities that burnt sootier, softer 'bituminous' coal rather than harder types like anthracite.* In 1854, Charles Dickens complained 'The great destruction of life from pulmonary disease is due to the fact that the soot which smudges the collars and chitterlings of our citizens, that ruins our finest paintings, that blackens our public buildings, that suffocates our country-born babies, that kills our plants, that fleeces our sheep of their whiteness, that blackens our faces, and buries our whole bodies in palls of fog, is also constantly passing into our lungs.' In 1868, *Atlantic Monthly* columnist James Parton memorably compared Pittsburgh to 'hell with the lid taken off', and a few decades later art critic John Ruskin would give a pair of lectures to the London Institution, *The Storm-Cloud of the Nineteenth Century*, describing the view of Manchester from his home in the Lake District as the 'devil's darkness'. Doctors started to tot up the deaths and publish the figures in medical journals, economists calculated how much money was spent on cleaning buildings and city smoke inspectors were employed to check and fine worst offenders. But the only changes they'd demand would be boiler design, possibly shifting to a different type of coal, or even natural gas if it was available, or move trains to electricity or oil. Otherwise they simply spluttered through it.

London was especially well known for its 'pea-souper' smogs, a reference to the mustard-coloured tinge that came from the type of coal commonly burnt in the area (it's a split-pea soup, not a

---

* As David Stradling describes in his book on air-quality protests in the US, *Smokestacks and Progressives* (1999), in 1902, a strike in the anthracite coalfields of eastern Pennsylvania meant New Yorkers had to turn to the more obviously polluting bituminous coal. Sales of Panama hats plummeted as there wasn't enough sunshine to need shade from, laundries reported a boost in business as clothes were dirtier and butchers complained smoke was spoiling meat as New Yorkers had a literal taste of what it was like to live in smokier cities like Pittsburgh. Stephen Mosley's book *The Chimney of the World* (2001) also offers an engrossing description of some of the earliest anti-smoke protests, which grew up around Manchester, and what it might have been like to live under those clouds.

bright-green one). The start of 1880 brought a particularly large serving and the *New York Times's* London sketch writer offered a vivid description. He described the city as having been plunged into 'a week of night' as 'a curtain of cloud, thick with the smoke of coal fires, has been drawn over London'. It made its way into every part of the city, taking the place of people in theatres, shutting off the electric lights at the British Museum and chasing postmen to deliver the mail to the wrong houses. He went on: 'It has tarnished our windows, sweated our china, poisoned some of our citizens to death; it has hurled skaters at each other, it has made the gas-lamps look silly, it has set pedestrians crying aloud in the streets lest they should be run over.' The traffic by the river had been suspended, but only after buses and cabs crashed and patients were left to die as doctors' carriages were smashed: 'It has plagued us at all times and all hours, in sitting, in standing, in sleeping and walking.'

Not long after, a new London Smoke Abatement Committee was set up, led by editor of the *British Medical Journal* Ernest Hart and open spaces' advocate Octavia Hill (who would later co-found the National Trust following a campaign against railway developments in the Lake District). They put on a large smoke abatement exhibition in South Kensington, which was visited by over a hundred thousand people and later moved to Manchester. This helped raise some discussion of the smoke problem in the press. *Nature*, for example, noted a wax company in west London had to keep moving further and further out of the city, because it needed sunlight to bleach its products and there just wasn't enough of it. Still, the focus of the exhibition was squarely on technologies that might help manage or reduce the smoke – a different boiler or less smoky fuel – rather than anything considering the root of the problem. It was a display of new gadgets and ideas very much in the World's Fair model, with prizes, medals and a name-drop in *The Times* for the best devices on display. More Ideal Home Exhibition than radical call for change.

★ ★ ★

In industrial cities nineteenth-century environmental campaigns were often led by middle- and upper-class women who laced their activism with an idea that to be clean was to be good, worrying not just about the physical health of the working class and all the soot they had to live in, but the moral corruption it might bring. Some even saw their environmental campaigns as an alternative to missionary work – instead of going abroad to save souls, they would live in poorer parts of industrial cities and do their good work there. This was an elite urban environmentalism that tended to maintain strict hierarchies, more about cleaning up the city to fit a particular image of what looked and smelt appropriate than working with communities to call for change. They'd pop in, wring their hands about how terrible it all was and leave again. It's perhaps no surprise little changed.

One environmental health activist who did want change was Ellen Swallow Richards. She'd been the first woman admitted to MIT and would go on to be the first woman to teach there too. From a relatively humble background, she'd worked in her father's shop to save enough to study at Vassar College in New York state. After graduation, she tried to get work as a chemist, but no one would hire a woman. One prospective employer offered to take her on if she paid them. Another laughed and told her to apply to MIT first – so she did. The MIT faculty were divided about whether to let her in and even when they did it was clearly defined as 'the Swallow Experiment', not a new policy that might open the door to more applications from women, and made her work in private in case she distracted the men. Initially, she was careful to always go along with what the men wanted. She knew where everything was in the lab and would happily darn her colleagues' clothes when they tore. One particular sceptic, professor of mining engineering Robert Richards, moved from vehemently disapproving of the experiment to proposing as soon as she graduated (she made him wait another two years, wanting to be certain he wouldn't stand in the way of her career). She slowly charmed MIT into accepting her idea of applying science to the home and in 1876 brought in the first cohort of female scientists to MIT, to study in a separate 'Women's Laboratory'.

Her lab at MIT would test ingredients for adulteration, exposing food fraud, identifying mahogany dust in cinnamon and arsenic in wallpapers. As she settled into her career, she became bolder, pushing the public health community to consider sanitation, ventilation and lack of infrastructure like fire escapes. She also became inspired by the ideas of ecology – the study of the relationships between organisms and their surroundings pioneered by thinkers like Alexander von Humboldt and finally given a name in 1866 by German naturalist Ernst Haeckel. Richards tweaked this idea of ecology for her highly practical science of everyday lives, announcing in a November 1892 lecture – which made the front page of the *Boston Daily Globe* – her new idea of 'oekology', a mix of the words 'ecology' and *oikos*, Greek for 'house'. A year later, some of this oekology would make its way into the Chicago World's Fair, with a model kitchen named after Count Rumford, which served 10,000 people lunch at 30c a pop. There, diners would be presented with special printed menus Richards had developed outlining to the general public for the first time the fat, carbohydrate and caloric values of the food they were eating.

And then, in 1910, a year before her death, Richards went a step further with the idea of 'euthenics'. Just as oekology had been a spin on ecology, this was a take on eugenics. And this is the bit that's sometimes left out of biographies of Ellen Swallow Richards, especially those who like to make her out as the grandmother of eco-feminism. As Swallow explained: 'Eugenics deals with race improvement through heredity. Euthenics deals with race improvement through environment.' She isn't, unlike some of the other supporters of eugenics at the time, talking explicitly about differences between races. But she doesn't hide her concerns about immigration, and it does reflect a belief – shared by other environmental reformers at the time – that the 'stock' of American life needed to be cleansed, not just of pollution in rivers and air, but of people. It's also been argued that her earlier work on 'scientifi cally prepared' menus was more about imposing a particular idea of 'American' food on immigrant communities than applying objective research. Richards' interest was nurture, not nature, noting that 'hope for the future' can be found in research by the immigration commission that children of immigrants, once settled in the US, have better 'head measurements' than their parents. But she sees eugenics and euthenics as part of the same project, with euthenics possibly playing the first stage, before eugenics follows.

The idea of euthenics would stick around even after Richards died. In May 1926, a spread in the *New York Times* introduced a new summer course in euthenics at Vassar. Birth control activist Margaret Sanger was a fan too, delivering a speech at the Vassar Institute of Euthenics a few months later, praising the Immigration Act the US had passed in 1924, but warning that 'While we close our gates to the so-called "undesirables" from other countries, we make no attempt to discourage or cut down the rapid multiplication of the unfit and undesirable at home.'* Eugenics was popular at the time, especially among a certain type of 'progressive' – people who wanted to improve the world and believed in a stronger role for government as opposed to a more *laissez-faire* approach, but were also concerned about the depletion of natural resources.

Back in 1798, philosopher Thomas Malthus had published an essay on population, responding to the more utopian ideas of his time – people like Mary Shelley's father William Godwin who were inspired by the American and French revolutions and shared Joseph Priestley's hopes for positive technological and social change. Any improvements in society, Malthus argued, would simply lead to a larger population as people wouldn't die from disease, hunger or cold, which in turn would mean more mouths to feed, and so at least some of the population dragged back into starvation and suffering. He'd go on to teach at the East India Company's college and when, in 1870, the viceroy dismissed famine in Bengal as simply a way to balance the population and that sending relief would only make it worse, it was Malthusian ideas he was drawing on. Roll on to the early-twentieth century and Malthus's arguments still haunted political debate, a few strands of thought also rolling in Jevons-inspired concern about limited coal reserves. For some progressives, eugenics was a way out of the problem posed by

---

* Sanger was openly a supporter of the eugenics movement, one of several birth control advocates who found a common cause in calls for population control (Marie Stopes was another). Unlike some of the more elitist eugenicists, Sanger herself came from a working-class background, largely looking for liberation rather than control and deplored Nazi sterilisation programmes in the mid 1930s. Still, she admired the basic idea, having earlier called for the sterilisation of 'mental defectives, epileptics, illiterates, paupers, unemployables, criminals, prostitutes and dope fiends', and remains controversial, especially in her work with Black American communities. For trivia fans, Sanger was also an inspiration for the character of *Wonder Woman*.

Malthus. Rather than cruelly expecting the excess poor to die off periodically of starvation or cold, we should limit the number of babies born in the first place. Wanting to apply what they saw as the leading biological science of the day, they figured it was only the right and progressive thing to do to encourage some inherited 'good' characteristics and discourage other 'bad' ones. The problems came in how they defined which people should be discouraged from giving birth and what method they picked for such discouragement. For all that eugenics appealed to people who genuinely wanted to improve the world, it also attracted those who simply liked the idea that they and their friends had been born better than everyone else or, worse, actively wanted to eliminate a whole group of human beings.

* * *

The people who really embedded eugenics into the environmental movement in this period – a particularly racist form of it too – were members of the Boone and Crockett Club. This had started as a way for rich men to share hunting stories and would end up building the basis of the conservation movement. Indeed, its members would go on to coin and popularise the term 'conservation', through their work with institutions like the Audubon Society, the Save the Redwoods League, Forestry Commission, Bronx Zoo, and several national parks and forests. This elite club would also support the growth of a science-smelling form of white supremacy and the passing of that 1924 Immigration Act that had so delighted Sanger. For several of their members, protecting the environment and white supremacy were part and parcel the same thing.

The initial idea had come at a dinner party in December 1887, hosted by 29-year-old Theodore Roosevelt. Two years before, his wife and mother had both died on the same day. Grieving, he'd retreated to a cattle ranch in North Dakota but, after losing much of his cattle investment in a difficult winter, he'd come home. Enjoying being back among friends, swapping tales of chasing bison and elk, he proposed they formalise a club for big game hunters. Membership was limited to an elite core of a hundred men who had killed and mounted at least three species of large North American game animals. Named after two iconic hunters of the American 'frontier', Daniel Boone and Davy Crockett, their first meeting was held in February 1888. The core of the Club was always hunting – killing animals – but from the get-go they had an interest in protecting wildlife too. After all, if the bison all died out, what would be left to chase?

The US had industrialised rapidly after the American Civil War, but for all the hope and sense of progress wrapped up in shows like the Chicago World's Fair, people were increasingly haunted by a sense it was all going to fall apart. Was this new modernity fragile? George Perkins Marsh, a former congressman who'd been involved in the founding of the Smithsonian, toured with lectures arguing humans could radically and unfavourably alter the fabric of nature. Like Humboldt before him, he warned that human destruction of the environment could cause climate change at a local level, as the draining of swamps and clearing of forests could lead to desertification, noting that London was a degree or two warmer than surrounding areas. Of particular influence to later environmentalists in the US was Marsh's 1864 book *Man and Nature*, which warned Americans might be in danger of running out of trees. This wouldn't just leave them wanting timber, but with land parched by the Sun in summer and flooded by rain in winter. Marsh urged what we might now call 'sustainable forestry' – an occasional pausing so 'nature ought to be allowed to reclothe them with a spontaneous growth of wood'.

While some Americans were gripped by a sense of impending doom that their society had degenerated morally and environmentally, some also looked at the ways train and steam ships sped up the colonisation of the interior of the US, some also started to feel as if the country had run out of frontier, mourning their particular idea of America as 'lost' in the process. Boone and Crockett were part of that in a way, nostalgically cosplaying Davy 'king of the wild frontier' Crockett because it spoke to their ideal of white American masculinity. The extinction of the passenger pigeon and the near extinction of other species such as bison exacerbated worries that America was, in a way, a dying land. Back in the 1810s, ornithologist and painter John Audubon had written about flocks of passenger pigeons in Kentucky, made up of at least 500 million birds, taking five hours to pass. A century later they were all gone. The American

Ornithologists Union offered rewards of $300 to anyone who might find a nesting pair, but in 1914, Martha the last passenger pigeon died in a Cincinnati zoo. Other North American species like the great auk, grizzly bear, caribou, moose, wild turkey and elk all looked threatened too. Blame was quickly cast on 'market hunters' – people who hunted, not for fun like Roosevelt and his buddies, but for food and profit. This in turn was blamed on immigrants and former slaves who, it was argued, just didn't have the refined understanding of man's interactions with nature that Boone and Crockett or their friends might.

As Roosevelt moved more into politics, two other Boone and Crockett members, George Bird Grinnell and Madison Grant, would take over the running of the club. As Jonathan Spiro describes in his excellent if sometimes disturbing 2009 biography of Grant, they transformed it from a place for elite hunters to swap stories over dinner into a much larger conservation organisation. Grant had graduated in law and set up offices in Wall St in the 1890s, across the street from his friend J. P. Morgan, but never really practised. His family money was more than enough to live on. He was an activist, though a quiet, lobbying type of campaigner who worked via powerful friends of powerful friends, inviting you for a drink, being the member of the right club. He'd keep Boone and Crockett in the same model, its membership still small and elitist even if the aim was large-scale political change. He was rich, he was old money, he was unrepentantly elitist and he was an unabashed, committed racist. He once wrote New York was becoming a 'sewer of races' and worried that people like him – people who could trace their families back to the first European settlers – were somehow as endangered as the redwoods and elks. Grant's book, *The Passing of the Great Race*, is often described as the 'bible' of scientific racism. Hitler certainly saw it as a sort of bible, saying so when he wrote personally to thank Grant for writing it.

Grant's racism didn't just influence the Nazis, it was woven through his environmental work. He once said to his friend and fellow Boone and Crockett member Henry Fairfield Osborn (director of the American Museum of Natural History and the man who named both the *Tyrannosaurus rex* and *Velociraptor*) that both conservation and eugenics were 'attempts to save as much as possible of the old America'. Grant actively identified with the large mammals and trees of America – he thought bison were under threat because of immigrant poachers and worried this would be his fate too. He similarly saw himself in the redwoods, the arboreal equivalent of his notion of the 'Nordic' race. He was less interested in smaller mammals and birds, but Grinnell was happy to pick this up, founding the first Audubon Society in the 1890s, naming it after the ornithologist* whose former home Grinnell had lived in as a child. Grant, Grinnell and Osborn would also work together to build the Bronx Zoo. Grant had a vision for a large 'zoo park' where people could escape modern America into a 'wild' image of the country's past.† It was ambitious, but Grant easily raised money via wealthy friends. Rockefeller gave a cool $1 million, and others like the Vanderbilts and J. P. Morgan signed up too; men who'd made fortunes from industrialisation of America and now wanted to build a way for people to escape it. Grant and Osborn would run monthly Galton Society meetings on eugenics, held at the American Museum of Natural History. The museum would host the second and third International Eugenics Congresses in 1921 and 1932. The exhibition at the second congress had been so popular the museum kept it on for another month and, along with Grant's book, helped influence the passing of the 1924 Immigration Act.

Running alongside the growth of the Boone and Crockett network was another club – this one based around protecting the Sierra Nevada, a mountain range out west. It had been formed in 1892 off

---

* The image Audubon created for himself also sat neatly in the Boone and Crockett model of a man of the past, full of adventure, stories and science. As his biographer Gregory Nobles notes he relied heavily on Black and Indigenous knowledge of American birds, although never acknowledged their contributions and was at pains to distinguish himself from Native and African Americans. He was also a slave-owner. Moreover, he relied heavily on Black and Indigenous knowledge of American birds, although never acknowledged their contributions and was at pains to distinguish himself from Native and African Americans. (It is also possible Audubon had African heritage himself, and this was a reason he put so much effort into distancing himself from people who were not white).

† It was also a place where in 1906 they'd end up displaying a man from Central Africa in the monkey cage, an incident the zoo has only very recently apologised for.

the back of campaigns to establish the world's first national park at Yellowstone. It was in part modelled on the Appalachian Mountain Club, a group founded by Boston academics in the 1870s for people to enjoy the White Mountains of New Hampshire in the alpinist mode Horace Bénédict de Saussure had pioneered in Europe a century before. The Sierra Club's first president was the Scottish American naturalist John Muir, who took an almost religious love of a sense of wilderness from the transcendentalists and fought not just to extend Yellowstone but establish many similar spaces around the country. In 1903 he famously took Roosevelt (who was president by this point) on a camping trip in Yosemite. Sometimes described as the most important camping trip in American history, it sparked the creation of five more national parks and 150 national forests. Roosevelt also had his Boone and Crockett friends lobbying for a similar expansion of national parks, but it suits various storytellers – then and now – to use this trip, the character of Muir and the backdrop of sequoia trees as the crucial moment.

There was an uneasy tension between wanting to create parks for Americans to visit to experience the natural wonders of their country and, at the same time, also maintain the landscape as somehow 'untouched'. For Muir and his followers, these parks had to be wild, but for that reason they also had to be controlled. Yellowstone had a host of geothermal features – the Old Faithful geyser for example – which would attract tourists from across the States, as people who could afford it both wanted to escape polluted cities and, because of the transport technologies on offer, could more easily do so. But these very people could get in the way of the romantic, transcendentalist feeling of being surrounded by things that were non-human. How could you play the romantic poet and 'wander lonely as a cloud' when there were other people in the way? Most offensively was the idealisation of these new parks as somehow uninhabited by humans in the first place, a retelling of a fantasy that white Americans had 'discovered' the continent. At first the white visitors to Yellowstone simply ignored Native Americans, acting as if they weren't there. A couple of wars and increased visitor numbers provoked a slightly less blinkered approach, and the last of Yellowstone's native inhabitants were removed by autumn 1879. Roosevelt, Grinnell and others at the

Boone and Crockett Club were part of this removal, protesting the destruction of nature by 'Indian hunting parties' and pushing the Department of the Interior to bring in the cavalry, establishing a principle that Native Americans must leave to 'preserve' nature, a pattern that'd follow in other parks in decades to come.

When Roosevelt became president, he put another Boone and Crockett man, Gifford Pinchot, in charge of forests. One of the eugenicists of the group, Pinchot had a strong faith in science as the way to improve the world, but a more mixed relationship with technology. Hugely influenced by the writings of Marsh, he felt an unbridled and rapid industrialisation had damaged the Earth and, adding a more Malthusian bent, worried that the benefits of modern life – like better access to warmth, food and medicine – had meant the survival of people who'd otherwise have died off. By the time Roosevelt left office, Pinchot had extended the forests from 45 million to 151 million acres. It was from Pinchot that we get the idea of a 'conservation' movement; in 1907 he coined the term 'conservation ethic' to describe the sort of work he and other Boone and Crockett types did to protect nature. Conservation was, for Pinchot, a form of management, and in terms of his work in forests it was to protect them as a renewable resource, just as the Boone and Crockett Club wanted to keep just enough bison to hunt without making the chase too easy or being in any way inconvenienced by the exuberance of nature.

* * *

Post-war, the eugenics of the conservation movement would be brushed over, even if their often rather fantastical (and racist) ideas of American wilderness stayed relatively unquestioned. We'll return to the environmental movement in later chapters and see how, in the 1960s and 1970s, green groups in the West would grow from the soil turned over by Grant, Roosevelt, Muir and their friends. This new generation of activists would be influenced by radiation fears of the post-war era too, as well as the new environmental problems the twentieth century was piling up. What's more, in places, they'd pick up tactics and ideas of the civil rights movement, itself heavily influenced by Gandhi's approach to non-violent direct action and other twentieth-century fights against colonialism. And yet, despite the influence of the civil rights movement, it took a while for the idea of environmental racism to emerge (that is, an awareness of the intersection of racial injustice and the impacts of environmental pollution). Even then, it didn't come from the mainstream environmental groups, but rather networks of low-income Black women mobilising over chemical pollution in Love Canal, to the east of Niagara Falls, in the late 1970s.

The mainstream environmental movement remained – and still remains in places – rather white. It also remained quite middle class, which didn't exactly help its relationship with the labour movement. By the mid- to late-twentieth century, union power had shifted from those early coal strikes of the nineteenth century. The rise of oil over coal disrupted power held by unions. Plus, as fossil fuels bedded in, parts of the labour movement also became strongly devoted to it. Mining unions have an interest in keeping coal burning long term, after all, even if their members would happily retrain to work elsewhere. So, despite the shared history of the Chartists critique of 'smokeocracy' and plenty of environmentalist labour organisers – César Estrada Chávez, for example – green NGOs and unions increasingly found themselves at loggerheads.

The green movement sometimes gets lumped together as a singular group. As the former British secretary for the environment Owen Patterson once memorably put it – 'the green blob'. However, it contains a diverse range of political ideologies, all with their own beliefs, cultures and histories shaping what they say and do. Its mixed heritage of civil rights and white supremacy mean parts of the environmentalist movement will occasionally clash on topics like population. But, if anything, activists downplay the mix of political views surrounding work, often preferring to hide behind pictures of polar bears and a loose sense of shared love of the environment to build the impression of a broad base. I don't say this to dismiss the motivations of their work or their power as a positive force for good (I work for an environmental NGO, I'm part of that green blob), just emphasise that there are a lot of different motivations at play and none of it is a simple game of goodies vs baddies. As we'll see in Chapter Ten, it'd be an organisation founded by the son of one of the Boone and Crockett eugenicists, Henry Fairfield Osborn Jr's Conservation Foundation, that would host what might be called the first conference on climate change in 1963. But we're getting ahead of ourselves slightly, as there's a lot more environmental destruction and a lot more science to happen first. For now, we need to catch up on the story of oil.

CHAPTER SEVEN

The Rise, Fall and Rise of Big Oil

The early rock oil business had been explosive – literally, when it came to some drilling sites. It had undulated between boom and bust and back to boom again, uncertain and rather ramshackle in places. A man named John D. Rockefeller would bring it under control. The empire he built, and the way it was eventually undone, would shape much of the modern oil industry. We'll never know, but possibly, if other characters had dominated in the oil industry's formative years, it might have turned out differently, responding to the climate crisis differently too.

Rockefeller was the son of a literal snake oil salesman, a point his later detractors would make much of. His father William had worked in lumber trading and built a reputation as a bit of a con man before reinventing himself as Dr William Levingston, selling herbal cures and elixirs including a popular remedy of the time – snake oil. This had its origins in a treatment Chinese labourers who'd come over to work on the railroads had used to ease sore joints after a hard day's work. They shared it with American workers, and soon quacks were claiming the cure-all effects of their own snake oils (though often made from rock oil). John D. Rockefeller built himself to be a very different man from his father. Serious and religious, he taught Sunday school and believed deeply in the oil business as a force for good. He rarely let emotion show and would often sit back to allow others to speak first, rather than letting anything of himself show. His famous philanthropy started as a small gift to his church, growing as his fortune did, and he started applying a business-like approach to charity, extending through medicine, education and sciences (including the science *de jour*, eugenics). Overall, he'd end up giving $550 million to different causes in his lifetime. In many ways, Rockefeller is one of the villains of this tale, but that's not how he would have seen himself.

Born in 1839 in rural New York state, young Rockefeller left home at 16 with dreams of making something big for himself. He started off working for a shipping firm in Cleveland, providing him a favourable spot to cash in on the economic growth that followed the end of the

American Civil War. As the rock oil business started to grow, he threw himself at the opportunity, pouring in any finance he could get his hands on to build a refinery. When this was profitable, he built a second. Keen to insulate his business from possible disruption from others, he bought land to grow timber to make his own barrels, and his own boats and railway cars to transport the barrels on. He also built up strong financial resources, so he didn't have to rely too heavily on outside financiers. This not only kept his business safe from the downturns of the early oil business, but let him take advantage of them, buying up resources when they were cheap. In 1867, Rockefeller employed Henry Flagler, who'd become one of his closest friends and help build the complex network of transport systems on which their oil empire would run. Flagler had made a small fortune distilling whisky but sold the business because his father disapproved of alcohol, only to lose the profits with imprudent investments in salt. In 1870, the two formed Standard Oil, otherwise known as SO, or 'Esso' if you say it out loud. The name 'Standard' was picked to reassure their customers, suggesting a 'standard quality' of oil that was less likely to explode or simply fail to light like their competitors'. Perhaps it also suited Rockefeller's approach to the oil business too – he wanted control and consistency; to be the standard choice.

He started buying out other companies, sometimes simply because they were going cheap, sometimes through persuasion, sometimes via other firms so people didn't realise it was him, sometimes by sweating them out (for example, putting them into a 'barrel famine' and cutting off their ability to transport oil) – and he was successful. By 1879, SO controlled 90 per cent of America's refining capacity. SO also dominated transportation, from barrels to trains to pipelines. Rockefeller was ruthless, building an empire that spread across the American oil industry. The press often portrayed the company as an octopus, its tentacles reaching out to suffocate any competitors. As Daniel Yergin argues in his detailed history of oil, *The Prize*,* Rockefeller saw all this as simply replacing chaos and volatility with stability, giving light, heat and movement to people who'd otherwise

---

* *The Prize* is a towering classic in the history of the oil industry, and a vital source in writing this book. It's one of those doorstops of a Pulitzer-winner, so you might want to look up the TV series first (though the book is very much worth the investment of your time if you're interested in the topic).

have much more limited and dangerous lives. 'Give the poor man his cheap light, gentlemen,' he'd tell colleagues.

* * *

When the first barrel of this 'Yankee invention' of rock oil arrived in the Russian capital of St Petersburg in 1862, it was eagerly pounced on by residents wanting to light up their long, winter evenings. Russia had oil of its own though, in Baku (today the capital of Azerbaijan, then annexed by the Russian Empire). The famous eternal fire at the Fire Temple there was a place of worship for Hindu and Zoroastrian faith, and back in the thirteenth century Marco Polo had talked of a spring near Baku that produced an oil not much use to cook with but that burnt well. A small-scale oil industry had developed as early as the 1820s. Output had initially been small, mainly based around hand-dug pits, but in the 1870s the Russian government opened up Baku oil to private investment and, applying the new American idea of drilling, wells soon started to shoot up. In 1873, a young chemist named Robert Nobel arrived in Baku. He was the eldest son of Immanuel Nobel, a Swedish engineer who had emigrated to Russia in the 1830s. Robert's younger brother Ludwig had built up the family armaments company and, finding himself with a large government rifle contract to fulfil, sent Robert down south with 25 million roubles to buy wood. But as Robert passed through Baku, he was distracted by the oil wells and, without stopping to consult Ludwig, spent the money on a small refinery instead. The Nobels had joined the oil business.

The first consignment of Nobel's Baku oil arrived in St Petersburg in October 1876. Ludwig had good connections with the government, and soon won the blessing of the viceroy of the Caucasus.[*] An industrial

---

[*] Robert soon resented Ludwig's intrusion into his find and went back to Sweden. Their brother Alfred stuck to the family business of the arms trade, inventing dynamite. When, over a decade later, in 1888, Ludwig died of a heart attack on holiday on the French Riviera, some newspapers confused the brothers and reported the death of Alfred, who was then put in the odd position of reading his own obituary while grieving the sudden death of his brother. Brooding over the negative way he was described (the 'dynamite king') he changed his will to establish prizes to honour the best of human endeavour, so the Nobel name would forever be associated with good.

leader in a similar model to Rockefeller, he set about analysing every aspect of the oil business, learning everything about the US experience. Like oil pioneers before and after, Ludwig Nobel invested heavily in research too, becoming the first oil company to have a permanent staff position for a petroleum geologist. Transportation was the biggest challenge, as it had been in the US, not least because their market was much further north, in St Petersburg. Oil was shipped in wooden barrels over the Caspian Sea and then up the Volga River, eventually reaching a rail line that could take it to markets in the north. Ludwig found an efficiency by building large tanks into ships that could be directly filled up with oil from a pipe (rather than split into barrels) – the first oil tanker. His *Zoroaster*, named after the founder of Zoroastrianism, set sail with its first cargo of oil on the Caspian Sea in 1878. Competitors to the Nobels started work on a railroad to take oil west of Baku too, but the price of oil dropped suddenly before they could complete it, and they ran out of money. They were bailed out by a French family who had bankrolled many railroads across Europe and had started investing in oil too: the Rothschilds. The railroad was completed in 1883, and in 1886 off the back of it the Rothschilds formed the Caspian and Black Sea Petroleum Company (aka Bnito). Soon, Baku oil was competing with SO in the European market and Rockefeller was wondering how far east his octopus tentacles might stretch, setting up a new international operation in London to keep an eye on things.

Despite all the new transport options, there was still more oil coming out of the ground than either the Rothschilds or Nobels could easily move, let alone sell, and they started eyeing up the Asian market too. The Rothschilds knew a London broker called Fred Lane (nicknamed Shady Lane because he'd work for so many people at once it was hard to tell whose interests he was really dealing in). He, in turn, knew a rising star in shipping, Marcus Samuel Jr.

★ ★ ★

The Samuel family lived in Upper East Smithfield, not far from the Tower of London. Samuel's father, Marcus Sr, was a descendant of Jewish immigrants from Holland and Bavaria who had settled in

London in the middle of the eighteenth century. He'd traded in pretty much anything going in and out of the nearby London docks, importing everything from ostrich feathers to bags of pepper, and exporting the first mechanical loom to Japan. For a while, he'd taken advantage of the fashion for exotically ornamental boxes decorated in shells, the sort you still see in English seaside towns. This is possibly why he was listed as a 'shell merchant' in the 1851 census and his shop on Houndsditch was known as the 'Shell Shop'. He built good relationships with Scottish businessmen who ran trading houses in Calcutta, Singapore, Bangkok, Manilla and Hong Kong, and it was sometimes said you could count Marcus Samuel's seven children by his trips to the Far East, each conceived on his return. When Marcus Samuel Sr died in 1870, the business was taken over by two of his sons, Marcus Jr and Samuel Samuel. Initially, they weren't oil men but rather, like their father, opportunistic traders working with whatever they could turn a profit on. Sam set up shop in Japan and the two firms, M. Samuel & Co in London and Samuel Samuel & Co in Yokohama, grew rich sending British tools of the industrial revolution out to Japan in exchange for rice, silk, copper, china and lacquer goods for European markets. The shells weren't forgotten either, at one point they were employing 40 women in special premises rented in Wapping to clean the shells and prepare them for boxes. Sugar, flour and tapioca also rolled through Marcus's end of the business in London, and he even dipped his toe in oil, buying small quantities of kerosene from Standard Oil and via Jardine Matheson & Co in 1886, for resale in Japan.

Between steam trains, ships and big canal projects like Suez, patterns of global trade were intensifying. Britain and the other European colonial powers were extracting enormous riches from the rest of the world, and they were selling goods back to the world too. The international back and forth that the early American whaling industry had been so wrapped up in had grown in size and complexity. There were raw goods, but also the sorts of manufactured tools and curios on which people like Matthew Boulton and Josiah Wedgwood had built their fortunes; all the goods so ostentatiously displayed at the Great Exhibition. And then there was the telegraph. Richard Edgeworth's very low-tech version of an 'optical telegraph' had relied on people seeing a sign raised five miles away and had struggled to catch on in cloudy Britain, but the basic idea came of age in the electrical era. An electric telegraph was first put to use on a bit of the Great Western Railway in the north of London and then, in May 1844, with Samuel Morse opening a telegraph line from the Supreme Court in the basement of the Capitol building in DC to Baltimore's Mount Clare Station. By 1845 Samuel Colt had a telegraph from Coney Island to Battery Park to send news of incoming ships to merchants. A few years later another line linked Chicago to New York, and the first telegraph messages between Europe and the Americas were sent in the summer of 1858, with cable from London to Bombay laid in 1870, and then on to Australia two years later. The world was getting smaller, more interconnected, and the Samuels cashed in on all of this new, faster, global back-and-forth.

Marcus was very much the ideas man. He always spoke very softly, so much so people sometimes had to strain to hear him, but he was a persuasive dealmaker, with high ambitions. He wanted more than just wealth, he wanted respect. He wanted the name of Samuel – his father's name, the name of a family of East End Jews – at the highest levels of British society. Sam, two years younger, was very much the loyal sidekick. Warm-hearted, generous, gregarious and always late. He loved silly riddles and would repeat them again and again (some for decades). Sam wasn't without ambition either though; he'd finish up his career in the House of Commons, Tory MP for Wandsworth, sparring with Winston Churchill. The family business had always been about seizing opportunities, shells or otherwise, and Shady Lane's suggestion of work with the Rothschilds was a big opportunity, if a risky one. The brothers had recently made a cool million by buying rice cheaply in huge quantities during a glut before selling it on at high prices during a drought. Marcus was planning a house move to upmarket Portland Place and had put his sons' names down for Eton. He was on a roll, but also had a lot to lose.

Marcus travelled to Baku with Lane in 1890. There he saw the tankers Ludwig Nobel had built up on the Caspian Sea, and started to plan a ship that would be specifically built to transport oil. This wouldn't be a tank retrofitted into the side of a boat, but a whole vessel designed for the purpose – a sort of floating bottle. How much more efficient would it be, how much further could it go? He continued his trip on to some other deals in Japan, selling his idea of a floating bottle to Scottish trading companies around the Far East as he went. Still, he made the Rothschilds wait for an answer, so he could talk it over with Sam first. There were risks in terms of technically building the thing, and also how Rockefeller would react. One sniff of Baku oil moving into Asia and the SO octopus would simply cut prices to push them out (rising prices in other parts of their network to subsidise the move). It was the Rockefeller way. Marcus figured he might be able to convince the Suez Canal Company to let his floating bottle through though, which would cut the voyage by 4,000 miles and with that greatly reduce the cost of the transportation. SO were stuck sending ships of their oil all the way around the southernmost tip of South Africa, having been refused access to Suez on safety grounds.

The brothers agreed to chance it and, inevitably, news got out. In the summer of 1891 the press was reporting rumours of a 'powerful group of financiers and merchants' hoping to take oil through Suez, fanning antisemitism with references to 'Hebrew influence' of the Rothschilds and Samuels. A London law firm launched an aggressive lobbying campaign against the enterprise too. It refused to reveal its client, even when the foreign secretary asked, but it reeked of Rockefeller. The Samuels had powerful allies too, however. What's more, they'd successfully started building their tanker in West Hartlepool. Named the *Murex*, a nod to their history in the shell trade, Lloyds of London rated it safe at the start of 1892 and, crucially, the Suez Canal soon followed. Meanwhile, Marcus and Sam were building storage tanks throughout Asia to receive the oil. They sent their nephews, Mark and Joseph Abrahams, to find new sites, paying them £5 a week, along with a stream of long-distance insults and interference. Mark bought a second-hand rickshaw to keep costs down, but was still told off by his uncles, who hounded him to keep busy selling coal while he was there to keep profits up while they were building this risky new venture.

On 22 July 1892 the *Murex* travelled to Batumi on the Black Sea where it could pick up Bnito's kerosene, and then on past Istanbul and through Suez and on to Singapore and Bangkok. It was a technological triumph that could carry 4,000 tonnes of oil quickly and safely, even in tough weather. It was also designed to be steam cleaned afterwards, so it could take goods back from Asia, including foodstuffs like rice that couldn't arrive back in Europe tasting of kerosene. Marcus Samuel wasn't going to waste a precious trip. There was one problem, however. The tin cans SO used to transport oil had become a prized product in some of the Asian ports, recycled to build everything from roofs to eggbeaters. Marcus wasn't about to be beaten by a bit of packaging and quickly chartered a ship filled with tin to the ports they were delivering to, instructing his contacts to make barrels in which to sell the oil. What colour do you want them, wrote one representative in Japan? 'Red!' Mark Abrahams replied without much thought. Soon, all the Samuel trading houses were also local tin container factories, their bright and shiny red cans, fresh from the factory, competing with Standard Oil's blue ones that had been battered on their trip all around the world. By the end of the following year, the Samuels had launched 10 more ships, all named after shells.

Back in the US, the SO octopus started to twitch. Using the cover of the 1893 World's Fair in Chicago, they opened up talks with Bnito. But the Rothschilds tried to use Marcus Samuel and his tanks as a bargaining tool, a role Marcus did not take well to. Rockefeller tried to take advantage of this rift with a generous offer to join the SO board but still, Marcus rejected it, wanting to retain control of his own business and keep it running out of London.

★ ★ ★

Asia had its own oil, and it wasn't long before people would start drilling there too. In 1890, the Dutch King agreed to support a project that had been trying to tap oil in Sumatra for about a decade, granting permission for it to be called the 'Royal' Dutch Company for the Working of Petroleum Wells in the Dutch Indies. In London, Marcus Samuel was looking for an Asian oil supply of his own too. He didn't want to be stuck as the Rothschilds' transport guy forever. He'd met an old Dutch mining engineer in Borneo who was obsessed with the idea of finding oil in the jungles of Kutei and sent his nephew Mark Abrahams out to investigate. As Abrahams passed through Singapore, the local SO agent was fast to cable back to New York, noting 'a Mr Abrahams, said to be a nephew of Marcus Samuel's' had arrived from London and, 'as

Mr Abrahams is the gentleman who started the Russian tank oil bases at Singapore and Penang … ' this trip might mean something. But this 'something' was to prove rather an amateur affair, at least in the short term. Abrahams had nothing more than a few weeks' training before he'd gone out and his uncles back home weren't much help either. The Samuels were in shipping, not drilling. Between disease, challenging weather conditions and accidents, the mortality rate among workers was very high. Abrahams' team struck oil in February 1897, and then the oil they found was too thick to be much use for lighting.

Still, Marcus was convinced the future of oil lay in transportation, not lighting, and he'd make something of this Kutei oil eventually. Electricity was starting to push oil out of the illumination market, but at the same time cars had come a long way from Erasmus Darwin's early musings of a steam-powered carriage. In 1899, Marcus had attended public trials of horseless carriages arranged by the Royal Automobile Club in London and had decided cars were going to be the next big thing. It wasn't clear, at that point, whether cars would be powered by electricity, alcohol, kerosene, steam or something else, but Marcus was willing to put a bet on oil playing a role. Some trains were already using oil, for routes through cities that didn't want to add to coal smoke pollution – he knew because he'd helped sell some of it to Russia. Even in coal-mad Britain, the Royal train had been converted to run on oil because it made for a cleaner ride. Marcus smelt a new opportunity.

★ ★ ★

Meanwhile, new fields were opening up in the US too. The SO octopus had managed to keep hold of pretty much every drop that ran out of Pennsylvania, buying up developments in Kansas too, but California and Texas were proving harder to grab. Rockefeller's attempt to quietly control the oil industry had started to attract attention and not everyone approved. A wave of popular progressive politics was asking questions about the so-called 'robber barons', and how exactly they'd managed to amass so much of the profit and control out of the industrial age. By the end of the 1870s, a Pennsylvania grand jury had indited Rockefeller for conspiracy to create a monopoly and tried to have him extradited from New York. Several other states followed with their own anti-monopoly suits. Henry Demarest Lloyd at the *Chicago Tribune* ran a series of campaigning editorials on SO, followed up in 1881 by a feature in *The Atlantic*, 'The story of a great monopoly'. It was so popular the edition went through seven printings. This didn't seem to have any immediate impact on SO's balance sheet and the various extradition attempts failed, but Rockefeller couldn't hide any more. When he testified at one case in Ohio, he stuck to his usual approach of being as unforthcoming as possible and was ridiculed by the press: 'John D. Rockefeller imitates a clam' ran the headline. He became increasingly ill and paranoid, suffering from alopecia and digestive problems, and would sleep with a gun by his bed.

It was the discovery of oil at Spindletop, Texas, that was to really remake the American oil industry, not just with the large quantity of oil that would gush out, but the new actors that it'd bring into the game too. Pattillo Higgins, a tall, burly, one-armed Texan with an appetite for violence, had first spotted the opportunity here. He'd previously worked as a logger, mechanic, gunsmith and draughtsman, and was rumoured to have once fought six men simultaneously. He'd then found religion and became a Baptist Sunday School teacher – and it was while out on a Sunday School trip to Spindletop Hill that he came across gas bubbling through springs. If you poked a lit stick at them, they'd ignite. The kids were amused and Higgins started to obsess about the oil that might be found under the hill. In 1892, he formed the Gladys City Oil, Gas & Manufacturing Co (named Gladys after one of the girls in the Sunday School) and went to work. At first, he found very little, but despite being told by his doctor he must be hallucinating, Higgins remained convinced he'd finally strike oil and poured everything he owned into developing the business. Finally, after recruiting Captain Anthony Lucas (a former navy engineer with experience prospecting salt domes like Spindletop) and some more experienced oil drillers from Pittsburgh, oil gushed out of a well in January 1901. First with a deafening roar of gas, then rock and finally oil, green and heavy, came rocketing out at an impressive rate, 75 barrels a day.

Prospectors rushed to cash in on this new gush of oil, just as they had 40 years before in Titusville. An acre of what only two years before had gone for less than $10, now went for as much as $900,000. The local town of Beaumont was said to have drunk half of all the whisky consumed in the state. It soon gained a reputation as 'Swindletop' as people flocked to cash in on all those other people flocking to cash in. A fortune teller named Madame la Monte made a pretty penny selling advice on where to find new gushers and a 'boy with the X-ray eyes' promised to be able to see through rock to spot the oil. But Rockefeller wasn't about to bring order here. Texas had been one of the states to bring an anti-monopoly suit against SO in the 1880s and Rockefeller wasn't welcome.

So new players emerged instead and the flow of oil from the Gulf Coast took enough of the overall market share that SO couldn't play their price-fixing games so easily. There was Texaco, for example, founded by Joseph Cullinan. He'd worked in SO's pipeline arm in Pennsylvania before setting up his own company selling oil equipment and was already working in Texas when Spindletop blew. Then there was the Pew family. They'd been in oil in Pennsylvania in the 1870s, even making use of the gas that came out of the ground too; the first company to supply a major city (Pittsburgh) with natural gas as a substitute for coal gas. They'd sold this project on to SO, but kept a small oil company, named Sun Oil, and saw an opportunity to build it in Texas. Spindletop is also where the Mellon family seriously committed to oil. Already an established bank, they'd flirted with oil in the 1880s but, like the Pews, sold it to SO. An early investor in Spindletop, they figured that to build an oil company at the turn of the century, they needed to take an integrated approach; producing the oil, as well as moving it on and selling it. In contrast, Rockefeller's approach had always been to control transport and leave the risk of production to others. By 1907, William Mellon had organised the family's various oil interests in the area into a company named Gulf Oil.

Another remaking of American oil would be kicked off by journalist Ida Tarbell. Growing up in Titusville, Tarbell had seen the boom and bust following Edwin Drake's discovery. Her father and brother were both in the oil industry and knew Rockefeller's quietly controlling approach all too well. She'd started working life as a teacher but switched to writing and moved to Paris in her mid 30s. Her life there had a certain bluestocking glamour to it; sharing an apartment with other women writers, interviewing luminaries including Louis Pasteur and Emile Zola, and hanging out with an Egyptian prince, all while she wrote a biography of the French revolutionary Madame Roland. While still in Paris, Tarbell wrote for *McClure's*, a literary and political monthly, first in pages previously filled by children's author Frances Hodgson Burnett and later taking up an editorial post once she'd moved back to New York. Tarbell apparently convinced the magazine's owner, Samuel Sidney McClure, to run an exposé on SO over a mudbath at an ancient spa in Italy (although that story fits the image of 'muckraking' journalists* just a bit too neatly not to take with a pinch of salt).

Tarbell's father pleaded with her to pick any other topic than Standard Oil – you can't win against the octopus, he warned. At a party in DC hosted by Alexander Graham Bell, one of Rockefeller's bankers took Tarbell to one side to say similar, adding a threat to the finances of the magazine for good measure. She curtly replied that the magazine's bank account made no difference to her – she was the journalist, not the owner. Via Mark Twain, she bagged an introduction to one of the SO directors, Henry H. Rogers, who remembered her father and was surprisingly candid. Tarbell sleuthed her way through Rockefeller's quiet mystery and soon had enough to go public. It was published as a series, starting in November 1902 and was immediately explosive, the talk of the nation. This publicity, in turn, brought her new stories, and the series ended up running for two years, published in book form in 1904 as *The History of the Standard Oil Company*. She followed this up the following year with a biography of Rockefeller, making him out to be a thoroughly amoral son of a snake oil dealer. He, in response reportedly called

---

* The term 'muckraking' refers to investigative, often reform-minded journalism that emerged at the start of the twentieth century in the US. The term itself didn't come until the practice was already reasonably well established. In 1906, Theodore Roosevelt complained some journalists seemed to prefer to wallow in the mud rather than support his policies and compared them to a character in *The Pilgrim's Progress* who could look nowhere but down to the rake he used to move muck.

her 'Miss Tar Barrel', complaining to a friend that, 'Things have changed since you and I were boys. The world is full of socialists and anarchists. Whenever a man succeeds remarkably in any particular line of business, they jump on him.'

The stories influenced the 1904 election, so much so that at one point Theodore Roosevelt found himself promising to return $100,000 in campaign contributions from SO directors. Eventually, in 1906, SO was told it had three years to divide itself up into smaller entities. The largest of these, about half of SO's value, was Standard Oil of New Jersey, which we today know as Exxon. The New York arm we now know as Mobil, the Californian one became Chevron and Standard Oil of Ohio became Sohio (later the American arm of BP). The octopus had been slain, but if anything, its arms were more powerful working independently. Rockefeller had been urged to sell his shares before the breakup, under the assumption they'd reached their peak. Wisely, he didn't. Most doubled in value. Some tripled.

★ ★ ★

Back in London, things weren't going exactly swimmingly for the Samuel brothers. Their work in oil had, at first, been extremely profitable. When they launched their new company, Shell Transport and Trading in October 1897, it floated at £1.8 million. Marcus had managed to build more of a name for himself too. He was immensely proud to be an Alderman of the City of London, and after the Shell tanker, SS *Pectan*, rescued a British warship that had run aground at the entrance to Suez, he was offered a knighthood. But he still felt like an outsider. If he wasn't excluded for his Jewishness, it'd be for his East End roots. Aristocratic Jewish families like the Rothschilds might use Shell tankers, but they wouldn't invite the Samuels to socialise with them and were less than enthusiastic about accepting Marcus's hospitality too. He still didn't have the status he craved.

They tried to get in on the Spindletop wave, only to be disappointed when too many wells were drilled too close together and the one due to supply Shell ran dry. All four of the large tankers Marcus had sent out to Texas were to be converted to transport cattle instead, destined for the east London meat trade. Twiddling his thumbs while Sam was visiting Mark Abrahams out in Borneo, Marcus took a gamble buying up vast quantities of oil wherever he could and investing in four more tankers in which to store it. However, the price of oil plummeted and it had to be sold at a loss. More problems came as the company's tanks were damaged in the Boxer Rebellion in China, and the Second Boer War in South Africa aggravated deals with the Dutch. Then one of Shell's tankers ran aground on Suez and when another tanker went to its aid, it caught fire. Throughout the turmoil, Marcus could be seen riding his horse, Duke, every day in Hyde Park. Another oil man noted that Marcus rode his horse like he rode his business, always looking as if he was about to fall off, but never quite doing so. He'd made a new friend in Vice-Admiral Sir John Fisher, and the two became increasingly convinced shipping should shift from coal-powered steam to oil. Between them, Fisher and Marcus cooked up a demonstration they hoped would convince the Admiralty, fitting up warship HMS *Hannibal* with an oil-burning engine. Sadly, this too would turn out to be a humiliating fiasco. A crowd of about a hundred gathered at Portsmouth Harbour to watch it sail. The ship started off burning Welsh coal, as normal, then shifted to oil and – at the point everyone was meant to be impressed by the smoothness and power of the engine – immediately became blanketed in thick, sooty smoke. It was yet another setback financially, and for Marcus it was also a blow to his dreams of being seen as contributing to British power, to be part of the great imperial machine. He got back to the office to find a venture in Australia he'd sunk money into was also in a bad way.

An alliance with the Royal Dutch company had been discussed, on and off, for years. Sometimes involving Rockefeller, or the Rothschilds, or both. Marcus had resisted, determined to keep his independence and remain a British company, but he wasn't really in a position to push back any more. A plan had been cooked up by Fred 'Shady' Lane that would build a new company out of the struggling Shell alongside Royal Dutch and the Rothschilds, all with equal power under the chairmanship of Sir Marcus. It would be a powerful setup, one that would combine Texan, Sumatran and Russian oil with markets across the world. But it would also include the exploitation of new oil fields in Romania and this last point was especially hard for Marcus to stomach. He had just been elected Lord Mayor of the City of London, and the ceremonial procession was planned to go through the East End, back where his father had started the business all those years ago and, more to the point, had recently become the home to many Romanian Jews taking refuge from persecution at home. Marcus felt uncomfortable working with Romania, but his complaints fell on deaf ears.

The first stage in what would become Royal Dutch Shell happened in June 1902, with the Asiatic Petroleum Company. Henri Deterding, who by then was running Royal Dutch, would move to London to help run things. Deterding and Marcus had never got on well. Deterding thought Marcus arrogant and a bad leader, strong on vision, maybe, but weak when it came to detail; in return, Marcus didn't trust Deterding. But for now, Marcus was busy with the pageantry of becoming Lord Mayor. The procession, in November 1902, was a grand affair, with carriages taking the Samuel family and various dignitaries through the East End, ending at the Guildhall for a large banquet. Deterding was not impressed, snobbishly telling a colleague it looked like a circus. But Marcus had a new slice of the status he'd so long yearned for. Behind all this, Shell's earnings were continuing to tumble. The only way out was full amalgamation and the Samuel brothers didn't have much choice other than to do it on Deterding's terms. Marcus procrastinated for a few months, but their position didn't get any better. Finally, the union was cemented in 1907 with the formation of a new company, Royal Dutch Shell. Shell was intentionally the last of those three words, reflecting its junior role. London was to remain its financial centre, with the technical side of the business in the Hague. Deterding would run things in London, settling into the role of English country squire with an estate in Norfolk. At first Marcus regarded himself as a failure, but after the initial disappointment – including a brief crisis where he bought himself a yacht that he had to abandon as it made his wife seasick – he managed to build a working relationship with Deterding. Marcus would continue to provide advice for the firm, and even after his death was referred to as 'the chairman'. What's more, for all the word Shell had been put at the end of the company's title, it's the bit of the name – and logo – that'd stick through time.

* * *

When the Wright brothers first took to the skies in their *Flyer* in December 1903, it ran on Standard Oil. It was clear from the get-go this would be an opportunity for the military and, behind that, a new growing market for the oil industry too. Within months the Wright brothers had been visited by head of ballooning at the British Army Colonel J. E. Capper and, by the end of 1906, they'd employed seasoned arms dealer Charles Flint – who went on to form IBM – to help them establish relationships with the American, French, German, Italian, Russian, Austrian and Japanese governments. It'd be the start of a new era of war and also a key new market for oil: transport. Today this market is one of the trickiest parts of the decarbonisation puzzle. As a 2020 Oxfam report on carbon inequality argues, it's largely due to flying and other high-carbon transport choices like SUVs that the richest 10 per cent of the world's population have substantially higher carbon footprints than everyone else.

The oil industry would find itself at the centre of some of the biggest political fights of the twentieth century, whether it inveigled itself there or not, explosively so in some cases. So much of the modern oil industry had started in Baku, but by the start of the twentieth century it was the hub of a lot more than oil tankers: it was also the base of 'Nina', a large, secret, cellar-based publisher that printed Lenin's newspaper *Iskra* ('Spark'). This was more than simply coincidence. Just as the comings and goings of whale oil in New Bedford had provided cover for the Underground Railroad, these new oil routes were perfect for smuggling propaganda through Europe. Baku was also the stomping ground of a young Joseph Stalin who led mass strikes among oil workers, including a prolonged one focused on the Rothchilds' interests that had spread several hundred miles along the railway line to Tbilisi. When, in January 1905, news reached Baku that soldiers from the Imperial Guard had fired on unarmed protesters petitioning the Tsar in St Petersburg (so-called 'Red Sunday'), strikes reignited there, as they did all over the Russian Empire. Local officials, fearful that Baku was especially susceptible to revolution, fuelled long-standing racial tensions in the area between the Muslim community and the Christian Armenians, arming one side agaisnt the other. As extreme violence spread through the area, exploding oil tankers literally fed the fire, their smoke blackening out the Sun for several days.

It was within this new context of the energy-hungry and newly politically volatile twentieth century that oil fields started to open up in the part of the world the British had dubbed – from their London-based view of the world – the 'Middle East'. Oil seepages and gas flares in the area were hardly a secret – you could trace stories of them back to the ancients. Geological surveys in the 1850s had pointed to the likelihood of substantial resources in Persia, and Baron George de Reuter (most famous for his news agency) had gained rights to extract what he could in 1872. This particular project had come to nothing, but that wasn't going to stop others giving it a go. British businessman William Knox D'Arcy had made a fortune in gold mining in Rockhampton, Australia, in the 1880s and wondered if he could replicate this success with the 'black gold' of Persia. Working on advice from Antoine Kitabgi Khan, a Persian general of Georgian descent, Knox D'Arcy sent a representative to Tehran to negotiate with the Shah. In May 1901, he got his deal, as the Shah granted Knox D'Arcy exclusive rights to search for, exploit, develop, render, transport and sell natural gas, petroleum and asphalt throughout the Persian Empire for the next 60 years – in exchange for £20,000 plus shares and royalty on profits. Knox D'Arcy hired George Reynolds, who'd previously drilled for oil in Sumatra, and started the long and difficult process of shipping equipment first to Basra, then 300 miles up the Tigris to Baghdad, where it was finally carried by people and animals over the Mesopotamian plains to the drilling site.

Conditions were hard and work was slow going. It seemed as doomed to failure as the previous attempts and Knox D'Arcy must have felt as if his money was simply evaporating in the heat. He needed more investment to continue, trying Standard Oil, the Rothschilds and even the catering company Joseph Lyons and Co (today better known for its Bakewell tarts), all to no avail. He'd find another route though via oil consultant Thomas Boverton Redwood. Always immaculately dressed, complete with orchid in his buttonhole, Redwood had a background in organic chemistry and had patented a valuable distillation process, becoming a key advisor to several oil companies. Perhaps most importantly, he had the ear of the British government on oil matters. The British Admiralty was coming around to the idea that it might need to think about this sticky alternative to coal, but was worried it'd have to rely on supplies from America, Russia or the Dutch East Indies. Unlike coal, Britain had very little oil of its own (North Sea oil and gas wouldn't be exploited until much later in the twentieth century). They tried tapping various parts of its Empire, but to little success. One of the world's major oil companies was part-British, as Marcus Samuel kept sticking his hand up to point out, but not everyone trusted them as quite British enough (some due to antisemitism directed at the Samuels, some simply because they had cut a deal with the Dutch).

Buoyed on by Marcus Samuel's old friend John Fisher, Redwood cooked up a syndicate which, assuming it found oil in Persia, would supply the Royal Navy with oil it felt it could trust. It'd combine Knox D'Arcy's outfit with a small Scottish-owned oil company in Burma, headed up by Lord Strathcona, an 84-year-old Canadian self-made millionaire who'd previously served on a Royal Commission to investigate the Second Boer War and was generally trusted as the right type of good egg, supportive of British interests. Knox D'Arcy had his funding at last and could try drilling at a new site. Finally, in May 1908 they struck a gusher – oil rose 50ft into the air, the accompanying gas cloud threatening to suffocate everyone around. It was clear they'd need a business to manage the flow of this oil into markets and the Admiralty wanted its role to be kept as quiet as possible. The syndicate became the Anglo-Persian Oil Company, today better known as BP.*

In 1911, Winston Churchill was appointed the First Lord of the Admiralty, the political head of the Royal Navy. Fisher and Marcus

---

* The British Petroleum Company was, before the First World War, an arm of Deutsche Bank, built to sell Romanian oil to the British market. When the British government declared war on Germany, it took control of the company, soon selling it on to Anglo-Persian, which later drew on the name when it wanted a rebrand in the mid 1950s.

Samuel took the idea of oil-powered ships to him again. Churchill wasn't impressed by Marcus's sales patter. Fisher had to apologise for his friend, explaining later to Churchill that Marcus was 'a good teapot', but 'began as a pedlar selling "sea" shells and now he has six million sterling of his own private money' so was a man to trust. Churchill was convinced by the central argument that oil was the future, even if he remained sceptical of Marcus and his seashells. He shared the Admiralty's old concern about having to rely on foreign supplies of oil, compared to British coal, but Churchill was equally worried that by banking on this British coal, they were ceding too much power to the unions. A strike by Welsh coal miners in 1910 had kicked off several years of intense industrial action across the UK, largely forgotten today but huge at the time and known as the 'Great Unrest'. Churchill had been home secretary briefly before moving to the Admiralty post and sent troops in against strikers led by transport unions in Liverpool. Churchill's choice was less a matter of a British or foreign fuel (or a technical question of coal vs oil-based devices), but rather to give power to the oil companies or the labour movement. He chose the former.

The new managing director of Anglo-Persian, Charles Greenway, started aggressively courting the navy for its contract, pushing an idea that his firm was run by trustworthy British gentlemen (with a less-than-subtle subtext that anything built by an East End Jew working with Dutch partners wasn't quite cricket). This resonated with Churchill and by June 1914 he was standing up in Parliament, proposing a bill where the government would invest £2.2 million in Anglo-Persian, acquiring 51 per cent of the stock. The MP for Wandsworth, Samuel Samuel, sat twitching in the backbenches as Churchill made a stream of sarcastic comments about Shell, especially the implication that it would charge inflated prices, which another MP later described as 'Jew-baiting'. Less than two weeks later, Austro-Hungarian Archduke Franz Ferdinand was shot in Sarajevo.

The war that followed – the Great War, the First World War – would prove to be one of oil-powered machines. Fisher and Churchill were prepared up to a point, but no one had quite expected the level or demand. Allies invested in cars, lorries and motorcycles to transport people and goods. Then there was the development of what was initially called a 'landship', but we'd know as a tank (and would, a few months after the war, be deployed by Churchill against striking workers in Glasgow). Most significantly, perhaps, this war was the serious start of airborne warfare too, with Zeppelins crossing the Channel to bomb mainland Britain on a cloud of Henry Cavendish's old 'inflammable air', hydrogen. The innovation that'd really stick, however, was the Wright brothers' plane, running on oil. Planes were intially used for reconnaissance, but then pilots started shooting each other with handguns. They were soon fitted with machine guns and before long they were vehicles for bombing too. And then there were the explosives. A few years before the war, chemists at Shell had realised the heavy oil they'd tapped in Kutei could be used to make toluol – the second 'T' in TNT – which otherwise, like mauve dye and creosote, could be extracted from coal tar. Ever the patriot, Marcus Samuel gave the Royal Navy first refusal, but they preferred to stick to supplies drawn from home-grown British coal gasworks. The German and French governments thought differently though, and Shell soon had enough orders to build a toluol plant in Rotterdam. When Britain's wartime supply of toluol from gasworks turned out to be less reliable than they'd hoped, Shell saved the day, dismantling its Rotterdam refinery piece by piece, shipping to London under the cover of night, before sending it on a specially cleared rail network to Portishead in Somerset where it was put, jigsaw-like, back together.

In the end, in the words of Lord Curzon, just before he became foreign secretary: 'The Allied cause had floated to victory upon a wave of oil.' Henry Bérenger, president of the French petrol committee, replied in agreement: 'As oil has been the blood of war, so it would be the blood of peace. At this hour, at the beginning of the peace, our civilian populations, our industries, our commerce, our farmers are all calling for more oil, always more oil, for more gasoline, always more gasoline.' It had got tight at times. Indeed, we have wartime fuel rationing to thank for the introduction of summer 'daylight savings' clock changes, designed to make more of natural light. American car owners had been forced to get used to 'gasoline-less Sundays', and many were worried this might be something they'd have to deal with long term. Just as Brits had worried steamships might suck all their coal from the ground in the 1860s, with American cars and other vehicles guzzling up more and more oil, they worried their wells might soon run dry.

Shell had suffered losses with the destruction of their oilfields in Romania (17 per cent of their production at the time), but this sacrifice – along with the clever reconstruction of the toluol refinery and steady supply of oil to British forces – was recognised. In thanks, Marcus was made Baron Bearsted of Maidstone in the 1921 birthday honours. Finally, he had the recognition from the British establishment he so craved.

CHAPTER EIGHT
Big Science

A student slightly late to their 9am lecture at Penn State on 8 December 1939 might have spotted a group of geology lab assistants methodically sprinkling a dark powder over a small part of the campus. It was a cold and foggy morning, and the ground was covered in a light dusting of snow. The researchers had carefully measured out a patch of the snow and were covering it with what looked, on closer inspection, to be coal dust. They were also sticking thermometers in the ground, taking careful notes of their readings. Looking up to the geology lab, their colleagues were placing coal-dust-covered ice cubes on a sunny window ledge too. A week later, they were at it again. This time there was a clear, cloudless sky, but it was still bitterly cold, probably colder than before. But they were still sprinkling the ground with coal dust, taking measurements and standing back to watch. Peering in for a closer look, you might have seen the patch sprinkled with coal dust had become criss-crossed with small channels, melting away faster than the fresh snow around it. The researchers seemed excited, like they'd found something. 'Best send some down to Byrd at the Antarctic then, eh? Let him have a go!'

Professor Helmut Landsberg, who was leading the experiment, wrote up the research the following spring for the *Bulletin of the American Meteorological Society*. The basic science wasn't new. Benjamin Franklin had done similar experiments, wrapping blocks of ice in dark fabric. Fresh white snow and ice reflect much of the Sun's rays; darken the surface and they'll absorb more of the heat, and so melt faster. It's why rooftops in sunnier parts of the world are sometimes painted white and it's a good idea to avoid wearing black in the summer. Today, the darkening of snow in the polar regions, caused by soot from wildfires, is something polar researchers study in detail. They wonder, for example, if soot from the Industrial Revolution may have caused the retreat of Alpine glaciers in the mid-nineteenth century, hastening the end of the 'little ice age', and is having a similar effect today.

But Landsberg wasn't working out of concern for the health of glaciers (quite the opposite in fact). He thought a more precise understanding of sooting ice might help people clear frozen roads so they'd be safer to drive on. Moreover, the same method might be applied at scale, opening up land in the northerly parts of the US so they could be better used for the growing of crops or timber. Landsberg wasn't about to stop there either: 'It is utopian to propose covering large ice fields each spring with dark dust,' he mused, 'but it would seem to be feasible at least to melt off considerable portions of glaciers.' Dust the Antarctic ice sheet with coal each summer to increase the rate of melting and, though it'd take a few years, eventually the glaciers would be gone. For Landsberg, at the time, this was an efficient use of the Sun's rays – it was his idea of solar power.

Picking up the paper, *Popular Mechanics* agreed – why burn tonnes of coal to melt ice when a sprinkling of soot would do? It might even save us from burning coal to keep warm if we could use it to make the climate milder. The *San Bernardino Sun* gave it a big illustrated splash: 'Sprinkle coal dust on snow, expose to Sun, and presto … no snow. Maybe we CAN do something about the weather.' Thick lines of text across the photos asked if we could: 'Melt glaciers and make polar valleys fertile?' or 'Make winter driving safer?' The weather, the cold in particular, was an inconvenience, but with enough ingenuity perhaps humans could beat it. The *Sun* also pointed to studies brought back from Antarctica, suggesting there might be great coal deposits to be found if only we could melt through the ice sheet first. They also speculated that if America didn't try modifying the climate with coal dust, the Russians would probably get there first, to make Siberia more habitable. 'Whatever one may think of the current Soviet government in a political way' they mused, 'it is not averse to trying new things.'

It seems ridiculous today, perhaps painfully so. Glaciers were, by this point, already observably melting due to carbon emissions from burning coal. And yet, here were scientists sprinkling coal dust on snow, wondering if they could speed up the process. They weren't the only ones, by a long way. By the end of the war, people were starting to think of atomic sources of heat too, and in December 1945 the then head of UNESCO Julian Huxley addressed an early arms control meeting in Madison Square Gardens with ideas of dissolving the polar ice caps with atomic bombs. This could well flood New York City, he warned, but might also possibly provide Britain with the climate of Spain.

As we'll see, running alongside the story of our accidental weirding of the weather via the greenhouse effect was research applying chemistry to make it rain, snow, or divert a storm. This would colour how people saw the climate crisis in the mid-twentieth century, although it would also give meteorological research a funding boost at times too. Sometimes this work was done for humanitarian reasons, to end a drought or protect a town from a hurricane; sometimes scientists were tasked with weaponising the weather. Either way, it's the old, Enlightenment idea of science – learn enough about nature and perhaps you can control it. Back at the start of the nineteenth century Jefferson had mused over the feasibility of climate engineering. In the 1840s, 'Storm King' James Espy had made headlines with his idea of setting fire to forests to emulate volcanoes and thus trigger rain. But scientists and their funders were about to find things weren't quite so simple.

* * *

To grasp how the scientific understanding of climate change grew in the second half of the twentieth century, we have to understand the changes to American science that happened in and around the Second World War, in particular the relationship with the military. It's the context within which modern climate change science flourished; it also laid the ground for politicians picking up the issue in 1970 and – as we'll see as we approach the end of the book – it also helps to explain the power of climate sceptics. And to understand all those changes, we need to meet Vannevar Bush.

Better known as Van, today Bush is remembered for his idea of the 'memex'. He introduced this in 1945 article for *The Atlantic*, 'How we may think', about post-war science and technology, and it's often cited as an influence for hypertext, the internet, the web and Wikipedia.

The problem, as Bush saw it, was that science had become so successful that it was easy to get lost in all its specialist branches. 'There is a growing mountain of research,' he wrote, but any investigator 'is staggered by the findings and conclusions of thousands of other workers – conclusions which he cannot find time to grasp, much less to remember, as they appear.' The solution, he mused, was some sort of device that would allow for the non-linear storage of information, with systems for speedy retrieval. He called it the memex and imagined it would look like a desk with two slanting translucent screens on top, along with a keyboard, and sets of buttons and levers. The illustration looks remarkably like an overgrown version of a 1990s PalmPilot, or possibly a Kindle.

In his *Atlantic* article, Bush gives the example of looking up the origin and properties of the bow and arrow: 'First he [the memex user] runs through an encyclopaedia, finds an interesting but sketchy article, leaves it projected. Next, in a history, he finds another pertinent item, and ties the two together. Thus he goes, building a trail of many items.' The user can leave comments, make side trails and notes as he goes, which will be stored. 'Several years later, his talk with a friend turns to the queer ways in which people resist innovations, even of vital interest,' Bush imagines. 'He has an example, in the fact that the outraged Europeans still failed to adopt the Turkish bow. In fact he has a trail on it. A touch brings up the code book. Tapping a few keys projects the head of the trail. A lever runs through it at will, stopping at interesting items, going off on side excursions. It is an interesting trail, pertinent to the discussion. So, he sets a reproducer in action, photographs the whole trail out, and passes it to his friend for insertion in his own memex.' The challenge of living in a world built on masses of diverse, specialised information that no one person alone can ever grasp in its entirety is one most of us born in the twentieth or twenty-first century can appreciate. But it was a challenge Bush had found earlier and more powerfully than most, leading the complex systems of US science funding during the Second World War, so its perhaps no surprise that he came up with the idea of the memex.

Bush had initially trained as an electrical engineer in the early 1920s and helped invent a new type of radio. Off the back of that he co-founded the American Appliance Company (now the defence contractor Raytheon) before becoming vice-president of MIT in the 1930s and later moving to DC to lead the Carnegie Institution's $33 million endowment for research funding (where, among other things, he pulled their support for eugenics). In 1938 he was appointed to the National Advisory Committee for Aeronautics, the predecessor of NASA, and it was from there he started drafting a proposal for a national government office for science. When the Germans invaded France in May 1940, Bush decided speed was of the essence and took the plan directly to the President. Roosevelt took just 10 minutes to approve the proposal. He made Bush head of the new Office of Scientific Research and Development (OSRD) and gave him a budget of $3 million a week to be spent on everything from radar to antibiotics and explosives, incubating early work on the atomic bomb.* By April 1944, *Time* magazine put Bush on their cover, dubbing him the 'general of physics'. Work on nuclear warfare would later open up opportunities for nuclear power, providing fresh competition for fossil fuels. For now, what's important to understand is what such work did to American science.

The OSRD was wound down once the war ended and Bush returned to his post at Carnegie, but in the last weeks of the war he prepared a report for Roosevelt – 'Science, the endless frontier' – distilling what he'd learnt in wartime science to an ambitious plan for federal support for science in peacetime. Bush wasn't the first to argue for large government grants for science, similar calls had been made in the 1920s by the UK National Union of Scientific Workers and US agriculture secretaries Arthur Hyde and Henry C. Wallace. Just before the war, Irish crystallographer John Desmond Bernal had published highly influential book *The Social Function of Science*, which argued that science is the most important thing humans do, and should be organised and funded to maximise its huge capacity to improve people's lives. There's a story that

---

* Part of Bush's job was choosing what not to fund too, fending off more preposterous schemes. At one point the British suggested chipping off a huge block of ice from the Arctic to use as an airstrip in the middle of the Atlantic, fitting it with a giant refrigeration plant to keep it frozen. Bush simply told Roosevelt it was 'bunk' and it was never spoken of again.

Bernal's interest in science policy was inspired after a delegation of Soviets gatecrashed a history of science conference at the Science Museum in 1931. Whether this is true or not, he was certainly a man of the left.* But the idea of government-funded science for society can be pulled to a range of ideological ends, and in Bush's hands you can see it more neatly shaped for an American version of progress than a Soviet one.

'The endless frontier' is very readable, as science policy reports go. It argues a need for science to help new products to fuel the economy, to underpin drug development and to help build new weapons (in peacetime as well as war). Rather than talking of 'pure' vs 'applied' science, instead Bush used the term 'basic science'; work that interested the scientists but would still please more pragmatically minded funders. The notion of basic science is more than slippery, not least because one way to get researchers interested in a topic is to let them know funding is available. In the 1980s, Reagan would cut 'applied' research into atmospheric science and solar energy, arguing 'basic research' better fitted a free-market philosophy.† Still, in 1945, it helped make the case for expanding the sort of work the state or military might fund. The work underpinning the bomb, after all, had once seemed entirely the games of the most abstract physicists. Bush's recommendations were slightly too rich for President Truman's

---

* It's sometimes said Bernal was never strictly a 'card carrying communist', having apparently dropped his actual membership card sometime in 1933 and never bothered to replace it. Bernal was also known for his pacifism, and his parties. After delegates at a 1950 international peace conference ended up stranded in London, Bernal invited them back to his flat. One of the delegates was Pablo Picasso, who drunkenly sketched a mural on the wall of Bernal's sitting room. When the building was demolished, Bernal gave the mural to the ICA. In 2007 it was sold to Wellcome for £250,000.

† In a 2020 essay looking back on 75 years of Bush's report, Roger Pielke Jr reapplies Patty Limerick's critique of language used to describe the American space programme to argue that the use of the 'frontier' metaphor romanticises the endeavour, working to avoid too many tricky questions being asked about the social responsibility of scientific enterprise. The idea of occupying an imaginary empty frontier works for Bush (just at it worked for the early conservation movement) to absolve the occupier of responsibility for forest fires, destruction of Indigenous societies, pollution of groundwater, etc.

taste, but the Office of Naval Research was more than happy to fill the gap while the White House sat on the idea, acting as a sort of surrogate for national science funding until, eventually, in 1950, an official National Science Foundation was established.

Today, we'd call the science that emerged post-war 'big science'. It's characterised by the scale of projects that, like the Manhattan Project to build the bomb, not only had large budgets but also a large number of people, often from a variety of disciplines, and sometimes physically large equipment too. The sort of complex project that would make science managers' heads ache and daydream about a device like the memex to help them keep track. The term 'big science' was coined by American physicist Alvin Weinberg, director of the Oak Ridge National Laboratory, in a 1961 article for the journal *Science*. He was calling it out because he wasn't exactly a fan. He memorably and provocatively compared big science projects to pyramids of Egypt and the Cathedral of Notre Dame: 'We build our monuments in the name of scientific truth, they built theirs in the name of religious truth.' It wasn't that Weinberg wanted a return to 'little' science, he wasn't against the idea of spending lots of government money, working at scale or applying science to social challenges. Indeed, he specifically mentions pollution of the atmosphere as one of several important issues he thought American scientists should be investigating. But he was concerned that this money seemed to come with both the military and journalists attached, and that science was being used in a game of national prestige.

* * *

Back to the climate part of the science story. Just before the outbreak of war, Swedish glaciologist Hans Ahlmann conducted research in Greenland, confirming something he'd been confident of for decades – rapid warming in the polar regions. His fieldwork was disrupted by the war so he gave a series of lectures in Oslo on the gradual retreat of glaciers, pulling in data from the South Pole too. Like Guy Callendar, Ahlmann saw this as largely a positive change, calling it *klimatförbättringen*, or 'climate embetterment'. The root cause of this embetterment came from nature, he argued, not anything like industrial carbon dioxide pollution. But he was clear it was happening.

Whether it was for good or bad, by 1947 the Pentagon was wondering if they should start measuring the thickness of Arctic ice themselves; if the climate was changing and the country did go to war with the Soviet Union up there, it made sense to be informed. Within a few years, all the American military services were funding studies of the Arctic, including sea ice, permafrost and the physical properties of snow. In May 1950, evidence for global warming was presented at a meeting of the American Meteorological Society and made its way into *Time* magazine. The write-up suggests some of the delegates were still unsure, with one especially enthusiastic researcher from Harvard going to the trouble of finding a thermometer from 1849 and checking it against modern instruments to show it was reliable. Still, it was hardly front-page news, tucked away in the middle of the magazine, at the end of the science section (squeezed into the hard left of the page by an advert for air conditioning, fittingly enough). Moreover, it wasn't clear what was causing this warming or even if it was long term. No one seemed overly worried. It was much more exciting to see if we could change the weather on purpose.

Callendar's wartime project for the Petroleum Warfare Department, Fido, had shown it was possible to fight back the fog if you burnt enough oil, and in 1942 a team at General Electric led by Nobel Prize-winner Irving Langmuir had presented Bush and other scientific dignitaries with a new way to produce clouds by sending hot lubricating oil at supersonic speeds through cold air. This had belched out thick smoke a mile wide and 10 miles long – a military-grade smoke screen. They'd made an artificial cloud, why not try modifying an existing one? This had been secret, but in 1946 they went public with tests to fertilise, or 'seed', clouds by throwing granulated dry ice into a four-mile stretch of stratocumulus over Massachusetts, which had appeared to make it lightly snow. In 1950, Langmuir made a splash with a *Science* article on experiments in New Mexico, injecting clouds with silver iodine to create rain. This had made the cover of *Time*, an illustration of a slightly glum looking Langmuir holding up an umbrella to protect him from the rain, the umbrella stick made from a string of pipettes.* Donald Duck had even got in on the action, selling 'rain seed' to a farmer (before getting drunk on power, over-seeding the clouds and turning them into a solid dome of ice). Scientists at the Weather Bureau had dismissed Langmuir's claims as grossly premature, and British MPs had joked in the House of Commons that 'if my honourable friend has any [rain] to dispose of, will he put some of it in my garden?' Behind this rhetorical back and forth of parliament, the British were also experimenting with cloud-seeding from balloons in Tanzania and using planes in the south-west of England. The Ministry of Defence had flatly denied any rumours that a disastrous flood in Devon in the summer of 1952 had been due to such experiments, but 50 years later the BBC found one of the pilots on the project, and he suggested it might well have been.

This was what, in the early 1950s, the phrase 'man-made climate change' meant to most meteorologists and policymakers – a deliberate play with the weather. In an April 1950 *Scientific American* feature on 'The changing climate', author and meteorologist George Kimble is keen to let readers know they are living through a particularly interesting climatic moment. Temperatures have clearly crept up gradually since the middle of the last century. Still, Kimble reassures, there's no reason to concern ourselves wondering why. It's certainly not due to all the heat we're producing from burning fossil fuels, he notes almost in passing, concluding it's probably something to do with the Sun. He's much more interested in Langmuir's research into seeding clouds. Callendar's fog-dispersal device gets a name check (though not Callendar himself), but carbon dioxide doesn't get a look in.

---

* Langmuir was an inspiration for the character of scientist Felix Hoenikker in Kurt Vonnegut's novel *Cat's Cradle* (1963), a seemingly emotionless man who cared nothing about the possible disastrous impacts of his work. Vonnegut's brother Bernard was a colleague of Langmuir's, is referred to in his 1950 cloud-seeding paper, and Kurt himself had worked briefly in PR for General Electric. In the 1970s, Kurt told an interviewer this was what got him into writing science fiction, there was no avoiding it there, 'since the General Electric Company was science fiction'. For more on the interactions between science and fiction in this area, see James Rodger Fleming's (2012) book on the history of attempts to manipulate the weather, *Fixing the Sky*.

However, by the time the *New York Times* weekend magazine would cover the newly warming climate in July 1953, carbon dioxide was back on the table, albeit only as one idea among several others. New York had just had its third wettest spring since records began in 1826, the author, Leonard Engel, informed his readers. And although the Weather Bureau suspected that the large number of tornadoes recorded was partly due to expanding its team of volunteer storm spotters, it was surprisingly high – three times as much as usual. This was not a one-off phenomenon, nor was it just an issue for the US. Rather, there's 'abundant evidence' that the world's weather is undergoing long-term change. Engel points to the Alps, Greenland and Alaska, where the winters aren't long enough to make up for the summer melting; the glaciers are retreating, falling back as they sweat off running water. In describing the possible reasons for this change, Engel was keen to stress this was not military experiments gone wrong. 'An atomic explosion is awesomely powerful but not powerful enough to open the heavens' and studies of test blasts have shown no impact on rain. He'd checked with Langmuir too, and there was no way the recent downpours in the East had anything to do with his older experiments in New Mexico. As the article's standfirst emphasised 'atmosphere, not atoms, is to blame'. But when it came to what exactly was messing with the atmosphere, there were still plenty of gaps in scientists' knowledge. Plus, Engel notes, this was the first climatic change they'd had a chance to examine first hand, so they were learning as they were going.

What was clear, Engel told his readers, was that the people of 1953 were living through a period of climate change, and that these changes would affect them deeply. He said it might bring new opportunities, but he also struck a warning note: a decline in rainfall could cause devastating drought, or the shrinking of the ice caps could lead to flooding. Above it all, the feature was illustrated with an imposing photo of clouds building before a tornado. Underlining the point, on the following page the magazine had compiled news and photos from *New York Times* correspondents across the country, summarising what was going on in other parts of the US. While Boston and Florida both reported booms in summer tourism, in Texas, all the news was of the drought. The Rio Grande ran dry for hundreds of miles, 'topsoil drifts over useless farm machinery', livestock had to be sold at ruinous prices and farmers were going bankrupt.

The anonymous 'some meteorologists' advocating the carbon dioxide theory of global warming might have been Callendar, but it's more likely to have been newer work that year by Gilbert Plass, which had already been reported by *Time* and *Popular Mechanics*. Plass had been doing postdoctoral research on infrared radiation for the Office of Naval Research at John Hopkins University, and became curious around the edges of what he was supposed to be doing. He happened upon some of the older work exploring ice ages in terms of carbon dioxide in the atmosphere – the nineteenth-century research from Fourier, Tyndall and Arrhenius we saw in Chapter Two as well as more research work from Callendar – and started digging further. He was aware of new lab work questioning Knut Ångström's assertion that carbon dioxide didn't really do much in terms of heat in the atmosphere. Crucially, a sabbatical at Michigan State University gave him access to a new electronic computer, meaning he could run detailed calculations. Before he'd finished the work, he got a new job at Lockheed in Southern California, researching heat-seeking missiles. But he kept up the climate change work as an evening project; a bit of light environmentalism as a break from the military-industrial complex.

★ ★ ★

Meteorology in mid-1950s America was a slightly different beast to the fellows of the Royal Meteorological Society Callendar had pitched to in 1938. As aviation had grown, it offered fresh demands for meteorologists. Just as the navy had long used and supported research into and around the oceans, any of the various strands of the military now taking to the sky wanted better data on what they'd be flying through. During the war, tens of thousands of Americans were trained up either as professorial meteorologists or as observers and technicians to support their work, largely from the 'big five' meteorology schools of UCLA, Chicago, Caltech, MIT and NYU. These were a new development themselves. America's first university department of meteorology, at MIT, had only been founded in 1928.

The founder of this first department – and architect of much of 1930s and 1940s American meteorology besides – was Carl-Gustaf Rossby. He'd grown up and studied in Stockholm and arrived in the US in 1926 with a fellowship to work at the Weather Bureau. After a series of run-ins with the Bureau's leadership (possibly due to a failed forecast for aviator Charles Lindbergh) Rossby left under something of a cloud. Still, by then, he'd developed other options. Rossby's ideas for how meteorology could help make flying safer impressed Daniel and Harry Guggenheim, who were eagerly spending some of their family mining fortune supporting aeronautical research. They appointed him chair of the Guggenheim Committee on Aeronautical Meteorology and then supported the building of a research group in MIT. There, Rossby built research and teaching programmes and made his way on to a number of national aeronautical committees. In 1931 he managed to pull together grants from the National Academy of Sciences, Guggenheim and MIT to buy a plane, allowing him to take meteorological measurement devices right up into and through the clouds (a big step up from Glaisher's ballooning trips). With projects like these, he started training a whole new generation of aviation-minded American meteorologists, many of whom would later be invaluable to the US war effort.

One especially notable Rossby-trained meteorologist was Harry Wexler. During the war, he'd become famous as the first meteorologist to fly a plane into a hurricane,[*] earning him the nickname 'Hurricane Harry'. He'd go on to lead science at the Weather Bureau, helping build the modern climate change science that emerged after the mid 1950s, something we'll pick up in the next chapter. What's most important now is the role he played in increasing meteorology's power through new technological methods of data collection and

---

[*] As Fleming reports in his 2016 book on Rossby, Wexler and other icons of modern meteorology, this had started as a joke when Wexler noticed a hurricane off the coast and, knowing it was his colleague Colonel Wood's afternoon to go flying, laughed 'why not fly into it?'. Wood took him at his word though, and they went up. On landing, Wexler's wife reported he looked a bit pale and needed a drink. Fleming also points out this wasn't the first flight into a hurricane, even if that was how the press reported it – that had actually been done the summer before by a Colonel Duckworth.

analysis. Meteorologists had been bursting out of our atmosphere into outer space for a while, even if they weren't yet orbiting. Back in the 1920s, Robert Goddard had developed rockets propelled with a mix of liquid oxygen and gasoline, and used them for weather photography. In the late 1940s, a new opportunity came via some tech taken from the Germans – the V2 rocket. These had been developed near the end of the war to target Allied cities in retaliation for the bombing of Berlin and other German cities (the 'V' stood for vengeance). Effectively the first space rocket, these would be launched up into the edge of space and as they re-entered the atmosphere they travelled at the speed of sound. You wouldn't hear it untill you were hit. Although there is no exact figure, estimates suggest that several thousand people were killed by V2. Far grimmer and less often spoken about were the tens of thousands of people who died building the rockets, in an underground factory by the Buchenwald concentration camp.

A repurposed V2 rocket had been launched in New Mexico in 1947, taking photographs of the Earth's cloud cover. But not everyone in meteorology was convinced by these space trips, they figured money could be better spent elsewhere. Initially Wexler brushed it off too but came around to the idea in the early 1950s, egged on by science-fiction writer Arthur C. Clarke. He presented his plans at a symposium at New York's Hayden Planetarium in May 1954, bringing with him a specially commissioned artist's impression of a satellite's eye view, 4,000 miles above Texas. The original image has been lost, with most copies in black and white, but historian James Rodger Fleming found a colour photo in the Library of Congress (and admits he whooped for joy when it fell out of an archive box). It had been inspired by V2 rocket photos, but the artist had clearly worked with Wexler to add a lot more from a scientifically informed imagination, with delicate clouds posed to swirl carefully around North America, all on a purple-blue sea. It's incredibly beautiful, possibly because it's clearly at least partly fantastical – a guess at what the Earth from space might look like. Over a decade before the 'Earthrise' photo from space, this was still science fiction.

If meteorologists were going to predict the atmosphere of the future, Wexler quite reasonably argued, they needed strong data on the atmosphere of the present and satellite tech offered a vital new tool to collect it. They had an exciting chance to go up into to skies, above where weather phenomena happened to look down on winds and rains below. Not a 'god's eye view' of Earth, but closer than scientists had managed before. Balloons and kites were all very well, but they still existed within the weather system. They also had to be launched from land, making it harder to study the atmosphere above oceans. Satellites would help researchers develop forecasting and understand longer-term climatic variations, he promised, but its primary use would be as a 'storm patrol.' You can only imagine how Beaufort or FitzRoy's eyes would have popped at the prospect of such a piece of kit. In 1955, it was announced that the US would launch a research satellite as part of the upcoming International Geophysical Year. Wexler was going to take the Weather Bureau into the space age.

It was all very well extending meteorological data collection into outer space, but you still needed to do something with all that new information. Even with techniques like Gilbert Waker's 'searchlight' and all the extra trained manpower the war had brought, meteorologists were craving more computer power. Back in the winter of 1917, British mathematician Lewis Fry Richardson had found himself driving a Quaker ambulance on the Western Front, having quit a job in a remote Meteorological Office station in Scotland, morally opposed to war but curious as to what was going on. He would sit in his spare moments, the mud and blood around him, distracting himself with dreams of a vastly improved mathematical forecasting system, calculating an imaginary weather forecast for 1910 (a 'hindcast' rather than a forecast), using just pencil, paper, slide rule and a table of logarithms. His predictions were not only incorrect, they were ridiculously so. Still, the complex system he worked out, written up in a 1922 book, was recognised as an impressive achievement. The book also proposed, slightly tongue-in-cheek, the idea for a 'forecast factory', a large circular hall with the walls painted with a map of the world. Inside would sit thousands of 'human computers and computresses' working through data for their region armed with pencil, paper, slide rules and logarithm tables, just as Richardson himself had done alone in the mud during the war. Outside, Richardson adds, would also be playing fields, houses, mountains and lakes for these computers and computresses as, he argued, 'those who compute the weather should breathe of it freely'.

A couple of decades later, this dream would start to take shape, but it would be electronic. The idea started with Vladimir K. Zworykin (most famous for work on television), director of research at the Radio Corporation of America who, in October 1945, wrote an outline for a weather computer project. This was passed along to John von Neumann who, after a career that included game theory, working on artificial intelligence with Alan Turing and the Manhattan Project, was heading up the Electronic Computer Project at the Institute for Advanced Study in Princeton. It also went to Rossby, who leapt at the opportunity to further modernise meteorology and nominated Wexler to lead from a meteorological point of view. Zworykin wasn't just pitching a project that could improve weather forecasts, it was about weather control too, including the longer-term changes of climate control. It's sometimes suggested that he was cynically trying to excite military funders with the prospect of unleashing storms on their enemies. However, he concludes with the hope that an international project of the necessary scope to properly compute the weather might well contribute to world peace 'by integrating the world interest in a common problem and turning scientific energy to peaceful pursuits.' We share the weather, after all.

They met with several military meteorologists at the start of January 1946, at the Weather Bureau's old DC HQ about a mile west of the White House. The meeting was supposed to be confidential, but the *New York Times* broke the story two days later (probably via a navy leak). The plans were still all strictly 'on paper', the *New York Times* reported, and it would take years of work and millions of dollars to even make a working model, let alone apply it. But, if this super-calculator could be built, it wouldn't just help scientists understand the weather better, it might help them control it, using a strategically targeted atomic bomb, for example, to divert hurricanes away from populated areas. The machine would be 12ft high, they guessed, 18ft wide and 20ft long. It'd use thousands of tubes, the interior a maze of wires and controls requiring 10–20 people to operate it. And they'd need 100 of these machines. But it could transform their relationship with the weather. The piece concludes with a Mark Twain quote, 'everyone talks about the weather but no one does anything about it', quipping that finally now maybe the time to 'do' something about the weather had come.

Undeterred by this unwanted press attention, the meteorology computing project received a green light, initially funded by the Office of Naval Research, later augmented by Air Force grants and supported by a large team of US Weather Bureau personnel. Von Neumann's wife, Klara, taught the team to code, and was in charge of hand-punching and managing each of the experiment's 100,000 punch cards (for which she got a small 'thanks' at the bottom of a research paper). After a slow start, Wexler recruited Jule Charney, a Rossby-trained mathematician-turned-meteorologist who gradually devised a model that roughly succeeded where Richardson had failed. To test it, they'd run hindcasts of past weather, just as Richardson had done. When Charney finally managed to get one to work in late 1953, he apparently phoned Wexler, who'd been sound asleep, in great excitement, declaring: 'Harry it's snowing like hell in Washington on 6 November 1952.' Still, for all this was a massive leap from Richardson sitting in the mud with a pencil and paper (or even his more utopian dream of a circular theatre of mathematicians), the computer was slow, making a chugging sound as it went. Part of Charney's challenge was to work out simple enough calculations for the computer to manage. It'd frequently break down too, eating up more time.

In London, researchers at Imperial College and the Met Office were also getting excited about the new electronic era of meteorological maths. But they had to make do with some borrowed time on the Lyons Electronic Office, Leo1 to its friends.* This wasn't as powerful as von Neumann's system in Princeton, so the Brits devised simpler calculations, causing much snobbish back and forth across the Atlantic. One of the Imperial meteorologists working on the project, Richard Scorer, called Charney's approach 'utterly useless', in return Charney

---

* Generally recognised as the world's first business computer, it ran out the J Lyons and Co offices in West London. It's the same company that had turned down a chance to invest in the company that had become BP a few decades before. Margaret Thatcher worked for them for a while, as a food chemist. In an alternative universe the world's run by a dictatorship of technologically advanced bakers.

dubbed Scorer 'a fool' in a letter to Rossby. The Americans had a much better relationship with Sweden. Rossby was now splitting his time between Stockholm and Chicago, and could supervise tests off a Swedish computer. For Rossby, international collaboration in this computing project was key, part of a larger ambition to put meteorology at the centre of international scientific collaboration. He'd have other opportunities to build on this ambition though, not least through this International Geophysical Year project the Americans were planning to build a satellite for.

★ ★ ★

All sorts of players and agendas ended up being involved, but the initial idea for an international year of geophysics (or IGY as it would soon be better known) came from scientists. In April 1950 a group of Earth scientists had met informally in a living room in Silver Spring, just north of DC. One of them was visiting from London and they were enjoying the chance for a bit of trans-Atlantic science chat. The conversation somehow ended up on the International Polar Years the World Meteorological Organization had run in 1882 and then again in 1932. Someone pointed out that with all the recent technological developments like computers and rockets, maybe it was time for something similar, but bigger – a massive coordinated, worldwide study of our planet. In the end it came out as a bumper year of 18 months, running between July 1957 and the end of December 1958 (timed because solar activity would be at a peak) with some 30,000 scientists, engineers and technicians taking part across 67 nations. When a science administrator in the US tried to make a rough estimate of the cost – hard with so many different national budgets involved – they thought it was around £500 million.

President Eisenhower kicked off the American end of proceedings on 30 June 1957, with a radio and television address promising 'the beginning of one of the great scientific adventures of our time'. He was keen to stress this wasn't just a scientific project about the planet we all lived on, but an endeavour in international collaboration. 'The most important result' would be not a new discovery but the 'demonstration of the ability of peoples of all nations to work together harmoniously for the common good'. Out of the mouth of a politician, it's hard to take this as anything more than rhetoric, and people were suspicious at the time. At a press conference in 1955, a journalist had asked Eisenhower whether the real reason for a recent mission to the Antarctic was to find new spots to keep nuclear weapons. The president replied flatly that such rumours were absolutely without foundation, they were purely doing logistics for an IGY scientific expedition that 'will be done under scientists and for the development and benefit of the world, nothing else'.

Britain's IGY launch happened a few weeks after Eisenhower's speech, but with slightly more pageantry – a special BBC TV show *The Restless Sphere*, fronted by Prince Philip. A report of the rehearsal in *The Times* a couple of weeks before the broadcast quoted a senior member of BBC staff promising 'the most complicated and massive programme of its kind ever attempted by BBC television'. The prince stood in a makeshift studio at the Royal Society offices in Burlington House, Central London, surrounded by maps and bits of scientific equipment, rubbing his thumbs with an awkward enthusiasm while he introduced a series of films that take us from the Antarctic to Japan, to Canada to what they referred to as the 'Belgian Congo', via the Swiss Alps and a high school rocket society in the US. We see fieldwork in all sorts of locations, as well as scientists' offices, lab work and, finally, a rocket launch. Environmental science is presented in the mode of a noble quest, an adventure of (wealthy, white, male) derring-do, exploration and wide-eyed discovery. The politics of the Cold War are implicit, but very much present, as is colonialism.

About half an hour in, there's an especially dramatic bit from a research station on top of a glacier in Switzerland talking about an international project on glaciers involving Swiss, French, Austrian, German and Danish scientists. 'Why bother?' the scientist on camera rhetorically asks, all the time wearing sunglasses and clutching a pickaxe. 'Well, for the last 50 years, the glaciers have been receding, and the high snow is melting. The levels of the ocean have been rising. If all the glaciers melted at once, the water would rise halfway up Nelson's column.' The central premise of the IGY wasn't fear, though – if anything, the ethos was incredibly hopeful, beautiful in places. People around the world were working together to understand something they shared – their home planet. As Prince Philip put it at the end of the BBC documentary, echoing Eisenhower: 'The IGY is the world studying itself, but it's also much more than that. It's a great experiment in world cooperation.'

And then, on 4 October, Russia literally put a rocket up it all, launching *Sputnik* from a site in the Soviet Socialist Republic of Kazakhstan. A simple metallic sphere, the design was meant to replicate a natural celestial body as much as possible, but with distinctive beeping radio signals the Russians knew could easily be picked up by 'ham' radio enthusiasts. The West was shocked, worried that in this big symbolic show of scientific and technical power between capitalism and communism they had been put in the shade. Their concerns were exacerbated when the Russians followed it up a month later with *Sputnik II*. This carried a living creature, Laika the dog. A stray mongrel from the streets of Moscow, she died within hours of the launch due to overheating. More to the political point, she was roughly the same weight and shape as a nuclear bomb. America's shock at *Sputnik* prompted the establishment of a new body in the Pentagon to better rationalise research – the Advanced Research Projects Agency or ARPA (which among other things would later produce the technical foundation for the internet) – and an update for that old committee on aeronautics Van Bush had sat on, a new National Aeronautics and Space Administration, better known as NASA.

Sputnik diverted attention from the idea of satellites for meteorology, at least temporarily. Still, all the fuss opened up more funding and in September 1958 Wexler gathered staff together to excitedly declare 'the Weather Bureau is about to enter the space age'. The following April, the first of 10 television and infrared observation satellites (TIROS) was launched. Solar-powered, it orbited the Earth every 99 minutes, 450 miles up, taking cloud photos on two miniature black-and-white TV cameras, returning them to a space centre named after that pioneer of space weather photography, Robert Goddard. After three weeks, *Tiros 1* had spun around the Earth over 300 times, sending back 7,000 cloud photos. It'd be the first of many satellites collecting data for atmospheric science, vital not just for weather forecasting, but our ability to track and understand climate change too. The press conference gave Wexler a chance to tell his favourite joke: oceanographers like to say 70 per cent of the world is covered by water; we meteorologists know 100 per cent of the Earth is covered by air.

★ ★ ★

It's healthy to be sceptical about the political motivations around IGY. And yet, at the same time it was one the most incredible science projects humanity has pulled off. The basic premise of the world working together to understand its shared home is gorgeous, even if multiple people found ways to take advantage. There was important polar research (including on climate change, as we'll pick up in the next chapter), which was later formalised as the 1961 Antarctic Treaty and agrees to keep Antarctica for peaceful scientific co-operation. Work during IGY also confirmed a worldwide system of underwater mountains and ridges that helped scientists better understand plate tectonics. The international centres established to manage all the data collected in IGY still live on today. And we got satellites. The US satellite plans didn't turn out as hoped, but when they did get into orbit, they detected the Van Allen radiation belt around the Earth (named after James Van Allen, in whose living room IGY had first been dreamt up). Plus, as historian Patrick McCray points out, for all that IGY was very much a matter of big, professional science, key work was undertaken by amateur enthusiasts too. Those networks that telegraphed in weather data to Joseph Henry at the Smithsonian in the mid-nineteenth century had a space age counterpart with Moonwatch, a programme involving thousands in tracking these new satellites. They were far from passive collectors of data either, creatively tinkering with their equipment, developing their own new techniques, and forming local and regional networks to share information both with each other and the broader public. It wasn't just an event for elites.

The spirit of international cooperation was clearly important to many of the scientists involved. Writing in *American Scientist*, director of the National Science Foundation Alan Waterman declared it 'the greatest single act of cooperation among the scientists of the world', hoping that this, not necessarily the science that came out of it, would be IGY's key legacy. Another of the American IGY organisers, Lloyd

Berkner, made a similar point, writing in *Science* of the 'genuine enthusiasm' researchers were coming to the project with. It was as if, 'Tired of war and dissension, men of all nations have turned to Mother Earth for a common effort on which all find it easy to agree.' Then again, Berkner was well known for being good at playing politics, and at one point also worked with the CIA to propose a system where scientists could act as spies, so possibly take that with a pinch of salt too.

You can see the appeal the cover of science gave politicians in this era. Study a bit of the Antarctic and you get to put your flag there (a point satirised by *Punch* at the time, with a cartoon of IGY delegates trying to teach Shakespeare to penguins). It didn't just have to be a matter of physically taking up space either, a nation's scientific knowledge of an environment could be used to stake a claim too. The UK was already making use of British research into falling whale populations around the Antarctic as a way to exert power over Chile in Argentina (save the whale, sure, but only if it helps make a colonial point). And yet, at the same time, this theatre of cooperation offered vital opportunities for geophysicists to work beyond national borders. The oceans and atmosphere don't stop at passport control, after all, so if the politicians were up for pretending they were friends, scientists could get some useful work done. Plus, there were plenty of IGY projects where political and scientific interests could work side by side. Trips to Greenland allowed scientists to study interesting chunks of ice, and for the US government officials who handled the logistics it was a chance to learn more about routes to the Soviet Union. When it came to oceanography, the navy got data they could use to better move their ships around, and in exchange, the scientists could use these ships to travel to bits of the planet they wanted to study.

In his book *Arming Mother Nature*, historian Jacob Darwin Hamblin invites us to see military involvement in IGY as part of a larger project of applying Earth sciences to the Cold War. Hamblin notes the buzzword of the IGY was 'synoptic'. On the surface, this was a simple desire to collectively amass as much information as possible, a chance to view our home planet together. All this data would collect a 'snapshot' of the Earth, which was fascinating for scientists, but it was also invaluable for military planners who would otherwise be limited to the data they could collect in their own regions or trusting data they managed to somehow wrangle from the other side. Moreover, having a greater understanding of the planet wasn't just good for the soul – or for the kinds of environmentalist values we often associate with Earth sciences today – but so we might manipulate it. Engineering had long had a relationship with the military, and the twentieth century had already seen chemistry, biology and atomic physics all applied to the tools of war – was it now the turn of environmental science? Might we not only weaponise the weather, but ecological systems, glacier melt and more?

Rossby died in August 1957, just after the launch of IGY, but not before he had set in motion a few more attempts to modernise meteorology. From the early 1950s, he'd worked with researchers across Scandinavia to build a project to measure carbon dioxide in several spots across Sweden, Denmark and Finland. They were talking to researchers at the Pineapple Research Institute in Hawaii too, cadged a lift to Spitsbergen on a Soviet icebreaker and discussed setting up a monitoring site somewhere in the Sahara. And all the time he was swapping letters with Wexler, back in DC, as well as Svante Arrhenius's grandson Gustaf, who was now working at the Scripps Institution of Oceanography in San Diego. They wondered if a larger carbon dioxide-tracking project might be worth running as part of the IGY. Because – along with a load of data, space dogs, systems for satellite tracking, the Antarctic Treaty, NASA, a dollop of political rhetoric and a set of commemorative stamps – IGY launched modern climate change science. In 1956, a century after Eunice Foote first placed a cylinder of carbon dioxide by a sunny window to see what would happen, science was about to get serious about climate change.

* * *

Six foot four inches, with a deep voice, Roger Revelle was literally an imposing figure in mid-twentieth-century Earth sciences – the self-appointed granddaddy of global-warming research. Initially he'd wanted to be a journalist, but was inspired by geology as an undergraduate and, once he'd taken his first research trip out on a boat, knew oceanography was for him. He'd married into a family of journalists though – an unusually rich family of journalists too. His wife Ellen had been named after her aunt Ellen Browning Scripps, who had made a fortune with a chain of newspapers and given much of it away to local science and education programmes in La Jolla, California. These included an oceanography research institution where we find Revelle, now employed as its director. Revelle was a highly entrepreneurial researcher, adept at making use of the many opportunities for federal 'basic science' funding Bush and his 'endless frontier' had opened up, and had an especially good relationship with the US Navy. The sort of researcher who was curious about a huge range of topics, collaborating with pretty much anyone who came near him, he always had a new project pitch in his back pocket. When Revelle first arrived at La Jolla as a fresh-faced PhD student in the early 1930s, the oceanography work there was reliant on Scripps family money and the odd bit of contract work, with just a small marine station, one lab and a single research boat. When he left in 1964, it had a larger fleet than Costa Rica.

As a PhD student, Revelle studied the way carbon dioxide was absorbed by the sea, concluding with his co-authors that only about half of the carbon dioxide emitted by humans was soaked up by the ocean, not 98 per cent as people had previously thought. This was back in the early 1930s though, during those decades when no one much thought carbon dioxide was going to be a warming problem in the atmosphere. For the time being, he was much more excited about the prospect of getting back on a boat, to discover something else new and exciting. He'd started working with the US Navy in 1935, first as a guest using their vessels for research, then as a reserve and, when war came, an officer, lieutenant and finally commander. He was in Manilla, attached to a group planning the invasion of Japan in the summer of 1945 when news of the atomic bomb landing on Hiroshima came through. He returned to DC, tasked with building post-war oceanographic research policy, including helping build the Office of Naval Research. In 1946 he led a study of the first post-war atomic test at Bikini Atoll – in fact, the idea of a scientific programme for the test was Revelle's. The research proved Charles Darwin's theories – from his trip with FitzRoy – that such atolls are sunken volcanic islands.

He returned to the Scripps Institution of Oceanography at the end of the 1940s and in 1956 had come back to the questions of carbon dioxide via a new technique that had recently been developed at the University of Chicago – carbon dating. This was primarily an archaeological project, a way to find out the age of the Shroud of Turin, for example, but had benefited from Air Force support as the military saw value in delicate new techniques for the measurement of radioactivity. Playing spot-the-Russian-bomb wasn't the only innovative spin-off of carbon dating though, and one of the group, Hans Suess, developed a collaboration with the National Park Service to investigate tree rings. Tracing radioactive carbon isotopes through the trees, he noticed that the rings that had formed in the past century or so had more of the type of carbon emitted from fossil fuels. He was tracing the history of carbon pollution right back to first coughs of the Industrial Revolution, written right there in cores of the tree trunks. Revelle had an eye on IGY funding and thought there'd be a bit of money available for anyone interested in chasing radioactivity through the ocean. He invited Seuss and another researcher, Harmon Craig (who'd been using carbon dating to explore the age of bits of seawater) to join the team at Scripps.

Using a mix of navy and Atomic Energy Commission grants, Revelle worked with Suess to trace carbon's movement from the air to sea. After some initial analysis, they concluded a typical molecule of atmospheric carbon dioxide would be absorbed into the oceans in a decade or so. At first this seemed to agree with some of the naysayers of the carbon dioxide theory of global warming – nature would just wash it all away for us. Revelle was preparing to submit a paper to a meteorology journal to this effect when he remembered something from his PhD work in the 1930s. The complex soup of chemicals in seawater does something chemists call 'buffering', which keeps it from getting too acidic. This meant that although oceans might initially take in our excess molecules of carbon, they'd spit a load of them back out again to protect themselves from getting too sour. From this, Revelle worked out that the ocean surface would only absorb about a ninth of the carbon dioxide his earlier calculations had predicted. It hadn't really sunk in at this point, but he quickly added a few lines at the end of the paper and sent it off for review. Reading over the paper, these lines stick out a bit. In fact, when historian Spencer Weart checked the original in the archives, he found it had been visibly taped on.

The paper also took the opportunity to lobby for funds for more research, ending with a call for more knowledge. It's a joke that scientists always say we need more research (or to put it another way, they want more funding), but it was true that if carbon dioxide was as important as Plass and Callendar suggested, we were going to need more data on where and how it might be found. The paper notes there might be an opportunity for some useful data collection on carbon dioxide as part of the upcoming IGY. In fact, Revelle had already been discussing the topic with Wexler at the Weather Bureau, who in turn was talking to Rossby back in Sweden about how IGY would be a chance to take carbon dioxide monitoring up a stage or two (as was Revelle's Scripps colleague Gustaf Arrhenius). They'd even recruited the perfect postdoc to the job – Charles D. Keeling, or Dave as he preferred to be known. We'll catch up with him in Chapter Ten.

CHAPTER NINE

A Carousel of Progress

The lightbulb didn't kill gas or oil any more than kerosene saved the whale. Oil found a new market in transport, the gas infrastructure moved to heating and electricity generators were happy to swallow any fossil fuel that would help them produce their sparks, even if hydro (and later wind, nuclear and solar) would take a slice of that pie too. By the 1960s, it was hard to find any part of US life untouched by this new, fossil energy regime. There was faster and cheaper travel, cheaper food, air conditioning as well as heating systems, and so many fun gadgets and miracle fabrics derived from oil.

At the New York City World's Fair in 1964, the Ford Motor Company invited you to take a ride on the Magic Skyway designed by Walt Disney, riding over a *Jetsons*-like 'City of Tomorrow' with towering metal spires and bubble-domed buildings. This even imagined the colonising of Antarctica, the logic being that because energy was so cheap and abundant in this tomorrow world, any place on Earth could be made habitable. General Electric also employed a bit of Disney magic, co-producing 'Progressland', a $17 million, three-storey pavilion and later inspiration for Epcot with an animatronic 'Carousel of Progress' at its centre. The audience here were rotated around six stages, showing stories of how the American family had, from the 1880s to 1960s, gradually been freed up to spend less time on chores and more with each other. Above it all, an audio-animatronic father figure acted as host and narrator, with jokes about 'a fellow named Tom Edison who's working on an idea for snap-on electric lights!' before the family finally arrive in their General Electric Gold Medallion Home, surrounded by the newest models of kitchen equipment. A specially composed song, 'There's a Great Big Beautiful Tomorrow' by the Sherman Brothers (better known for the Mary Poppins score), promised with a jingle, 'There's a great big beautiful tomorrow, shining at the end of every day.'

This was all very much in the model of the bright, gleaming electrical light of the 1893 Chicago World's Fair, just more so. The 1964 World's Fair was especially lavish in its use of energy, with even more lavish use implied for the future. Electric companies were keen to show off nuclear power as a way to such energy abundance, but fossil fuels were still core to the event. Gas provided 80 per cent of the cooling, something the gas companies proudly advertised to counter all the images of an atomic future being pushed. Even if nuclear power hadn't been part of the Progressland story (and it very much was, with a 'manmade sun' thermonuclear fission explosion as part of the show), the fair was a key event in the Cold War, articulating not just an image of the future, but how this was better than any communist visions. Lenin had been an electrical utopian in the early days of the USSR, proclaiming there would never be full communism until full electrification had been achieved. But stories of scarcity in communist nations was a mainstay of American Cold War rhetoric, and the kitchen gadgets made for a useful proxy for other tech developments, just as images of spaceships did elsewhere.

In the summer of 1959 there had even been a televised debate between Nixon and Khrushchev in a General Electric model kitchen, part of an exhibition of American goods held in a park in Moscow. The leaders had been led towards the Pepsi exhibit when an enterprising young press agent shouted: 'This way to the typical American house!' Nixon remarked: 'Would it not be better to compete in the relative merit of washing machines than in the strength of rockets?' General Electric had provided him plenty to show off with too – a built-in oven, washer-dryer and combination fridge-freezer. It was all done up in canary yellow, but GE was at pains to stress that the freedom-loving Americans could pick turquoise or pink if they so preferred. Khrushchev was less than impressed, however. Nixon: 'I want to show you this kitchen. It is like those of our houses in California', pointing to a dishwasher. Khrushchev brushed it off: 'We have such things', to which Nixon replied: 'This is our newest model. This is the kind which is built in thousands of units for direct installations in the houses. In America, we like to make life easier for women.' Again,

Khrushchev shrugged it off: 'Your capitalistic attitude toward women does not occur under communism.'

* * *

Electricity had long promised some freedom from household chores. Indeed, there was a spate of electro-utopian feminist novels in the mid-nineteenth century imagining women's work both done by machine and socialised in the process.* Roll on to 1913, and Edison was writing in *Good Housekeeping* that the woman of the future would be less of a domestic labourer and more of an engineer, 'with the greatest of all handmaidens, electricity, at her service' (lest you take this as some sort of feminist statement, he also thought that would mean a housewife's brain might finally equal her husband's). Having pretty much won lighting from oil and gas, the electricity industry wanted to grow its domestic market in other parts of the home. General Electric bought smaller domestic appliance companies like Hotpoint, and substantially increased its advertising budget, its messages spread across newspapers, magazines and billboards as well as via door-to-door selling, contests and parades.

Electric companies built electrical model 'homes of the future' that toured America. Westinghouse's 'House of Tomorrow', complete with an electric garage door-opener was soon countered by General Electric's 'Home of the Future' competition for architects, with $21,000 prize money up for grabs. Winning designs were published in magazines like *McCall's* and *Good Housekeeping*, and several were built as model homes, including one displayed at the Rockefeller Center in 1935. They were popular attractions – sometimes 10 per cent of a town's population would visit them – and showcased a mix

---

* So much so they were satirised by conservative writers. In Anna Bowman's *The Republic of the Future, or, Socialism a Reality* (1887) housework only took two hours a day and was performed by machines, children were cared for in state-run nurseries and food was delivered through electric 'culinary conduits' (note: this was meant to be a bad thing). For this story and more on what might today have been seen as 19th century visions of a 'fully automated luxury communism', see David Nye's excellent (1990) book, *Electrifying America*.

of electrical devices that were by the 1920s becoming more common (fans, irons and vacuum cleaners) along with those that were still only really for the very rich (like fridges, washing machines or ovens). The marketing rhetoric drew on the early ideas of electric liberation but, at the same time, would also often suggest electricity as a replacement for a servant, appealing to middle-class aspirations to be rich enough to have 'help'.

By the 1930s, manufacturers decided it was time to push fridges in a big way to the American public. An international ice transportation industry had grown in the middle of the nineteenth century, with ice magnate Frederic Tudor building a multinational industry shipping 150,000 tonnes of ice out of Boston harbour each year. Henry Thoreau mentions the ice extraction industry around Concord in *Walden*, musing: 'The sweltering inhabitants of Charleston and New Orleans, of Madras and Bombay and Calcutta, drink at my well.' Between 1830 and 1880 dozens of patents were filed for mechanical refrigerating machines, and by the end of the century you might expect a brewery or meat packer to have a refrigeration plant. There were even moving refrigeration machines on trains and boats. This, along with other developments in food preservation, like canning, changed how and what we could eat considerably, detaching us from seasons and increasing the miles food travelled between farm and plate. Still, you wouldn't have a fridge in your home. The popularity of iceboxes (cabinets that used ice and insulating materials to keep food cold), suggested there could be a market for domestic refrigeration devices, and electric companies were keen to pour chunks of their R&D budgets into developing something sellable.

The first domestic refrigerator to go into large-scale production was the Kelvinator in 1918. It was expensive and broke down regularly, but the basic idea seemed popular enough that by 1923 it had over 50 competitors. Ruth Schwartz Cowan, in her classic study 'How the refrigerator got its hum', traces the history of these, alongside others developed around the same time by the gas industry. Gas fridges, though slightly later to market, were quieter (they didn't hum) and broke down less often. But, Cowan argues, gas didn't have the R&D or marketing budgets the electric companies were willing to risk and found it hard to catch up with the head start the electric companies already had in this market. General Electric really did go all out in selling its (humming) fridges too. There were actors dressed as swashbuckling pirates in storerooms, special exhibition railroad cars toured the country, complete with their own jazz bands, and animated puppets danced in dealers' windows. In 1928, one of the fridges was even sent on a submarine journey to the North Pole with Robert Ripley (of *Believe It or Not* fame), and in 1935 General Electric launched the first Technicolor commercial, an hour-long film with Hollywood stars of the day, blending comedy, romance and a complete electric kitchen.

Despite the early promises, as the electrical assault on the home worked out, women did as much housework as before, it was just different work. And, although people now went shopping rather than growing their own food or making their own clothes, there was something self-sufficient about a lot of the new domestic devices. You stayed at home to do your washing, rather than it being a community activity, just as the middle classes started to buy cars rather than sharing public transport, which further isolated women at home. And so, energy use grew not as a tool for cooperation in the way some of the Victorian utopians imagined, but as a way to close family units off from one another. The shared networks of gas and electricity, like those for water and communications, were running all this, but buried, hidden from view. Electricity still retained something of its early promise of a healthier, happier, easier life for all, possibly because of mystical associations with lightning, maybe because it let you keep the smells and soot of oil, gas and coal at arm's length, or maybe just because they were cannier about investing in PR. And yet as electricity became more mainstream, like gas before it, the spectacle faded. It became normal, buried under pavements and plastered behind walls, forgotten or taken for granted. The flick of a switch became something everyday, not a reason for a parade.

★ ★ ★

Oil would find a new, massive market with transport, but electricity gave it a good run for its money, at least at first. Electric transport had come a long way from 1851 and Charles Grafton Page's smashed battery cells, broken in the bumpy train ride out of DC. Siemens had presented a model electric railway at the Great Industrial Exposition of Berlin in 1879, and an 1881 International Exposition of Electricity in Paris featured an electric tricycle from Gustave Trouvé (the same man who made the electric jewellery). By 1883, there was an electric tramcar working through parts of London; Magnus Volk had launched a seafront electric train in Brighton, which you can still ride today;[*] Siemens was running a line at Mödling near Vienna; and you could take a hydro-powered tram to visit the Giant's Causeway in Northern Ireland. By the end of the decade, thanks to a system built by former Edison man Frank J. Sprague, electric streetcars were criss-crossing cities across the US and Europe. Indeed, the first volts down the line from Niagara to Buffalo were for the streetcars, lighting came later. Soon you could even take an electric train through a tunnel that ran under the Thames, ferrying commuters from the City of London down to Faraday's birthplace at Elephant and Castle.

In December 1894, chemist Pedro Salom drove his 'Electrobat' car up to the doors of the Franklin Institute in Philadelphia and gave a talk on the history of automobiles, presenting his own work as the conclusion. It was clean and safe, he argued. What's more, as an electric car rather than one based on steam or an internal combustion engine, it was quiet. It didn't shake or get too hot, and it didn't belch out fumes. It was the future. By the spring of 1897, the Electric Vehicle Company was running a cab service in Manhattan with 13 of these Electrobats and a central station where batteries could be swapped. As the fleet grew, they converted a skating rink on Broadway into a new battery-swapping station, and opened offices in Chicago, Boston and Newport, with agents in

---

[*] In 1896 Volk opened an extension to this route on rails under the sea, the passengers held above water in a carriage on stilts that soon became known locally as 'daddy long legs', its gangly, beast-like image often appearing on postcards. This part of the railway closed in 1901 and all that remains of it today are some of the concrete blocks that supported the tracks, but both the Brighton museums and the National Railway Museum, York, hold archive photos.

San Francisco, Mexico City and Paris. The London Science Museum has a brilliant example of one of the old British electric cabs – it looks just like something you'd expect Sherlock Holmes to step out of into the London smog. The electric car's main competition, at first, were the steam cars made by identical twins the Stanley brothers who'd produced a popular and remarkably light steam engine. It needed refilling with water every 20–30 miles or so, but the existing infrastructure for horses allowed for that (until they were shut down due to a foot-and-mouth outbreak). The batteries of electric cars made them heavy and could be slow to recharge, but electrics had fewer parts so broke less often and were easier to fix – and best of all, they produced no smoke (at least at the point of use, the battery still needed charging, often using coal-powered electricity).

But most people didn't drive at all and in the early days cars were generally seen as the toys of rich boy racers. Women preferred electric cars, apparently. Or at least they were sometimes marketed to the ladies and this has been given as a reason why men avoided them. It's also that cars were sold to Americans with the idea of going for long drives into the countryside, and electric cars couldn't handle the range. The Electric Vehicle Company in particular seems to have been badly managed too. By 1902 the regional offices closed, and although they kept a cab service in and around Central Park for a few more years, they went bankrupt at the end of 1907. Meanwhile, Henry Ford, who'd built his first automobile at home while working nights at the Edison Illuminating Company, focused on selling to a mass market beyond the super-rich enthusiast. In 1900, there'd been nearly 4,200 cars in the US and they cost at least $1,000 – two years' working-class wages. In 1908, Ford's Model T sold for $850, which was already low, but by 1916 it was $360 and his annual sales had exploded from 6,000 to 600,000 cars. This model, in the early days, applied a hybrid 'flex-fuel' system that ran on either gasoline (which if you lived in the city you could buy at a newly built filling station) or alcohol (which if you lived in the countryside, you might make yourself).

The rise of the car would transform the streets, not just in terms of the emergence of curbside gas stations, motels, or 'drive-in' theatres or eateries, but the design of towns, how we interacted with space around us and who we allowed to take control. As with the refrigerator getting its hum, none of this was inevitable and a fair bit of it was down to public relations. Until the 1920s cars were, at best, uninvited guests on American streets. Back when only a few people had access to cars, they were largely seen as toys for the rich, not the necessity to modern life that they'd later be considered. In 1906, Woodrow Wilson, then president of Princeton University, said: 'Nothing has spread socialistic feeling in this country more than the use of the automobile, a picture of the arrogance of wealth.' As car use grew there was greater social acceptability of driving, but it still took a concerted effort from the motoring community to dominate the streetscape in the way they did. But within a decade, cultural norms had changed profoundly and the car was king.

In the four years after the First World War more Americans were killed in automobile accidents than had died in battle in France. As Peter Norton notes in his fascinating (2009) book *Fighting Traffic*, this fact did not escape public attention.* The press was full of stories of the innocent victims of these reckless drivers, and public monuments to those killed by cars were built alongside ones grieving the war dead. But the motoring community organised, using street-safety campaigns as a way to promote the need for people to behave better around cars, not the other way around. And so, the image of the reckless driver faded into one of reckless pedestrians. The term 'jaywalker' was key to this. It had started simply as a retort to epithets directed at cars, a counterattack to 'joyrider'. The 'jay' in jaywalking came from a derogatory term for someone out of place in the city; the implication being that a jaywalker was someone who did not know how to move around in modern life. 'Chauffeurs assert with some bitterness,' noted the *Chicago Tribune* in 1909, 'that their "joy riding" would harm nobody if there were not so much jay walking.' The term wasn't picked up uncritically. When, in 1915, New York City police commissioner used it to describe someone crossing the road mid-block, the *New York Times* protested, calling the it 'highly opprobrious' and

---

* If you've ever looked at a bit of a town and wondered why cars have the amount of space they do, Norton's book will help explain. A core source for this chapter, I cannot recommend this book enough.

'a truly shocking name'. But the car companies knew the term had power. Boy Scouts were recruited to issue cards to offenders, explaining that they weren't simply crossing the road, but jaywalking. In New York City in 1923, 600,000 cards were distributed; an all-caps banner proclaimed 'DO YOU KNOW YOU ARE GUILTY', before explaining what jaywalking was and that you must 'CROSS ONLY AT CROSSINGS'. In Cleveland, crowds of jaywalkers (actors) taught parade-goers to disapprove of the practice. In San Francisco pedestrians found themselves pulled into mocked-up outdoor courtrooms where crowds of onlookers watched as they were lectured on the perils of jaywalking. Pedestrians learnt to believe they were at fault, not drivers, and it was their responsibility to let the cars take space.

The fabric of towns changed too, adding sometimes literal concrete to the cultural shifts the idea of jaywalking had established. The death of the 'walking city' is often discussed as a technological victory – it certainly represented people being able to move further, faster, with an extended breadth of movement. But there are different ways to extend movement, and as many cities chose to build for the car, aspects of social life were lost, sometimes literally paved over, as people simply drove through town. Roads didn't just mean more space for cars – and so more air pollution, more carbon emissions, more cracking open the Earth for oil – but also more concrete. Due to the way it is made, concrete has incredibly high carbon emissions (if the modern cement industry were a country, it would be the third-largest carbon polluter in the world). Land also had to be cleared to build roads, which could mean clearing people and the destruction of natural carbon sinks like forests. In several places, road building concreted in not just carbon emissions but social inequality too. For example, it's often said Robert Moses – a mid-twentieth-century public official and president of the 1964 World's Fair – had especially low overpasses built on Long Island so buses could not pass under them, thereby cutting off Black working classes who relied on public transport. Whether this is true or not (you can read up the arguments online for yourself), it's hardly the only example. After the toppling of the statue of slave trader Edward Colston in Bristol in June 2020, the *LA Times* ran an op-ed suggesting the monument to racism in their city most due to come down was the freeways. The construction of these not only bulldozed Black communities, but left an enduring legacy of both segregation and air pollution (which, as in many other cities around the world, disproportionately affects ethnic minority groups).

★ ★ ★

As oil companies organised themselves around this new transport market, they invested in new logos that were easily identifiable while driving, and new marketing schemes designed to build relationships with individual motorists. Shell sent its female customers Valentine's Day cards and had poet John Betjeman edit guides to English counties that motorists might visit. The oil companies would also find themselves having to invest in new places to extract oil and form new alliances to control it.

William Knox D'Arcy's deal with Persia was only the start of the story of oil in the area. Plans to tap Mesopotamia (modern day Iraq, along with parts of Iran, Syria and Turkey) had started not long after oil had been successfully drilled in Persia, with a deal put together by Armenian oil expert Calouste Gulbenkian. The son of an oil dealer, Gulbenkian had grown up with the business and studied petroleum engineering at King's College London before publishing a book on Baku. Escaping Armenian persecution in Turkey at the turn of the century, he ended up as an oil sales rep in London, working with the Samuel brothers and Henri Deterding at Shell. Gulbenkian and Deterding had even discussed the Knox D'Arcy's concession, but passed it over as too much of a gamble. Annoyed he'd missed this chance, Gulbenkian later adopted the maxim 'never give up an oil concession', which would guide him for the rest of his life. He was notorious for being doggedly persistent in everything he put his hand to and suspicious of everyone.[*]

---

[*] He built a prestigious art collection later in life, and he'd always have two or three experts appraise anything before buying. Determined to live longer than one of his grandfathers who'd made it to 106, he employed two sets of doctors so he could always have a second opinion. Though one of his doctors also advised that he should always have a mistress no older than 18, for the good of his health, so he was perhaps selective about picking them in the first place (he died aged 86, in case you were wondering).

By 1907, anti-Armenian feeling in Turkey had waned enough for Gulbenkian to go back home. He convinced Marcus Samuel to let him set up an Istanbul office and it was from this base that he built the Turkish Petroleum Company, which secured a concession for Mesopotamian oil. Gulbenkian's deal, signed in 1914 (the day Franz Ferdinand was shot, in fact) saw half of Mesopotamian oil go to Anglo-Persian, with 22.5 per cent each to Royal Dutch Shell and Deutsche Bank, and the remainder for Gulbenkian himself. The nickname 'Mr 5 Per Cent' would stick. War got in the way and changed the political setup slightly, but the oil was far from forgotten. France got Deutsche Bank's share in the Turkish Petroleum Company as war damages and set up the Compagnie Française des Pétroles (today's Total). American oil interests were also keen to get in on the act and leant on the British until they had a seat at the table. The official argument for this was that the US was smarting from having to introduce daylight savings policies and gasoline-less Sundays during the war so, worried scarcity might become the norm, went searching for more oil. Arguably the American oil majors were also worried about a potentially big new competitor; they either wanted a slice of that action, or at least to slow down its extraction as much as possible so they could still keep charging a high price for American oil.

The US side was led by Walter Teagle, the man running the largest remaining tentacle of Rockefeller's SO octopus, Standard Oil of New Jersey. Teagle was, like Gulbenkian, the son of an oil man who'd gone on to study the subject at university. His mother also came from oil. Back in the 1860s her father had co-founded an oil-refining company with Rockefeller, but sold his half before things got big. Not long after Walter graduated, the Teagle family business was swallowed by SO. He went with it, and rose rapidly in the larger company, becoming president of Standard Oil of New Jersey in 1917, aged 39. He had Rockefeller's old rolltop desk installed in his office and could play on a sense of continuity with the old days – his grandfather had sort of co-founded the business, after all – but in many ways he was all about change. Teagle was very aware that he'd inherited a company that was primarily about the refining, storage, movement and selling of oil, not drilling – Rockefeller having always left that to other, smaller companies. But Teagle knew that to be a player in the twentieth-century oil business, you needed to tap your own crude, declaring in 1920, at the 50th anniversary of SO that the company was now 'interested in every [oil] producing area no matter in what country it is situated'.

Negotiations dragged on throughout the 1920s. Daniel Yergin characterises this part of the story as Gulbenkian fighting with Teagle over his 5 per cent. For political theorist Timothy Mitchell, it was part of a larger narrative of oil companies at the time delaying the exploitation of new oil fields as much as possible, to avoid too much competition with their existing reserves. Stories about running out of oil might sound like a threat to the oil industry, but they can be useful PR if you want to keep prices up. Either way, while Teagle and Gulbenkian fought it out, Anglo-Persian, Royal Dutch and the American interests eventually sent a geological expedition in 1925 and started drilling in April 1927. A few months later, on 15 October at 3am, at a drilling site in Kirkuk a little north of Baghdad, a huge roar went up. Soon the area was drenched in oil and noxious gas, flowing at 95,000 barrels a day. The deal-makers needed to get a move on. Finally, in July 1928, six years after Teagle had first sailed to London to try to make a deal and 14 years after the plan had initially been brokered by Gulbenkian, the so-called Red Line Agreement was signed. This divided up Mesopotamian oil between Anglo-Persian, Shell, Compagnie Française des Pétroles and the Near East Development Corporation (the Americans), along with, of course, 5 per cent to Gulbenkian. It took its name from the thick red line Gulbenkian drew on a map so everyone was clear on the boundaries. It was the area he'd built the agreement around, the lines of the empire he'd grown up in, the area within which they'd agreed to work together.

American oil interests did not stop there. They tried Persia, seeing if any of the oil regions not already covered by the deal with the Brits might be available, and were initially met with enthusiasm. American partners seemed like a welcome alternative to the imperial interests of the European powers, but they were soon to conclude the Americans were even 'more British than the British' (not a compliment). The red lines of Gulbenkian's agreement were constraining too; if one signatory made a deal they had to work with the others. One of the smaller SO tentacles, SOCAL (Standard Oil of California, today's Chevron), made deals first with Bahrain and then Saudi Arabia. Jack Philby, a British convert to Islam and close advisor to the Saudi king (and father of 'Cambridge Five' spy Kim Philby) had spotted the opportunity, and played British and American oil interests off each other. Once drilling started in the early 1930s, SOCAL found itself short on capital and marketing outlets, many of which were controlled by SO of New Jersey, but found a partner in another American firm that wasn't bound by the red lines – Texaco. It was rapidly expanding and had plenty of eager customers – including Franco's Spain (more on that in a bit).

Back in Baku, Deterding had bought out the Rothschilds in 1912 (or rather, he bought them out with shares, so they owned a lot of Shell), and Teagle made a deal with the Nobels in 1917, arranging to pay $11.5 million via a Swiss bank account. Both would regret it. The Soviets simply claimed the oil and flooded the European market with low prices, similar to the ways Rockefeller had flushed out his competitors back in the 1880s. Teagle tried working with Deterding to build a united front, but they fell out, Deterding launching a price war of his own along with a public press campaign claiming Standard Oil was collaborating with communists. Between this problem, a new boom in Texas and new extraction sites opening up in Venezuela and Mexico, Teagle began to worry whether overproduction (not a lack of oil) might send the industry into crisis. He started to run 'big three' secret meetings with Deterding and Sir John Cadman, who'd taken over as chairman of Anglo-Persian. This culminated in September 1928 with a meeting at Achnacarry Castle in the Scottish Highlands, which also included William Mellon, head of Gulf, and a couple of Teagle's SO colleagues (plus two of Deterding's nieces who got bored and put treacle in the oil men's beds). Ostensibly they were there for a weekend of shooting and fishing, but they were cooking up a deal, basically an oil cartel allowing them more control over prices. They couldn't have complete control, because it didn't include Russian oil, but would be a strong force against the Soviets and any other interests that might pop up.

By the mid 1930s, Deterding's behaviour had become increasingly autocratic, his interventions at Shell HQ described by one colleague as 'thunderstorms'. His anti-communist sentiment was ever-more elaborate too. When he openly started supporting Hitler, the secret service tapped his phone. Deterding's shift through anti-communism to fascism is sometimes put down to the influence of his wives. He'd married the daughter of a Tsarist general in 1924 (Gulbenkian's ex-girlfriend in fact, a point they'd fallen out over), divorcing her a little over a decade later to marry his German secretary. But he'd already been badly burnt by the Russian Revolution, and Hitler was far from the first controversial dictator he had dealt with. There was Mussolini and Franco, for a start. When it came to Juan Vicente Gómez in Venezuela, Shell's official line said Gomez caused a 'severe moral quandary' for the company, which it tried to get around by building a refinery not in Venezuela itself, but the island of Curaçao 50 miles away. (Glossing over the point that this had been done on Gomez's request, because he thought he could avoid union organising if the refinery was offshore). Deterding himself, however, seemed to be under no quandary, describing the government under Gomez as having 'consistently insisted on fair play'.

It was possibly Mexican oil that tipped Deterding over the edge. British businessman Weetman Pearson had built much of the Mexican oil industry at the start of the century. Having already built the Blackwall Tunnel under the Thames and several more under New York's East River, he had been invited to oversee some industrial projects by the Mexican government. So the story goes, he missed a rail connection in Texas not long after the discovery at Spindletop in 1901. Forced to spend the night, he saw the area, as he put it, 'wild with the oil craze'. Inspired, he bought up concessions for likely land in Mexico. £5 million worth of investment later, he'd found huge reserves, founding El Águila (Mexican Eagle) in 1908. Mexican oil would become a critical source for the US during the First World War, but a coup in 1913 made the political situation more than sticky for Pearson. So, when in October 1918 Deterding approached him with an offer to buy a chunk of Mexican Eagle and take over its management, he was very happy to agree (today Pearson PLC is better known for publishing, media and education, it used to own the

Economist). It'd be an investment Deterding would later regret, when Lázaro Cárdenas was elected president in 1936, he swiftly nationalised the oil industry, declaring 18 March 1938 Oil Expropriation Day. Deterding started to see communist plots everywhere.

Shell HQ paid a PR man in London to keep any whiff of Deterding's increasingly controversial political views out of the press, threatening papers with withdrawing advertising if they ran a hit piece. This apparently worked for at least one British newspaper, but was harder on the Continent. There was a brief run on Shell shares in France and 'Death to Deterding' graffiti started to appear in Amsterdam. When it finally came out in 1935 that Deterding had offered Germany a year's supply of oil – that is, a military reserve – on credit, he was forced to resign, retiring to Germany. (When he died, a few months before Britain declared war, Hitler sent flowers to the funeral.)

Deterding wasn't the only oil man of the 1930s taken down by his association with the Nazis. In 1942 Teagle found himself in an embarrassing deal with German chemical firm IG Farben (the company that manufactured the poison gas used in camps). He'd set this up back in the 1920s as a way of sharing patents, swapping information on leaded petrol in exchange for German research on synthetic rubber. He was only officially guilty of breaking American anti-trust laws and paid a $50,000 fine when caught, but Teagle had been rather politically naive, wilfully so perhaps, as it was easier for business for him to turn a blind eye. Shares suffered and Teagle resigned at the end of the year. There's also the case of Torkild 'Cap' Rieber at Texaco, who gave Franco's nationalists oil on credit. This was illegal under US neutrality laws, so he snuck tankers out under the pretence they were travelling to Belgium or the Netherlands, issuing captains with sealed orders to be opened at sea redirecting them to nationalist ports in Spain. When the FBI caught wind of this, Texaco was issued a fine, but supplies continued, and through contacts in Spain he started to deal with Italy and Germany too, weaving his tankers through neutral ports to avoid embargos. Things eventually came to a head in 1940 when Rieber financed a German lawyer to work out of the Chrysler Building in New York City lobbying American businesses against supplying arms to the UK, and it was discovered that another German he'd employed was a spy. The head of British intelligence in New York passed the story to the *New York Herald*, Texaco shares briefly slumped, Rieber was asked to resign and the company started sponsoring broadcasts of the Met Opera to improve their public image (an arrangement that didn't end until 2003).

Twentieth-century global politics was a lot more than simply world wars, fascism and communism, and it was in the context of eroding colonial control the oil business started to open up in Nigeria. Oil seepages had long been known about on the west coast of Africa and bitumen could be found washed up by the ocean. In November 1905, the Nigerian Bitumen Corporation had been founded by a British businessman using Canadian and Galician crew, backed up with a loan from British government. The project's manager Frank Drader wrote home to his wife excited that the lagoon he was working near became polluted with oil after drilling: 'There was so much oil at our wharf here that the Doctor got all covered last night when he went in swimming.' The project was a failure financially, but left some political legacy, as the Southern Nigeria colonial government figured it should pass some oil regulations. Modelled on those drawn up for Trinidad and its pitch lake, these gave Britain a monopoly over all oil exploration in Nigeria and insisted any oil concessions had to go to British companies with British directors. In the 1920s, both the D'Arcy Exploration Corporation (i.e. Anglo-Persian wanting to diversify to compete with Shell and the Americans) and the Whitehall Petroleum Corporation (Pearson looking to invest some of the money he'd got from Deterding for Mexican Eagle) explored again, but decided it wasn't worth it.

D'Arcy revived its interest in the 1930s after a new geological report gave fresh advice – and this time it did so in partnership with Shell. The war delayed things. There was plenty of appetite for oil, but it was hard to get the staff and equipment needed. When work started up again post-war, it was very much mid-twentieth-century oil exploration, different from the early days of Anglo-Persian or Shell's first drilling project in Borneo. Sites were connected by helicopters, plus the growing calls for independence were getting harder to ignore, with the activities of British oil companies given particular attention. Gradually protests against Shell quelled down, possibly as nationalist leaders saw the benefit of oil in a future independent Nigeria (or the oil leaders saw benefit in cosying up to nationalist leaders). In preparation for independence, Shell formed a Nigeria-based company in the 1950s and started to make some attempts at Nigerianisation. This was slow though – by 1954 the company listed only three Nigerians as occupying senior-level positions out of a total workforce of more than 2,000. Eventually, in March 1958, the first shipment of Nigerian crude arrived at Rotterdam. Tensions between people of Nigeria and Shell were far from gone though. Indeed, in the 1990s, it would cause the company its largest human rights scandal after Ken Saro-Wiwa and eight other activists from the Ogoni region of Nigeria (where Shell had started drilling in 1958) were executed. In 2009, Shell agreed to an out-of-court settlement of $15.5 million to the victims' families. Make of that what you will.

* * *

Oil in particular had found a tight little spot for itself at the centre of the global economy, but there was a growing sense that fossil fuel's days were numbered. General Electric's 1964 Carousel of Progress was all about the future of nuclear power (even if the gas industry was keen to stress its role in actually powering the event). As we'll see in the last few chapters of this book, one of the reasons people didn't, at first, worry too much about what fossil fuels might do to the climate was because they figured this phase of human history – where we did something as weird as power ourselves by exploding the buried remains of ancient bugs and trees – was very much a temporary state of affairs. Nuclear received most of the shiny happy electrical future hype, but renewables got a look in too. After all, the first really big electrical project had been hydro, at Niagara, and wind and solar were about to start to at least talk about catching up too.

Sometimes dubbed 'the Danish Edison' Poul la Cour had got into electricity in the 1870s working on telegraphs at the Danish meteorological office. Inspired by the more utopian hopes for electrical power of the day, he wanted to ensure this bright new future was available to people across Denmark, not just in the cities. He was awarded a research grant from the Danish parliament and in 1891 used wind power to light up the high school where he was a teacher. His designs soon spread throughout Denmark, and la Cour was canny to set up training courses, a professional Society of Wind Electricians and a journal. Over in the US, brothers Joe and Marcellus Jacobs had grown up among screaming winds in the north-east corner of Montana, and built a small wind industry aimed at the off-grid rural market. They had little formal engineering training but liked to tinker. After playing around with a surplus First World War aircraft, adding propellers to sleds to get through the snow, they used this knowledge to build a wind generator using bits from an old Model T and the fan blade of an old wind-powered water pump. They started selling these to local farmers, setting up the Jacobs Wind Electric Company in 1928, eventually opening a factory in Minneapolis in 1932. Admiral Byrd took a Jacobs turbine on one of his trips to the South Pole and 22 years later visitors found it still spinning, even in the harshest of environments.

In Russia, the early Soviets were determined to make the most of the promise of electrification and Lenin instructed the Academy of Sciences to research wind power. In 1931 a 100ft wind turbine was constructed near Balaklava, overlooking the Black Sea. It moved around a circular rack so it could be positioned to face directly in to the wind and had an output of 100kW. To put that in some context, the turbines the Jacobs brothers were pushing out at their factory in Minnesota around the same time were just 1.5–3kW. There were bigger ones, but still, this Balaklava project was of a whole new magnitude and the Russians had started plans for a giant machine 50 times larger. In the end, Soviet wind power research fell by the wayside during the war and wasn't picked up again. No one knows what happened to the Balaklava turbine – there are some reports that it was used until it was destroyed in 1942, possibly doubling as an observation post for a while. An American entrepreneur named Palmer Putnam was inspired by the Russian talk of big wind and figured he could do better. Ten times better than the Balaklava turbine – the first megawatt. An MIT-trained geologist, Putnam had no particular background in generating electricity, but he had ambition and he was charming. Plus, he was from a well-connected family of high flyers – his mother was the first dean of Barnard, his father was a celebrated American Civil War veteran who'd gone on to run a prestigious publishing company and his cousin was an arctic explorer who married Amelia Earhart – and soon managed to raise cash.

Before long, Putnam's 1,250kW turbine was taking shape on top of a mountain in Vermont with the slightly unfortunate name of 'Grandpa's Knob'. It was a huge undertaking. The turbine's tower came from a bridge-builder in Pennsylvania, where the blades were also built, but then everything was put together in Ohio before being shipped to Vermont. It was too heavy for local roads, so bridges had to be temporarily reinforced. It took 10 rather hair-raising trips and that just took it to the bottom of the mountain. Reaching the top was another 2,000ft trek. Plus, there was no road, so they had to build their own. Finally, it was all there and assembled. On 19 October 1941 it fed electricity into the grid. But a bearing broke in 1943 and by then American engineers had other troubles to be dealing with. For another two years, Putnam's turbine just sat there, on top of Grandpa's Knob, while the world went to war. Finally, in 1945, someone found time to fix the bearing and it was switched back on. It promptly broke again. After 1,100 hours of operation, a blade fell off, sailing 750ft through the night, knocking Putnam off his feet. Speedily dubbed 'the blade that failed', congressional hearings in 1951 cited it as a reason to write off the tech altogether. Putnam himself turned away from wind too, arguing instead for nuclear and solar. Today, a phone tower stands on the top of Grandpa's Knob. Still, the company that bankrolled the project put the patents in the public domain and got Putnam to write a book detailing everything that happened. When, in the 1970s, people came back to the idea of big wind, it'd help give them a head start.*

---

* You can read more about Putnam and other American wind pioneers in Alexis Madrigal's (2011) ace book on the history of green tech in the US, *Powering the Dream*. It's also got some great stories on tidal, which I haven't had space for in this book but is fascinating, as well as solar, electric cars and more.

Dreams of solar power hadn't ended with Augustin Mouchot and his steam-solar printing press either. In the 1900s, pioneering solar-preneur Frank Shuman started to build a power plant in his backyard in Philadelphia. Strips of blackened iron pipes covered with glass were filled with a liquid that had a low boiling point; the Sun would heat these pipes, turn the liquid to vapour and this would power a steam engine. It was clever, you concentrate the power of the Sun with the glass and blackened pipes and cut down the amount of work they have to do by picking a liquid that turns to gas more easily than water. Still, it needed a lot of space. He figured that if he could establish solar power in countries where coal was expensive but the Sun was plentiful, he'd be able to scale it up, and tried a project in Maadi, a suburb of Cairo. Sadly, despite some favourable coverage in the *New York Times*, the First World War largely put paid to the idea. Still, on a smaller scale, solar hot water heaters were produced; an insulated box of pipes of water that could sit on the roof, using the Sun's heat to warm your bath. In June 1979, 32 such 'solar thermal panels' would make their way to the roof of the White House, as Jimmy Carter used them to make a point about alternative energy (Reagan would later take them down).[*]

In the 1930s, George Keck, an architect working on a dodecahedron glass 'House of Tomorrow' for the 1934 World's Fair in Chicago, noticed that despite the freezing weather outside and the furnace not yet installed, workers inside the building were dripping with sweat. The principle was pretty straightforward, it wasn't all that different from how Joseph Paxton had managed to grow Amazonian lilies in the UK back in the early nineteenth century. Keck started building homes with south-facing windows, dubbed 'solar homes' by the press, popular with customers keen to cut heating costs post-war. In 1947, the glass company Thermopane published a book entitled *Your Solar House* with 49 examples of different solar homes, each by a

---

[*] As a *Nature* editorial quipped in 2016, you can trace a history of American energy policy through solar panels at the White House. Carter put some up, Reagan took them down, Obama put some more up. Meanwhile, during George W. Bush's time in office, a few were quietly installed in the White House gardens, seemingly without the president's explicit knowledge or interest.

different architect, designed with different states in mind. Whole parts of Chicago and New York suburbs were built with solar home principles, and prefab designs meant components could be cheaply turned out in a factory. It could go awry, with bad planning, either losing heat at night through the windows or overheating, and they were still some way from a modern super-energy-efficient 'passivhaus'. And at the same time, they had to compete with the same sort of stories of uncomplicated energy abundance electricity companies were pushing in the World's Fair Progressland display; an image of the future where you can flick a switch for air con or heating, not needing to worry about how much electricity or gas it used, or even where that energy was coming from at all.

The real gee-whiz excitement for solar power was the promise of photovoltaics (solar PV); using the Sun to produce electricity and doing that without the need to turn a turbine. There'd been experiments with this since the mid-nineteenth century, but no one had managed to build anything that could really compete with steam, wind or hydro. An accidental discovery at Bell Labs in 1940 would change this. Russell Ohl was playing with some silicon samples and noticed one with a crack in it. Impurities had built up on either side of the crack, one side positively charged, the other negative. When he shone light on this odd little broken and dirty sample, a current would flow. Ohl had inadvertently made a positive-negative junction, the basis of the modern solar cell. It would take a lot more work before it was anywhere near efficient to use, but in April 1954, Bell Labs proudly presented their new solar cells to the world, using a strip of them to run first a toy Ferris wheel and then a radio transmitter that could broadcast music. Excited to get on board with the solar hype, the *New York Times* stuck the story on their front page. The new device, made from the same ingredient as sand, 'may mark the beginning of a new era, leading eventually to the realisation of one of mankind's most cherished dreams – the harnessing of the almost limitless energy of the Sun'.

It was all still hype though at this point. The first major order came from a hearing aid company, the idea being that small cells could be mounted on spectacles. But they went bankrupt before paying, so at first solar PV was largely left to toys. Solar cells had been developed by a telephone company and one early use was to power telephone lines in rural off-grid areas. But the panels got covered in bird droppings so had to be cleaned every week and the silicon transistor (based on similar tech) soon meant the lines didn't need so much electricity anyway, so the cells weren't needed. The army was excited though. As soon as it heard about the Bell Labs breakthrough it sent its lead power researcher to find out more. He was immediately smitten by the idea and quickly got a team together to work out applications. But after months of searching, they could only find one project where solar power might be viable: a top-secret plan, codenamed Operation Lunch Box, to launch a satellite. This wasn't entirely original; Arthur C. Clarke had already written about space stations running on solar. It simply made more sense to use energy direct from the Sun in space, rather than bringing a store of fossil fuels. As we saw in the last chapter, the US would be beaten to launch a satellite by the Russians – but in March 1958 *Vanguard 1*, the first solar-powered satellite, went into space as part of the IGY science programme. It's still up there, if you want to give it a wave, holding the record for being in space longer than any other human-made object. It'd be a while before solar made sense anywhere other than space, but ultimately the work that would go into powering satellites would bring down the cost of PV panels on Earth.

* * *

In his social history of American energy, *Consuming Power*, David Nye notes that the average American household in 1970 used more energy than a whole town in the Colonial period. A single TV at the time used as much energy in an hour as a team of horses could provide in a week. And yet, on the whole, this abundance wasn't seen as remarkable. If anything, it was a vindication for the American way of life, the success of capitalism. With the growth of unions and better labour laws, the average American working day decreased from 12 hours a day in the middle of the nineteenth century to eight in the 1930s At first this trend seemed likely to continue but, in the end, no excess of leisure ever emerged. People spent money in the free time they had and on new goods that now seemed necessary to navigate a modern life, and so maintained long working hours to pay for them. Consumption, not relaxation, would become the pattern of aspiration for twentieth-century life. Economies would soon become reliant on this pattern too, further yoking themselves to the fossil fuels that produced and shipped goods around the world. The marketing techniques people like Josiah Wedgwood had pioneered in the eighteenth century – celebrity endorsement, careful segmentation of different audiences to sell to, illustrated catalogues – became ever-more elaborate and the advertising industry boomed, all lit up in electric signs. The first department stores had started to pop up from the nineteenth century onwards, with malls joining them from the middle of the twentieth century. Shopping became a leisure activity in itself and shopping sites tourist attractions.

Plastics played a big part in this shiny new version of the American dream – along with whizzy new electrical devices, cars and concrete – and it's worth sketching some of their history here. People had started to research ways to improve naturally occurring plastics in the mid- nineteenth century. In the 1840s, Charles Goodyear patented processes for strengthening rubber (from trees), 'vulcanising' it with sulphur. Charles Macintosh had worked out how to use a by-product of the coal gas industry to make waterproof raincoats a few decades before, and submarine telegraph cable-makers would create their own special recipes for treating gutta percha (indeed, it was from playing with such mixes that led Willoughby Smith to make an accidental discovery that helped build the first solar cells). In 1862, artist turned chemist Alexander Parkes displayed his new material Parkesine at a World's Fair held in the space left by the Great Exhibition in London. Made from gun cotton, Parkes sold this new material as a cheap replacement for ivory or mother-of-pearl. Back in the eighteenth century, manufacturers like Matthew Boulton had made a fortune with new techniques that offered the sort of goods that were once only available to the super rich to a larger market, and Parkesine was just another stage of this. Parkes opened a factory in Hackney Wick, east London in 1866, but wasn't a great businessman – reports that some of his products exploded didn't help – filing for bankruptcy after just two years.

A couple of years later, American inventor John Wesley Hyatt entered a competition offering $10,000 for an alternative to ivory to make billiard balls, utilising an improved version of Parkes's idea. He added camphor to the mix and gave it the name 'celluloid'. Working with his brother Isaiah, he also developed a process of 'blow moulding' to produce hollow tubes of celluloid, paving the way for mass production of cheap toys and ornaments. Another of the advantages of celluloid was that it could be mixed with dyes, including mottled shades, allowing the Hyatts to produce not just artificial ivory but coral and tortoiseshell too. The next big plastics breakthrough came in 1907, when Belgian-born American chemist Leo Baekeland patented a manmade alternative to shellac – a resin secreted by the female lac bug that could be used for electrical insulating – which he named Bakelite. This soon found a range of applications – it was marketed as 'the material of a thousand uses' – and was joined by a host of new plastics in the 1930s and 1940s. Nylon offered a sort of synthetic silk, useful for both parachutes and women's stockings, and Plexiglass fed the burgeoning aviation industry.

Plastics could be durable, one of their appeals (as well as their tragedy) is how long they last, but the growth of plastic goods intensified a growing culture of disposability. The mass production of cloth and paper in the nineteenth century coupled with the emergence of germ theory had fostered a culture of disposables; products that might be used once and thrown away. This model suited companies that wanted to keep selling to customers and the number of disposable products grew in the twentieth century, from the first disposable razors in 1903 to nappies in 1961. Styrofoam, first developed for flotation devices in the war, found a new market as single-use coffee cups and take-away food containers. Colourful, squeezable polythene bottles soon became the norm for the packaging of a range of household goods, from washing-up liquid to shampoo to ketchup. The durability of the plastic was part of its attraction. A plastic straw, for example, doesn't get soggy while you use it. But it also meant they didn't rot so easily once they'd been used. Along with the simple growth in how much stuff people were consuming, the persistence of plastics meant the waste piled up.

The first calls to curtail disposables came in 1953, from dairy farmers wanting to protect their animals from ingesting stray glass. As Heather Rogers describes in her illuminating (2005) book *Gone Tomorrow*, the packaging industry mobilised fast and within months had launched the non-profit organisation Keep America Beautiful. Modelled on the sort of beatification organisations that had been running since the nineteenth century, it ran off a similar playbook as the jaywalking campaigns – shift the blame. The organisation printed pamphlets for schools to encourage 'lasting acceptance of good outdoor manners' and recruited a range of local community groups. Within a few years it was active in 32 states, with membership over 70 million, enjoying active support of four federal departments. The focus was quite squarely on the behaviour of individual consumers, not the existence of packaging in the first place. As one executive from the American Can Company put it, 'Packages don't litter, people do'; or, from a 1963 education film narrated by Ronald Reagan, 'Trash only becomes trash after it has first served a useful purpose. It becomes litter only after people thoughtlessly discard it.' It's a persuasive line, not least because it offers the individual, on the face of things, some apparent agency. But, ultimately, it's a fudge to avoid changing a system that probably never should have been allowed to establish itself in the first place. It's an approach that the fossil fuel industry would make use of later too, most notably with BP's promotion of carbon footprints as part of its 'Beyond Petroleum' PR push in the early 2000s.

* * *

The packaging industry had been quick to avoid blame, but by the mid 1960s, the sort of world Disney presented in its 'Carousel of Progress' was starting to lose its shine. In his 1965 State of the Union address, Lyndon B. Johnson celebrated what he called the nation's 'flourishing progress', but invited the American people to consider what they wanted to do with this flourish, who'd benefit and how they might protect themselves from any negative side-effects. 'We worked for two centuries to climb this peak of prosperity. But we are only at the beginning of the road to the Great Society,' which, he said, 'asks not how much, but how good; not only how to create wealth but how to use it; not only how fast we are going, but where we are headed.' This, he promised, would shape his policies on education, health, crime provision and the arts, but also work towards ending 'the poisoning of our rivers and the air that we breathe'.

Johnson followed this up a month later with a special statement on the preservation of natural beauty to Congress, which warned of 'the storm of modern change' as 'modern technology, which has added much to our lives can also have a darker side. Its uncontrolled waste products are menacing the world we live in, our enjoyment and our health … the same society which receives the rewards of technology, must, as a cooperating whole, take responsibility for control.' There was a sense that consumption had got to a point where the country was saturated with pollution – 'skeletons of discarded cars litter the countryside', rivers were 'over-burdened with waste'.

Johnson's speech-writers were perhaps reacting to a larger cultural shift, something pollster Daniel Yankelovich termed 'the new naturalism'. In a series of studies of college students in the late 1960s and early 1970s, Yankelovich discovered a widespread conviction that everything artificial was bad, while everything 'natural' was good. Ideas like these had a history – one that flowed through conservation, and back to the transcendentalists and Romantic poets – but also reflected a more recent rebellion against the world their parents built and a desire to build new cultural values. There's a scene in the 1967 film *The Graduate* that illustrates this well. A friend of the Dustin Hoffman character's parents takes him aside and says he has one word of advice for the young man: plastics. The idea of plastics, once at the forefront of the sort of shiny futures on display at World's Fairs, was now used by the film to denote everything the Hoffman character despises about the world he's invited to join: cheap, sterile, ugly, mass-manufactured and all too easily disposed of.

The great utopian promise of fossil-fuelled abundance was starting to show its age, or possibly had just got a little too big for its boots. It wasn't just heat and light and transport, but also broken toasters, not having as much free time as you'd been promised and, as Johnson's speech-writer put it so vividly, skeletons of discarded cars littering the countryside. As we'll see in the next chapter, this would set the scene for a growth spurt in the environmental movement and – at first more in science and politics than with activists – a growing concern about all this carbon dioxide we were adding to the atmosphere too.

CHAPTER TEN

Growing Concern

In the late 1950s, Frank Capra (most famous for the 1946 film *It's a Wonderful Life*) found himself disenchanted by Hollywood and turned his hand to a short series of TV science specials. They were made with AT&T's advertising company, on the encouragement of Van Bush who was on the AT&T board and keen to see them push a public science programme. It featured a double act of 'Mr Fiction Writer', a clean-cut man who acts a little like the audience's surrogate, asking questions and helping to explain the science; and a more awkward, balding 'Dr Research', offering the scientific expertise and a host of technical devices. The setup was that they were writing a science show together, with the help of a 'magic screen' that allowed them to interact with a cohort of cartoon characters. A rich score carries the narrative from serious drama to more light-hearted humour, and the cartoons help explain some of the more abstract parts of the science as well as adding character. After shows on the Sun, blood and cosmic rays, in 1958 they turned to the of weather and in this episode's last 10 minutes, it issues a warning that burning fossil fuels could lead to a rise in sea levels.

Titled *The Unchained Goddess*, the cartoon characters here were led by the weather goddess Meteora. She's painted as emotional, throwing temper-tantrum tornadoes, acting as a contrast to video footage of military meteorologists as heroic adventurer scientists, flying into the eyes of hurricanes. 'How clever,' she declares, flirtatiously, when informed that the Weather Bureau give hurricanes girls' names. 'Maybe you'll name one after me some day Mr Scientist?' she adds, batting her eyelashes. 'Even in my vilest moods, you still understand me!' Meteora gushes. 'We've hardly begun,' Dr Research reassures her, but 'man is coming closer and closer.' They cut to a large computer – a 'giant, electronic weatherman' – powered by punch-cards and a chalk board of complex equations, before moving to a shot of the 'couple of human mathematical wizards' who built the machine, John von Neumann and Jule Charney.

'Can you do anything to control my moods, Mr Scientist? My fits of temper? Ooo, sometimes I scare myself!' 'We've already started', he replies, showing footage of cloud-seeding experiments and Callendar's Fido fog-clearing project, promising they'll be more to come. 'The possibilities are endless, the unanswered questions fascinating. No wonder more and more young students are turning to meteorology.' It's important we learn more too, he adds, before we start messing with deliberate weather modification. And here's where the climate change bit comes in: 'Even now, man may be unwittingly changing the world's climate through the waste products of his civilisation.' Due to the six billion tonnes of carbon dioxide that factories and cars emit each year, the atmosphere seems to be getting warmer. 'This is bad?' Mr Fiction Writer raises an eyebrow, playing his role as the audience's proxy. 'Well it's been calculated that a few degrees' rise in the Earth's temperature would melt the polar ice caps.' Cue dramatic music and shots of falling ice smashing into large waves before a map of a flooded south-east coast of the US fades to a cartoon of tourists in glass-bottomed boats viewing buildings of Mississippi drowned under 150ft of tropical water.

This meteorology episode didn't get especially good audiences at the time and received largely negative reviews. It was the last of these shows Capra would make; he wasn't enjoying it as much as he'd hoped and AT&T were starting to balk at the cost. Still, Capra would later estimate the films ended up with an audience of 200 million over the next few decades, largely through showings in schools. Today the clip about climate change gets passed around as an example of how far back climate concern goes: 'Look! We knew! We all knew! Frank Capra knew!' Although that is true, it's worth appreciating the way Capra talks about climate change in this film. It's only a small section at the end, used as an example of the many questions that remain unanswered about the skies, after presenters have fizzed with excitement over some 'Buck Rogers' kit the army were pulling together to study hurricanes, wondering whether storms could be steered out of our way with a strategically placed oil slick. Moreover, the programme concludes with a big Hollywood reassuring ending, appealing to science laced with religion: God has given us reason to ask and answer big questions, and we'll use it so we can live hand in hand with nature.[*]

The key message seems to be not to worry, the science guys will work it out in the end, those 'weather wizards' with their clever machines and courageous friends in the armed forces. If anything, Capra uses images of melting ice caps simply for a bit of dramatic effect, a momentary shift to minor key before the big rousing finish. The cartoon character playing the goddess of meteorology clasps her hands to her (generously drawn) chest in delight and proposes to Dr Research. Everyone chuckles as she rumbles up a storm outside when he explains he's already married. The curtain falls.

* * *

It wasn't as if Plass had published his paper, Revelle followed up with Suess and now everyone was terrified that carbon dioxide emissions would dangerously warm the Earth. Still, for all that Revelle had pulled in IGY money for carbon monitoring, with the topic now mainstream enough to get a mention in a Frank Capra show, Plass still felt the issue was ignored by the scientific community. 'You used to get up and give a talk to the American Meteorological Society or whatever,' he later recalled, 'and maybe find five or ten people in the audience.' And so Plass dropped the subject. When he returned to academia after work at Lockheed and a stint at the labs at Ford, he pursued other research topics.

Even for Revelle, the issue of carbon dioxide in the atmosphere wasn't something to be too concerned about, at least for now. It wasn't that he saw global warming as a good thing. He wasn't talking in terms of Hans Ahlmann's idea of climate 'embetterment' or suggesting, as that 1912 *Popular Mechanics* piece had done, that our descendants might thank us for their changed weather. He knew the impacts of climate change could happen soon too. He was looking at decades,

---

[*] As James Gilbert argues in his 2008 book on science and religion in American culture, a big motivation for Capra in this whole project was to combine scientific and religious messages for the American public.

not the centuries Svante Arrhenius or Charles van Hise had talked about. Still, it was all far enough away that he figured people would do something before it got bad. When *Time* magazine interviewed Revelle in 1956 it warned that the burning of fossil fuels could have a 'violent' effect on the Earth's temperature in the next 50 years, but that Revelle and his colleagues weren't about to issue a warning against this catastrophe just yet, they simply intended to calmly keep watching and recording what was going on.

In his 1957 paper with Seuss, Revelle wrote that 'human beings are now carrying out a large-scale geophysical experiment of a kind that could not have happened in the past nor be reproduced in the future'. Looking at it today, it sounds like he means we were taking a risk, something dangerous, like a Hammer horror Dr Frankenstein would cook up. But for Revelle, it was more that this was interesting, this weird thing we were doing to the atmosphere right now. We should study it, but there was no reason to be scared. 'Roger wasn't alarmed at all', one colleague (Svante Arrhenius's grandson, Gustaf) would later say, 'he liked great geophysical experiments.' A similar sentiment was echoed by Harry Wexler in a 1957 article on IGY, arguing that the current rate of fossil fuel emissions made it 'a particularly interesting time in man's occupation of this planet', but it wasn't going to last. Indeed, this was one of the reasons it was so important to get in now with carbon dioxide monitoring during the IGY. We'd all swap over to nuclear energy soon enough, he imagined.* The fossil fuel age would be over and with it the chance to study carbon dioxide's impact on the atmosphere would be gone.

Because these carbon dioxide-monitoring plans were part of the big international science show-and-tell of IGY, Revelle was invited

---

* Nuclear power is one of the topics I decided to only touch on in this book. It's been written about extensively elsewhere, and I'm yet to be convinced it's a central character of our story, even if it certainly is a context and crops up here and there (e.g. as above, with nuclear hype setting a scene for complacency over fossil fuels). It's a fascinating story, though; some places to start reading up on it include Spencer Weart's (1989) *Nuclear Fears* and the sections on nuclear in Richard Rhodes' (2018) *Energy: A Human History*.

to testify to Congress. After introducing himself, the Scripps Institution and his work at the Bikini tests, he mentions an aspect of the oceanographic programme 'I thought you gentleman would be interested in', a combination of meteorology and oceanography. 'We are burning, as you know, quite a bit of coal and oil and natural gas. The rate at which we are burning this is increasing very rapidly' and it is 'producing tremendous quantities of carbon dioxide in the air'. It's striking how, back then, Revelle framed the issue as an interesting intellectual opportunity, reusing the experiment line and stressing it was temporary: 'Here we are making perhaps the greatest geophysical experiment in history, an experiment which could not be made in the past because we didn't have an industrial civilisation and which will be impossible to make in the future because all the fossil fuels will be gone.' He does play one political card, tapping into the Cold War fears of his audience, warning that if the Russian coastline changed due to the melting Arctic, the Soviets might take the USA's spot as the world's premier maritime nation. But aside from this, it's framed as a matter for scientists to interest themselves in, not really of much political concern.

★ ★ ★

By October 1959, Shell seemed concerned enough about the new attention to place an article on the topic in *New Scientist*. In a three-page feature, Shell scientist Dr M. A. Matthews is at pains to let us know that carbon dioxide is only 0.03 per cent of the atmosphere. Moreover, the bulk of that is from natural sources, not the burning of fossil fuels, and the oceans are swallowing it up anyway, as are plants. 'Man's efforts in burning large quantities of fossil fuels are inevitably small compared with the magnitude of Nature's carbon cycles.' It's an understandable spin on the topic, one many believed in at the time; who are we puny humans to believe we might have the power to change the global climate?

Matthews seems to have taken particular umbrage at a warning given by Edward Teller (aka the 'father of the hydrogen bomb') at a 1957 meeting of the American Chemical Society: continue to burn coal and oil at the rate we were, and the polar icecaps would melt. Or possibly Shell was more concerned that if Teller was talking about this issue, so might other people. Either way, Teller wasn't about to shut up, repeating the message only a few months after Matthews' article was published, at a special symposium held in New York City organised by the American Petroleum Institute (API) and the Columbia Graduate School of Business to celebrate the centenary of Edwin Drake and William Smith first drilling oil in Titusville. Teller, a strong advocate for nuclear technology and rarely afraid of upsetting his audience, had turned up to the oil industry's own 100th birthday party with a warning that if they kept burning, the Empire State Building would soon be under water. Teller was a bit of an outlier, but it was the start of something.

It wasn't just Teller who was telling the API about climate change, its own scientists were too. The oil industry had been battling concerns about smog for decades, creating the Smoke and Fumes Committee in 1946 to help coordinate a PR response, first around LA and then nationally. They knew the oil industry was going to be blamed for the bulk of air pollution, and so invested in scientific research to help fight back against what they saw as overly hasty environmental regulation that could seriously damage their bottom line. They started to commission research at Stanford Research Institute in northern California, and the Franklin Institute in Philadelphia, including a mass spectrometer known as 'Silent Sam, the Smog Detective'. Although the bulk of this work was on smog, after the new work on carbon dioxide emissions in the 1940s, they started to explore climate change too. Not long after Revelle and Suess's 1957 research was published, researchers at Humble Oil (then part of Standard Oil of New Jersey, now Exxon) followed up with a similar paper.[*] It disagrees with Revelle and Suess on how fast the ocean spat carbon dioxide back out again; according to these results

---

[*] You can read a lot of these climate papers online easily enough yourself. Both 'Smoke and Fumes', from the Center for International Environmental Law, or the Climate Investigations Center's 'Climate Files' have amassed early examples of the oil industry's interest in climate change, with sources and contextual commentary.

we'd have rather more time before climate change started to bite. Still, they agreed on the basic problem.

★ ★ ★

One of the reasons the carbon dioxide problem didn't just get filed away again was this new monitoring project Revelle and Wexler had fundraised for, and the dogged work of the postdoc they hired to run it, Dave Keeling. His PhD had been on carbon, but more toward the end of things that might be useful to plastic manufacturers than anything to do with industrial emissions. He'd spent a lot of grad school reading up on geology though, rather than polymers, and started to itch for a research project that'd get him out of the lab. He craved trees and mountains, not a desk. In 1953, he'd moved to CalTech to do a postdoc in geochemistry. The funding was from the Atomic Energy Commission, and he was meant to be looking at ways to extract uranium from rock. But after one look at what he later described as the 'ear-piercing, dust-belching rock crusher' that lived in the geology building basement, he requested a different project. His supervisor agreed to a study that involved taking carbon measurements outdoors, first checking rivers and groundwaters' relationship with the air around them, soon focusing just on the air itself.

Keeling developed a device had read about in a 1916 paper, laboriously refining his technique to get tighter results. He tested it first on the geology building roof and then in the cleaner air out by the sea; he found much less 'noise' in his results than the Scandinavian project Rossby had led. He'd take samples every few hours, noticing that although carbon dioxide levels would undulate with plant growth between the start and end of the day, there was a remarkably consistent baseline in the afternoon, wherever he'd set up his station. He repeated the experiments up and down the Pacific coast, in the Hoh Rainforest west of Seattle and in high mountain forests in Arizona too; again and again, the readings were consistent. He knew the experiment didn't necessarily need such a thorough sampling strategy, but he was having fun. He enjoyed designing, building and testing equipment, and the fact it meant he could spend more time in Big Sur State Park rather than going back to the office didn't hurt either (even if he did have to get out of his sleeping bag several times a night).

Wexler heard about this young researcher pulling dazzlingly good carbon dioxide readings from the west coast air and invited him to DC – Keeling's first ride on a plane. Wexler was busy, he'd been appointed chief scientist for the IGY Antarctic mission and had plenty of other projects to be getting on with. He quickly looked over Keeling's data and started telling him about an exciting new observation station the Bureau was building at the Mauna Loa volcano in Hawaii. Surrounded by thousands of miles of clean ocean with a high volcanic peak, it was the perfect place to study the atmosphere. It also happened to be a good place to test high-altitude military equipment too, so the navy had been good enough to help them build a road (assisted by local prison labour). Wexler was keen to establish a continuous programme measuring carbon dioxide at Mauna Loa and offered Keeling the job of running it. Before rushing to his next meeting, he showed Keeling the office he would work in, in DC. It was a long way from this exciting Hawaii volcano with its pristine air; a rather grim-looking basement of the Naval Observatory where the only other work was cloud-seeding experiments.

Revelle, meanwhile, had been plotting a network of carbon dioxide sensors spread across the Earth, which would collectively offer a single snapshot of carbon dioxide in the atmosphere. He figured it would be a nice, iconic global project for the IGY and offer a baseline for measuring future carbon emissions; perhaps someone might apply for funding to repeat it in a decade or two to see what had changed. He also tried to recruit Keeling. Keeling preferred Wexler's idea of continuous measurement, but the idea of working at Scripps in La Jolla was much more exciting than that grim DC basement. In the end, a sort of fudge of a double job was agreed: Keeling would work from Scripps running Revelle's global network of sensors, but he'd also receive funding from the Weather Bureau setting up continuous measurement on Mauna Loa and Antarctica. Keeling was determined to do everything as thoroughly as possible and lobbied for IGY funds to buy some expensive tech. 'Keeling's a peculiar guy,' Revelle later remarked, not necessarily as a compliment. 'He wants to measure carbon dioxide in his belly. And he wants to measure it with the greatest precision and the greatest accuracy he possibly can.'

Still, Keeling got results. Stalking down problems like contamination from machines in Antarctica or the volcanic vents on Mauna Loa with obsessive attention to detail, he produced remarkably accurate and stable data. One of the first things he'd spotted was that the results seemed to undulate with the seasons, with carbon dioxide levels rising to a maximum in May before dropping to a low point in October. This pattern repeated itself in 1958, 1959, 1960 and so on. It was plants in the northern hemisphere absorbing the carbon dioxide during their spring and summer, before dying back in their autumn. He was watching the planet breathe. Plot Keeling's results on a graph and you'll see it zigzag up and down with the seasons. But keep plotting through several years and another pattern's clear under the rhythm of the seasons: the carbon dioxide levels were steadily creeping up. For that reason, the graph of Mauna Loa's data is known as the Keeling Curve. This was all quite clear even from the initial few years' worth of data, and in May 1960 Keeling published his era-defining finding: the carbon dioxide level in the atmosphere had detectably risen; as Seuss and Revelle had warned, the oceans weren't swallowing those emissions up for us. It was even covered by the *New York Times*, although they were seemingly more interested in what it said about a theory linking meteor showers to rain. Plus, *The Times* reports, according to latest Weather Bureau figures it looks as if, after 50 years of gradual warming, the world's climate could be cooling.

* * *

People might still have been finding ways to brush off any concern over carbon dioxide pollution, but environmental concern was on the rise. The Second World War had changed the tone of Western environmentalism slightly, but it had never gone away. Julian Huxley (grandson of Tyndall's friend Thomas Henry), for example, was a committed eugenicist, but his was a slightly different flavour of eugenics than the sort promoted by characters like Madison Grant. He'd been a strident critic of the Nazis before the war, publicly declaring any racial take on eugenics as pseudo-science. Still, he found it hard to shake the idea that it was better for some people to breed than others. For example, in his role at London Zoo during the 1930s and 1940s, he supported Jewish refugees, but it was driven by his desire to save the brains (including a sexologist from Berlin who wanted to study the handprints of animals and compare them with the handprints of 'mental defectives'). After the war, Huxley was appointed the first director general of UNESCO, the UN body set up to promote peace through education and the sharing of science and culture. It was meant to be a six-year term, but between Huxley's stand on birth control and his openly left-wing politics, it was cut to just two. Still, he managed to build concerns about overpopulation into UNESCO's 1947 founding manifesto, arguing there is an optimum population size of the world and that 'man's blind reproductive urges' should be controlled. He was also key to the 1950 UNESCO statement on race, a scientific and moral condemnation of racism, and highly influential in setting up the International Union for Conservation of Nature – the group that draws up the annual 'Red List' of threatened species.

Jump to 1960, and Huxley got back from a UNESCO trip to 10 Central and East African countries, shocked and angered by the destruction of wildlife he saw. Being the opinionated and well-connected man he was, he turned this anger into a trio of articles for the *Observer*, warning that between hunting and the destruction of habitats, much of the region's wildlife could disappear within the next 20 years. The header of the first essay cried: 'Millions of wild animals have already disappeared from Africa this century. Does the wildlife of the continent now face extinction – threatened by increases in population and the growth of industry in the emergent nations?' Businessman Victor Stolan read it and, sitting in his home in South Kensington, penned a letter to Huxley, arguing they needed to set up an international organisation to raise funds to combat this problem. Huxley replied, fixing him up with Max Nicholson, who had worked on the IUCN, and they took advice from ad man Guy Mountfort, as well as Godfrey Rockefeller (grandson of J. D.'s brother) and ornithologist Peter Scott (son of Scott 'of the Antarctic'). This would become the World Wildlife Fund, the iconic charity with a black-and-white panda logo, launched at the Royal Society of Arts, London, in September 1961, with Prince

Philip the first president of the British appeal.* This was an update on the conservation movement of the Boone and Crockett Club set, but was still arguably in the same mould. Huxley had a distaste for Nazis, sure, but he had little problem picking up a British colonialist stance that looked at the continent of Africa and saw something that had to be saved and controlled by rich white people.

A couple of popular science books – *Our Plundered Planet* and *Road to Survival*, by Henry Fairfield Osborn Jr and William Vogt respectively – had called for greater environmental protection, both published just after the war. Both repackaged concerns about overpopulation for a new generation, the racism dialled down a little (but only a little), with the focus broadened to questions of the global population at large, rather than simply what was happening at home. They also both emphasised soil health, drawing attention not just to the charismatic megafauna and megaflora of bison and redwoods, but plain old dirt under our feet. And both implied population problems and scarcity of resources had been causes of the war, concerned that mistakes like these could be made again. A feature in *Time* magazine in November 1948 declared: 'After more than a century of intermittent haunting, the ghost of a gloomy British clergyman, Thomas Robert Malthus, was on the rampage', as both books were selling like hotcakes to glowing reviews.

Osborn Jr was a man of some zoological whimsy. He'd once turned up to a talk at the New York Chamber of Commerce with a skunk, a lemur, a hedgehog and 'a frisky, uninhibited monkey'. As a child, he'd kept a pet baby alligator who he'd take to bed with him. But he also had a catastrophic streak, and his experience captaining a field artillery unit in France during the First World War had convinced him humans were inherently destructive. In New York Society, he was known to request an extremely dry martini before regaling his companions with tales of man's corruption of nature. Similarly, *Our Plundered Planet* warned that as people unravelled forests, grasslands, soils, water and other animal life, the Earth was on course to 'become as dead as the

---

* Along the way, they dropped Stolan, something Stolan was apparently less than happy about, and the *Observer* puts this down to snobbishness around his work as a hotelier and his status as a Czechoslovakian refugee.

moon'. People must temper their demands and use less of the Earth before it was too late: 'Parts of the Earth, once living and productive, have died at the hands of man. Others are now dying.'

In 1948, Osborn Jr set up the Conservation Foundation, a sort of scientific think tank on conservation issues, which aimed to be different from earlier conservation groups in that it took an interest in what might be seen as more abstract scientific ideas like pesticides or population growth. Crucial to our story, in March 1963 the Foundation brought together a group of scientists to consider the implications of rising carbon dioxide in the atmosphere; the first scientific conference on climate change. Keeling and Plass were both part of the assembled expertise, but sat alongside biologists, with British conservationist and vice-president of the foundation Sir Frank Fraser Darling in the chair. This meeting took the carbon dioxide problem out of the more analytical, rather military-orientated field of geophysics it had grown up in and explored it as a question of ecology. Compared with earlier discussions of the topic, the different attitude to risk expressed in the report is striking. The recommendations and sense of alarm were still pretty meagre by today's standards – they mainly call for more research – but they do frame carbon dioxide emissions as a clear matter of concern. Moreover, noting 'the most alarming thing about the increase of CO_2 is how little is actually known about it', they see the uncertainty surrounding the issue a reason for more caution, not a reason to simply sit back and see what happens. Darling would, in 1969, expound on this in a series of lectures for the BBC, warning that we could already see the impact of warming oceans on fishing stocks, and although the bigger danger of the polar ice caps melting some time in the future might seem remote, that's no excuse to ignore it. He said: 'So often I've heard it said posterity must look after itself. I can think of a no more callous viewpoint.'

★ ★ ★

By the time of this 1963 conference, another writer was raising awareness about environmental issues: Rachel Carson. Her book *Silent Spring* had started to cause controversy even before it was published, via a series of essays in the *New Yorker* in the summer of 1962. Carson had been interested in pesticides, DDT in particular, for decades. She'd tried to get *Charlotte's Web* author E. B. White to write an essay about it and when he declined, she decided to have a go herself. Initially, the idea was to produce a collection of essays on pesticides written by several authors, with Carson as editor, but as she started researching she decided to do the whole book herself. The *New Yorker* essays were published in three segments in June 1962 and immediately people took notice. The controversy raged through op-eds and letters to editors to the floor of Congress. Prince Philip got hold of a few advance copies of the book and started handing them out to British politicians, including Christopher Soames, son-in-law of Winston Churchill and minister for agriculture, fisheries and food.

Carson and her publisher were especially good at tapping into the needs and worries of suburban America. This meant the chatter didn't just happen in newspapers and between politicians, it was in book clubs and gardening groups – and it kept on going, right into 1963 and after. The chemical industry mobilised a $250,000 PR 'war chest' to fight back, but little could be done to quieten the debate. Carson was seriously ill with breast cancer by this point and limited her personal appearances. However, she managed a few select public appearances before her death in April 1964, including a CBS television special on the book. Carson was slightly wary of television. When her earlier book *The Sea Around Us* was turned into a film by RKO, she had been annoyed at the ways the filmmakers had produced a climax with the Earth drowning as polar ice melted, arguing this sudden catastrophe simply wasn't scientific. But Carson trusted the CBS journalists and appreciated it was a powerful platform. When the show aired in April 1963 it ended up with an estimated audience of between 10–15 million. Sponsors had pulled their commercials when they found out about the topic, but this just made the show look even more important and trustworthy. She perhaps should have remained wary of television, as the film pitted her against a stiff, white-coated chemical industry expert, setting a frame of Carson as anti-science and anti-industry, which would later be used against her. Still, its impact was huge, further mobilising a call for change.

One of the reasons the book sparked in the way it did was because interest in environmental issues was already on the rise. Arguably, they hadn't ever really gone away, just shifted with the times. The women's groups of the nineteenth century hadn't stopped, and although they sometimes still saw themselves as doing 'civic mothering' or 'municipal housekeeping' they were getting more strident in their calls for clean air and water. Although there was a fair amount of sexism used to pour scorn on Carson (and still is), women's organisations helped make *Silent Spring* both a bestseller and a political force, a power that Carson herself cultivated while she was alive. Some of these women's groups had been involved in the anti-nuclear protests – the 1961 Women Strike for Peace, for example – and were up for a bit more of a political fight. It's also sometimes argued that concerns over the bomb helped open the public's mind to more abstract, seemingly invisible environmental threats like pesticide use (and climate change). Plus, with the growth of suburban populations and car use, outdoor activity was on the rise. The car companies had been very keen to sell their products on the promise of quick and easy trips to 'get out into nature'. People had taken them up on this and were deepening their environmental concern as a result. In 1964, Lyndon B. Johnson signed the Wilderness Act, a massive extension of what John Muir and the Boone and Crockett set had started, protecting over 9 million acres of federal land under a legal definition of wilderness.

As we saw in the last chapter, Johnson also referred to environmental issues in his 1965 State of the Union, following it up a month later with a special statement on the preservation of natural beauty to Congress. He was far from the first president to be interested in conservation, but he would be the first to talk about the climate crisis. His statement on the preservation of natural beauty extended familiar calls to clean up the air and rivers with a reference to the carbon dioxide problem: 'Air pollution is no longer confined to isolated places. This generation has altered the composition of the atmosphere on a global scale.' This, he notes, is in part through radioactive materials (this was the atomic 1960s after all), but also 'a steady increase in carbon dioxide from the burning of fossil fuels'. It was a blink-and-you-miss-it mention, but the carbon dioxide problem had reached the White House. The PSAC report 'Restoring the quality of our environment' that followed later that year included a section on carbon dioxide alongside the more visible, more politically charged issues like air and water pollution. A 23-page appendix on carbon dioxide from a subcommittee chaired by Revelle repeats the 'vast geophysical experiment' line, but this time warning that we could expect to see a marked change in climate by the year 2000. If we kept burning at expected rates, sea levels would rise about 4ft every decade and we could already see species of fish moving as ocean waters warmed.

Presidential interest understandably got the oil industry worried. In 2016, researchers at the Center for International Environmental Law unearthed a 1968 report prepared for the API by the Stanford Research Institute, 'The Robinson report', named after its lead author Elmer Robinson. This makes it clear carbon dioxide emissions are now outstripping nature's ability to breathe it back in. Moreover, it echoed the new, more worried mood about the impacts of the greenhouse effect, repeating the PSAC warning that sea levels could rise 4ft a decade. They conclude that although scientists are still unsure about much when it comes to the long-term impact of fossil fuel pollution, 'there seems to be no doubt that the potential damage to our environment could be severe'.

A little like the Capra film, today this report sometimes gets passed around with jeers that, 'They knew! The oil industry knew.' Of course they knew. They had some of the best geophysicists in the world. The oil industry had been investing in science since Benjamin Silliman Jr's report on Pennsylvanian 'rock oil' back in 1855 and hadn't stopped there, simply diversifying from chemistry into geology, engineering, oceanography and, when it suited them, atmospheric physics. Sometimes fossil fuel companies and their defenders get painted as 'anti-science'. In truth they've run on science and always have done. They're just strategic about how they use it.

* * *

For some, the solution to the new environmental problems was simply better tech. Today, when people call a project a 'technofix' they tend to mean it pejoratively, especially in climate contexts; pouring resources into a sticking plaster of an engineering 'solution' rather than addressing the root causes of a problem. However, the man who coined the term loved the idea. It was the same man who coined the term 'big science', Alvin Weinberg. It was 1967, and Weinberg had been impressed by Ralph Nader's campaign for automobile safety and the various promises of the contraceptive pill. He was also a fan of the ways atomic weapons had grown so large that the Cold War was stuck in a position of mutually assured destruction. For Weinberg, a technological fix was often preferable to social engineering. He would later go on to attack trends of what he saw as technological pessimism, describing much of his career as that of a 'technological optimist' and 'technological fixer', ideas he'd later apply to climate policy. Technofixes for climate change were already being talked about, well before Weinberg gave them a name. The 1963 Conservation Foundation report, for example, had referred to ideas of burying carbon dioxide, but dismissed it as expensive and easier to simply replace fossil fuels as a source of power. The 1965 PSAC report also suggests changing the Earth's albedo (how much it reflects light) to battle the greenhouse effect. It notes that spreading reflective particles on the ocean could cost $500 million a year but, considering the possible impacts of climate change, that might be necessary.

We should be careful of falling into Weinberg's binary of pro or anti-technology, not least because it has been weaponised by spin doctors trying to avoid regulation, dubbing anyone questioning the application of a technology as an out and out luddite. Plus, the world's simply more interesting than that. Stewart Brand's *Whole Earth Catalog* offers a good example of the new cultures of alternative technology that were emerging at the time. The 'whole Earth' had come from an earlier project by Brand. In 1966, he'd launched a public campaign to have NASA release the then-rumoured satellite photo of Earth as seen from space; the first image of the 'whole' Earth, commonly known as 'Earthrise'. Famously he'd planned this while on an LSD trip on the roof of his house in North Beach, San Francisco, convinced the image would bring people of the world together with a shared sense of their home. The catalogue featured essays and articles, but the idea was primarily to propagate tools for change – 'tools' being anything from masonry equipment or shears to maps or courses – in contrast to the new left's focus on political power. There was an element of seizing the tools of production about it, though it wouldn't be accurate to call it Marxist (today Brand describes himself as 'post-libertarian').* If you wanted to know where to buy a windmill or learn beekeeping, this was the place to go.

Beyond Brand, the more hippie end of late 1960s counterculture added a new layer of theatre to environmental activism. A group in New York, for example, sprayed black mist and passed out blackened flowers at a 'soot-in' in front of the Consolidated Edison Building. The new left of the time was less interested in the environment, initially preferring to focus on social and economic issues, but discovered a political critique similar to what the Chartists had called the 'smokeocracy' after the 1969 Santa Barbara oil spill. As tens of thousands of barrels of crude spilled out of a Union Oil rig in the Dos Cuadras oil field, just weeks after Richard Nixon's inauguration, the media swarmed in to give it blow-by-blow coverage and thousands of people took part in demonstrations against 'big oil'. Again, this was not new, Ida Tarbell had fought the size of oil companies back at the start of the century, but it was a resurfacing of anger that would be built on in years to come.

Concerns about population continued into the 1960s and 1970s, with the publication in 1968 of *The Population Bomb*, by Paul Ehrlich and his (uncredited) wife Anne. Ehrlich had been egged on by David Brower who had been discussing population over the last decade as

---

* Brand would, in the 1980s, found the 'Whole Earth 'Lectronic Link', or WELL as it was better known, which applied some of the similar principles and was highly influential in early online culture. He also famously declared 'information wants to be free' at an early hackers conference in 1984 and today remains a popular thinker among the Silicon Valley set. Steve Jobs described the *Whole Earth Catalog* as 'Google in paperback form, 35 years before Google came along' and Jeff Bezos gave $42 million to Brand's 'Long Now' project to encourage the long-term future of the planet.

director of the Sierra Club,* and the book caused instant controversy. For John Maddox, editor at *Nature*, Ehrich was an irrational 'prophet of doom', part of a larger problem that had been kicked off by Carson and also included concerns over pesticides, damage to the ozone and worries about the melting ice caps. Calling for a return to optimism around science, Maddox argued it was all unnecessary hand-wringing, a 'wave of gloomy speculation' that will eventually 'be considered to have been a wave of fashion' from privileged people in the West who, finding they no longer had the threat of nuclear war hanging over them, were looking for something new to get all Chicken Little about. People were simply enjoying worrying about some distant calamity to avoid sorting more immediate problems like air pollution. It was lazy of Maddox to lump a wide range of environmental worries together in that way, but the idea of environmentalists raising irrational concerns would be a characterisation that would stick, no matter how unfair or dangerous it might be (and arguably he was entirely right about people avoiding dealing with more immediate problems).

The biggest environmental event of the period, however, was Earth Day, first held on 22 April 1970. This had started as a proposal for a nationwide 'teach-in' on the environment, from Wisconsin senator Gaylord Nelson. To lead the project, Nelson hired 25-year-old former student activist Denis Hayes. Hayes's activism was forged in anti-war campaigns and civil rights, not conservation, but that didn't mean he wasn't committed. He'd read Ehrlich and Carson, but possibly more to the point came from a working-class background in an industrial town that had been heavily impacted by pollution. When April rolled around, 1,500 colleges ran teach-ins, as did a further 10,000 schools. There were also events in churches and temples, in parks, and protests in front of businesses and government offices. In some places, events lasted the whole week, not just the

---

* Discussions of what the Sierra Club called the 'population explosion' varied from Frank Fraser Darling critiquing the excesses of the American economic model to UCLA herpetologist Raymond Cowles arguing South Africa's wildlife would only be preserved as long as white rule remained. For more on this (as well as the books by Osborn and Vogt mentioned previously) see Thomas Robertson's *The Malthusian Moment* (2012).

single day, and it's been estimated 10 per cent of the American population were involved in some way. Like any big protest, it was a chance for people to realise they were part of a larger – huge – movement, but also an opportunity for them to network and learn from each other. Earth Day turned thousands of participants into committed environmentalists, its local hubs providing leadership training for people wanting to take their environmental activism further. At the Earth Day rally in DC, Hayes told the assembled crowd: 'They are talking about filters on smokestacks while we are challenging corporate irresponsibility. They are bursting with pride about plans for totally inadequate municipal sewage treatment plants; we are challenging the ethics of a society that, with only 6 per cent of the world's population, accounts for more than half of the world's annual consumption of raw materials.' The environment movement was rediscovering structural critique.

As the 1960s turned into the 1970s, other new environmental groups emerged, slightly less tied to the Boone and Crockett model. Friends of the Earth had an explicit break with the past, as it was founded by Brower after he was sacked by the Sierra Club. Greenpeace's origins were slightly more circuitous. Greenpeace was, initially, just the name of a boat, funded from a benefit concert for a Don't Make a Wave Committee against nuclear weapons testing near the Alaskan island of Amchitka. People had been worried the nuclear explosion would trigger a tsunami, hence 'Don't Make a Wave' and, frustrated that the established environmental groups like the Sierra Club weren't doing enough, they decided to take matters into their own hands with a protest. They built some media interest and a community of activists, and off the back of that they built a global movement named after the boat. There's a story that at one of the early planning meetings, one of the members flashed a peace sign with his fingers and another replied, offhand: 'Make it a green peace.' When they tried to put both words on the 25c badges they were selling for the fundraiser, there wasn't room, so they cut the space and merged them: 'Green Peace' became 'Greenpeace'. They'd later shift their focus from Alaska to French nuclear testing at the Mururoa Atoll, and then also start campaigning on whaling, eventually including a wide range of environmental issues. No one much in these new groups was talking about climate change though. That would come later.

* * *

Meanwhile, Keeling's carbon-monitoring project had been steadily clocking up more and more data points, curving ever upwards as fossil fuel emissions increased. What had started out as a temporary job for Keeling would turn into a life-long career. It'd be a family business too, as his son Ralph would become director when Dave died suddenly of a heart attack in 2005. First though, they'd have to get over the initial hurdle of the IGY funding winding down. Hans Suess had written to Revelle arguing the project should be continued, even offering to move some of his own research funds to help, and between them they found $10,000 to keep the monitoring project going a while longer. This bought Keeling enough time to get a grant from the National Science Foundation, whose budget had been boosted by *Sputnik* anxieties. The Weather Bureau helped out too, as did Scripps.

Support for the Keeling Curve would undulate over the years, not dissimilar to the zigzagging up and down in its graph. After Congress went on a cutting spree in the early 1960s the Antarctic monitoring had to be dropped, and the Mauna Loa Observatory was under threat too. In the spring of 1964, a delicate instrument broke down and, missing a full-time technician, it had to shut briefly. Keeling's research muddled through for a few more months before the 1965 PSAC report was published, which had made a point of arguing a carbon dioxide-monitoring programme should be continued for several more decades (perhaps no surprise as Keeling was on the carbon dioxide panel and it was chaired by Revelle).

People were starting to worry about climate change, but they were worried about lots of other things too. Keeling himself wrote in 1970 that 'the increase in CO_2 is of no special concern to our immediate well-being. The rise in CO_2 is proceeding so slowly that most of us today will, very likely, live out our lives without perceiving that a problem may exist.' He was more worried about smog. There were rising temperatures and rising levels of carbon dioxide, he noted, but also 'rising numbers of college degrees, rising steel production, rising costs of television programming and broadcasting, high-rising apartments, rising numbers of marriages, relatively more rapidly rising numbers of divorces, rising employment, and rising unemployment. At the same time, we have diminishing natural resources, diminishing distraction-free time, diminishing farmland around cities, diminishing virgin lands in the distant countryside.' Still, he concluded with a note of caution on short-term thinking: 'Have you noticed that practically all master plans do not project beyond the year 2000 AD? Our college students, however, today expect, or at least nourish the hope, to live beyond that date.'

Another problem for Keeling's research was that long-term monitoring just wasn't what most science-funding bodies saw as exciting. Revelle's initial idea of a single snapshot of data collection was the way science was (and is) usually funded – a single project rather than long-term commitment. The longer carbon-monitoring project doesn't fit the patterns of most scientific funding applications, where you are asked to find something new; instead it just keeps ticktocking away. To make things harder, the work was expensive, so for a body to take ownership was rather a commitment. As a grants reviewer who reluctantly gave Keeling money in late 1979 put it: 'CO_2 monitoring is like motherhood ... It does appear, however, that the former is even more expensive.' It didn't help that Keeling sometimes rubbed people up the wrong way with his dogged determination to, as Revelle had so memorably put it, 'measure carbon dioxide in his belly', coming across as a demanding know-it-all. Plus, this kind of data on how we're messing up the planet isn't exactly the sort of evidence everyone wants to be reminded of. It's immensely useful, but also rather vulnerable to a politician's knife.

There was plenty more to be studying when it came to climate change too, aside from monitoring carbon from the top of a giant volcano. Researchers were increasingly interested in how much further they could go back in time, whether they could piece together a picture of carbon dioxide levels from before Keeling had started to track it. The bottom of the ocean, for example, holds all sorts of stories if you know how to unravel it. Back when Revelle had been studying the after-effects of atomic bomb testing in the 1940s, he'd taken cores of coral to work out the history of the Bikini Atoll.

He had also hired Gustaf Arrhenius, who'd been working with new technologies to draw cores of deep-sea sediment in Sweden, to help him build on this work in the Marshall Islands. From the 1950s onwards, a new generation of chemically minded ocean researchers would apply techniques from nuclear science, like carbon dating, to samples of deep sea ooze to work out, by proxy, past climates. Dig deep enough and you could find fragments of sea creatures who'd lived thousands of years ago. Similar cores taken from ice offered further prizes, you could find ancient pollen, dusts from centuries' old volcanic eruptions or the fingerprint of Roman smelting, all caught in snow buried deep under centuries of ice.

For Danish meteorologist Willi Dansgaard, the real prizes were bubbles of air, artefacts of past atmospheres, persevered in deeply buried ice. He'd spent time in Greenland just after the war and yearned to go back. When the chance to hitch a ride on a seal- hunting boat came up in 1958, he jumped at it. It wasn't the most comfortable trip, his berth had him squeezed into the side of the boat meaning his sleep was always soaked with seawater, but he enjoyed watching the dolphins follow them. When they arrived, Dansgaard and his colleague set up a makeshift lab on the foredeck, naming it *Boblebua* ('home of the bubbles'). The skipper crossed himself at what this new structure might do to the seaworthiness of the boat. Ice was floating around them, but they didn't want to just fish it out of the sea, as they didn't know where it came from. Instead, they inched themselves close to the glaciers and waited for a decent-sized chunk to fall off, which they'd then further chop up on deck and melt down, filling thousands of plastic bottles with samples to analyse back home. Sometimes they'd ram the boat into smaller icebergs, which after some practise proved reasonably efficient (as long as someone held on to the lab's glassware).

In 1964, Dansgaard was able to take a second bubble expedition, this time as guests of the US Army, to a place called Camp Century. The US military had been preparing for a possible war with the Soviets in the Arctic since the late 1940s, and in 1958 built Camp Century as part of a Cold Regions Research and Engineering Laboratory. This strictly isolated research centre housed up to 250 men in 32 buildings, and Dansgaard describes it as fitted out like a modern town, with a radio station, shops, laundry, hospital, barbershop, fitness centre, cinema, church and library, all powered by a nuclear reactor buried 300ft below the living quarters. The mess hall had a generous bartender who charged 25c a drink, no matter the size (if you ordered a pint glass full of gin, it might go up to 30c).[*] Dansgaard wandered around the camp and got whiff of a new deep ice-core drilling project they were testing nearby. This used a special thermo-drill, which melted the ice while preserving the core it was extracting, and Dansgaard knew that if only he could get access to some of these cores, he'd be able to travel much further back in bubble-time than anyone had managed before. He wrote to a colleague in the US who he knew was handling some of the cores and asked if he could get a look-in. Eventually in 1967, under much secrecy, Dansgaard had cores to study stretching back 100,000 years, roughly 50 years a metre.

And then there was chaos. Or more specifically chaos theory. Edward Lorenz was one of those mathematically trained meteorologists who'd worked as a weather forecaster during the Second World War before moving on to academic work in one of the centres Carl-Gustaf Rossby had pioneered in the 1930s. He'd landed a professorship at MIT in 1955 and was lucky enough to be given a 'personal computer' in his own office, almost unheard of at the time. One day in 1961, he'd been running simple models of the atmosphere based on a series of equations Fourier had first developed back in the 1830s. Lorenz decided to run one of them a second time, fed it into the machine and left it running while he went for a coffee to escape the noise of the machine. When he came back, to his surprise, he found the weather patterns it had predicted were entirely different. Looking again at what he'd done, he noticed a very subtle difference between the figures he'd fed in each time; just a 1 per cent shift, to do with rounding up decimal places to save space. He'd go on to give an influential speech at the 1972 meeting of the American Association for the Advancement of

---

[*] Dansgaard's diary from the time also recalls a poster on the mess room wall, with advice of what to do in case of a nuclear attack: '1. Open windows and doors. 2. Loosen tie and belt. 3. Sit down on the floor with the head bent down between the legs. 4. And then – kiss your ass goodbye.' He has an entertaining autobiography you can look up on the Niels Bohr Institute website.

Science on this topic, with the now famous question: 'Does the flap of a butterfly's wings in Brazil set off a tornado in Texas?'*

Although the idea of the butterfly effect wouldn't go mainstream for a while, it excited climate scientists at an August 1965 conference on the causes of climate change held at the relatively newly established National Center for Atmospheric Research in Boulder. Lorenz gave the opening address and it reflected a sense at the conference that the climate was a complex system, you couldn't just say the Sun or the ocean or one chemical or another caused a problem, you had to look at its interactions with everything else. The idea that the atmosphere is chaotic and unpredictable is a talking point that's since been used by sceptics as a reason to ignore climate models, and it's one of those times when the distinction between specific weather events and the longer trends of climate is key. You can't predict exactly where and when a tornado would hit, for example, but you can know reliably when the Texan tornado season is. Climate scientists would later talk of climate change 'weighing the dice' towards particular types of extreme weather event, or as one climate scientist put in the late 1990s, the climate is like 'an angry beast' and as we pour greenhouse gases into the atmosphere, 'we are poking it with sticks'.

* * *

Another group applying computers to the problems of the day was the Club of Rome, an international group of scientists, industrialists, politicians and campaigners founded by Italian industrialist Aurelio Peccei and British scientist Alexander King. At the core of the Club of Rome's approach was Peccei's sense that the big problems of mankind in isolation – be they poverty, health, environmental degradation – could never be solved if you looked at them in insulation, instead they

---

* The butterfly here is sometimes traced back to a 1952 Ray Bradbury story where a time-traveller on a trip to hunt a T-Rex accidentally crushes a butterfly and then returns home to the year 2055 to discover changes in language and politics. Lorenz later said he'd initially planned to use the image of a seagull as an illustration before a colleague suggested a butterfly would have more impact (and then Brazil was picked for alliteration). The term 'butterfly effect' wouldn't be talked about until James Gleick's *Chaos: Making A New Science* (1987).

must be understood as non-linear interrelated systems. With some funding from the Volkswagen Foundation, the club commissioned a group of MIT researchers to collect data on development, pollution, population and resource depletion, build a computer model and watch it play out from 1900–2100. The predictions, published as *Limits to Growth*, were pessimistic, arguing that a business–as–usual approach would lead to overshoot of resources and social collapse after the middle of the twenty-first century. The basic premise was simply that the Earth was finite; nothing new or popular at the time.* Plenty of people criticised this focus, stressing it didn't take into account the ways technology and culture can change how we use resources. Others distrusted the computer modelling. *Foreign Affairs* laughed it off: 'The computer that printed out W★O★L★F★.' Still, released a month before the UN Stockholm Conference on the Human Environment, *Limits to Growth* sold more than 7 million copies in 30 languages. It only gives carbon dioxide one paragraph in its 200 pages, but its frame of a brewing crisis of industrial capitalism would heavily influence much of the climate debate.

Also planned for the run-up to the 1972 UN Conference was an international workshop on Man's Impact on Climate held for three weeks in July 1971, again near Stockholm. Just as the Conservation Foundation's 1963 meeting had brought a new take on the issue of carbon dioxide via its conservationist framing, this involved geophysicists, but wasn't led by them. Instead, it was spear-headed by MIT professor of management, Carroll Wilson. Although this didn't generate new knowledge, the event deepened participants' awareness of the issue, and helped build political engagement too. The book that came out of this workshop – *Inadvertent Climate Modification* (inadvertent as opposed to deliberately covering glaciers with coal dust) – was labelled 'required reading' for all delegates to the UN Conference the following

---

* Also popular around the time was the idea of the Earth as a spaceship, floating in space with only the finite resources we brought with us (it is a rather lonely idea of a ship, hardly the Millennium Falcon, or the ISS). US Ambassador Adlai Stevenson had used it in a speech at the UN: 'We travel together, passengers on a little spaceship, dependent on its vulnerable reserves of air and soil.' It was later used by British economists Barbara Ward, in a 1965 book *Spaceship Earth*, and then in 1968 by Buckminster Fuller, with his *Operating Manual for Spaceship Earth*.

summer. Wilson also addressed the UN directly, noting that it was easier to define the problem than say what to do, but warning that the effects of climate change could already be seen in the Arctic and work to avoid the worst consequences would take some 30–50 years.

When the UN Conference on the Human Environment did come around, it had a packed agenda. There were the more traditional conservation concerns like deforestation and endangered species, as well as increasing worries about less visible pollution from industrial or military chemicals and radiation. Moreover, there were knotty questions about how this would all play out geopolitically; what it meant in terms of conflict and cooperation between nations. Maurice Strong, the Canadian oil man leading the conference, framed it as a great moment in history (which, his desire for rhetoric notwithstanding, it was): 'Man today stands, at one of those critical moments, of change in human thought and progress. He has suddenly perceived the unitary nature of his planet, and he has become aware – quite suddenly – that he is all but reconstructing it, in some cases, purposely, in some inadvertently, but in the aggregate with startling speed and effect.' At the core of this, he argued, was science and technology, the science and technology that took us into space to give us that Earthrise photo, but also the science and technology that had given us agent orange (a herbicide used as a weapon by the US during the Vietnam war).

Alongside the formal UN business there was a semi-official fringe, with an Environmental Forum funded by the Swedish government including exhibitions, films, slideshows, panel discussions, lectures and workshops. Population was hotly debated here, with a much more critical take than at the mainstream UN conference. Ehrlich along with Edward Goldsmith,* founding editor and publisher of *The Ecologist*, advocated population control, backed up by a delegation from Friends of the Earth. They were met with what one attendee

---

* Goldsmith too had released a pre-Stockholm publication, *A Blueprint for Survival*. It recommended a shift to smaller, decentralised and to some extent de-industrialised communities, partly for environmental reasons, but also moral and psychological ones. Goldsmith was well networked and pulled in an impressive number of eminent scientists to sign his blueprint – Julian Huxley and Frank Fraser Darling among them – but received plenty of criticism from scientists (and others) too. John Maddox's *Nature article* was especially scathing.

described as a 'severe battering' though, not least from a vocal delegation of young scientists calling themselves the 'Oi Committee' who didn't think population was an environmental issue. It was far from the only topic on debate though. A 'People's Forum' – developed by local radical groups – ran sessions on discussing the Vietnam War, ecocide and the environment of work. There were also witness testimonies of survivors of agent orange and a Japanese mercury poisoning disaster. A heatwave boosted a sense of carnival, with an alternative technology group putting on 'alternative sightseeing' tours of the city and an 'Olympic Games of Pollution'. The modern art museum had solar and wind tech on display and, at the more theatrical end of things, the American hippie collective, the Hog Farm Commune, ran a 'whale-in' (as opposed to a teach-in) including a brief poem ostensibly written by a whale.

Back at the official end of the proceedings, the Soviet bloc boycotted the whole event in protest at the exclusion of East Germany, Beijing took every opportunity to criticise the United States, and Japan was fervently trying to avoid a moratorium on whaling. Indian Prime Minister Indira Gandhi famously argued that 'poverty is the worst form of pollution', challenging the way the Western conservation movement had framed the issue thus far. The hosts, Sweden, also caused some controversy when Prime Minister Olof Palme described US actions in Vietnam as 'ecocide'. Still, they founded the United Nations Environment Programme, to be headquartered in Nairobi, Kenya, which would later help build some of the key infrastructure for international climate negotiations. And, after much argument, a declaration was signed, agreeing basic principles under which the world's nations might work together on environmental issues. This included the basic tenant that humankind was affecting the environment on an unprecedented scale, that protection of the environment was key to the enjoyment of basic human rights and – following Indira Gandhi's provocation – that 'In the developing countries most of the environmental problems are caused by under-development' (i.e. the West shouldn't think it can find environmental salvation by keeping the rest of the world poor).

CHAPTER ELEVEN

Crisis Point

In August 1974, the CIA produced a study on 'climatological research as it pertains to intelligence problems'. The diagnosis was dramatic. It warns of the emergence of a new climatic era, leading to political unrest and mass migration (which, in turn, it argued would cause more unrest). The main concern was food, aware that political stability is so easily disrupted by scarcity in 'the new climatic era'. This new era wasn't necessarily one of hotter temperatures; the CIA had heard from scientists warning of a global cooling as well as warming. But which direction the thermometer was travelling in wasn't their immediate concern, it was the impacts. The little ice age had brought drought, famine and political unrest, and so could these new climatic changes.

'The climate change began in 1960,' the report's first page informs us, 'but no one including the climatologists recognised it.' The crop failures in the Soviet Union and India in the early 1960s had just been put down to standard unlucky weather. The US shipped grain to India and the Soviets killed off livestock to eat, 'and Premier Nikita Khrushchev was quietly deposed.' But, they argue the world 'quietly ignored' this warning, continuing to race ahead with a growing world population and massive investments in energy, technology and medicine. Meanwhile, climate change continued to roll on, shifting to a collection of West African countries just below the Sahara. People in Mauretania, Senegal, Mali, Burkina Faso, Niger and Chad 'became the first victims of the climate change' the report argues, but their suffering was masked by other struggles, and simply that the rest of the world wasn't paying attention. As the impacts of climate change started to spread to other parts of the world, the early 1970s saw reports of droughts, crop failures, floods and heavy rain from Burma, Pakistan, North Korea, Costa Rica, Honduras, Japan, Manila, Ecuador, USSR, China, India and US. But few people seemed willing to see a pattern: 'The headlines from around the world told a story still not fully understood or one we don't want to face.'

The claim no one was paying attention was somewhat unfair. As we've seen, some scientists had been talking about this issue for a while. It had been in the papers, referred to by the president and chuckled over in a Frank Capra television special. Earlier that year, American secretary of state Henry Kissinger had addressed the UN under a banner of applying science to 'the problems which science has helped to create', including his worry that the poorest nations were now threatened with 'the possibility of climatic changes in the monsoon belt and perhaps throughout the world'. Still, the CIA report had a point that climate change wasn't getting the attention it could have and there was a lack of urgency in discussions. There was no large public outcry, nor did anyone seem to be trying to generate one. For all that Frank Fraser Darling and others at the Conservation Foundation had warned about the dangers of sitting back and waiting to see what would happen, no one seemed ready to stand up and get started.

One of the things the report gets wrong is that it predicts this sleepy, sometimes wilful ignorance of the problem would stop pretty soon, as climatic changes start to bite. What's more, the CIA report predicts that when countries do try to fight back against climate change, this will manifest itself not against the use of fossil fuels, but rather the technofix of climate modification (cloud seeding and the like). 'Thus, any country could pursue a climate modification course highly detrimental to adjacent nations in order to ensure its own economic, political, or social survival.' This reads today as slightly ageing sci-fi, but it was a reasonably widespread analysis of the problem at the time and arguably a pretty reasonable guess (plus we may yet see it come to pass). When in 1975 anthropologist Margaret Mead pulled together a multidisciplinary group to develop a social science perspective on climate change, her report similarly expects weather modification techs to be at the heart of political fights over climate change, suggesting developing countries should be given support to set up their own weather modification programmes. Something else that Mead and the CIA agreed on was the lack of systems to digest the economic and political impacts of the changes, and ways to alert policymakers to concerns. As we'll see, that would change, although it would take another decade and a half.

The CIA report was initially prepared as a classified working paper. Still, it ended up in the *New York Times* a few years later. By this point, February 1977, the problem of burning fossil fuels was seen more through the lens of the oil crisis at home rather than famine abroad. The climate crisis might still feel remote, the *New York Times* muses, but as Americans feel the difficulties of unusual weather combined with shortages of oil, perhaps this might unlock some change? It reports that both energy and climate experts share the hope 'that the current crisis is severe enough and close enough to home to encourage the interest and planning required to deal with these long-range issues before the problems get too much worse'. As we'll see, the oil crisis would shake things up a bit, but if anything debate on climate change in the last third of the twentieth century would be characterised as much by delay as concern, not least because of something the CIA report seems to have missed – fightback from the fossil fuel industries.

* * *

The oil industry had moved on from the days of the chairmen of Shell and Standard Oil cooking up deals over a spot of pheasant shooting in the Scottish Highlands. In September 1960, an Organization of the Petroleum Exporting Countries (OPEC) had been founded at a meeting in Baghdad. It was initially led by five oil rich countries – Iran, Iraq, Kuwait, Saudi Arabia and Venezuela – though it currently has 13 member states; a cartel of sorts, but one to get around the cartels the oil companies were already dealing in (including the colonial politics they'd been built in). OPEC started hiking up prices in the early 1970s and for America this exposed a rather larger problem. The number of oil rigs had been declining steadily since the mid 1950s and crude oil production had peaked in 1970. In 1971 Nixon warned that the US was entering a period of increased energy demand, but also shorter supplies, and by 1972 the word 'crisis' was commonly used to describe the issue. In January 1973, *Newsweek* ran 'The energy crisis' as its cover story. The nuclear industry saw an opportunity to push an argument it'd been making for years, offering itself as the solution. Others in the energy business complained about excessive environmental regulations when it came to exploiting fossil fuels, bristling at the ways protests from both Native Americans and conservationists had delayed a giant oil pipeline being built through Alaska.

In April 1973, Nixon addressed the nation, warning Americans to prepare themselves for energy shortages and increases in energy prices. 'As America has become more prosperous and more heavily industrialised, our demands for energy have soared.' It had 6 per cent of the world's population and yet consumed almost a third of the energy: 'If recent trends continue unchecked, we could face a genuine energy crisis. But that crisis can and should be averted, for we have the capacity and the resources to meet our energy needs if only we take the proper steps – and take them now.' These steps largely involved drilling and digging. Natural gas was, in the president's words, 'America's premium fuel', the least-polluting option, but coal and oil were still on the table too. He cut regulations, opening up the Outer Continental Shelf (submerged lands off the coast of the US) for the oil and gas industry, tripling the acreage leased for drilling there by the end of the decade. The legal barriers to that pipeline in Alaska were removed, offered as at least a part-solution to the problem. Also called for an expansion of nuclear, hoping it might be bubbling up a quarter of the nation's electricity by 1985, half by the year 2000.

Running alongside OPEC was a group of Arab oil-rich states (the Organization of Arab Petroleum Exporting Countries, or OAPEC). In October 1973, this new group decided to flex some of its political muscle, issuing an oil embargo on nations supporting Israel in the Yom Kippur War. For the first time in decades, Americans couldn't simply have as much fuel as they wanted. Finding themselves in long queues at the gas station, they were horrified. Nixon announced $10-billion research and development programme Project Independence. When it came to any ideas that Americans might cut their use of energy, the White House turned off floodlights on the Lincoln Memorial at 10pm and twisted its thermostat a few notches to the left, asking citizens to follow. But energy efficiency was largely left for individuals and local communities to decide for themselves. As historian of American energy David Nye notes, in the end there was little effort made to shift energy use in the 1970s. In fact, electricity use increased 50 per cent over the decade, which was less than the 100 per cent increase of the 1960s, but still growth in a period of apparent scarcity.

On the whole, solutions to the oil crisis (and, on the rare occasion it might catch a ride, climate change too) were on the technofix end of things, with nuclear the key winner. Indeed, one of the many people pushing nuclear as a way out of both reliance on foreign oil and growing worries about fossil fuel pollution was Professor technofix himself, Alvin Weinberg. In the 1970s, the number of reactors increased from 15 to 74, producing 10 per cent of US electricity by 1980. However, at the same time, there was a growth of anti-nuclear feeling, exacerbated by an accident at Three Mile Island in Pennsylvania and the disaster film *The China Syndrome* at the end of the decade. Although the fossil fuel industries might point to their technically advanced methods for extraction as a shovel-ready technofix to oil shortages, controversies around strip-mining exacerbated coal's already rather grubby reputation and – between smog pollution and high-profile oil spills in 1969 and 1971 – it was hard for oil to play itself as the progressive way forward (although gas would try this game). Westinghouse promoted monorails for transport and heat pumps for homes, but neither made their way to the mainstream. Solar and wind captured the public's attention, but they weren't ready to compete economically.

The new political, economic and cultural context of the 1970s brought some expansion of the debate over how we might do tech well, wriggling out of Weinberg's binary of tech optimists vs the tech pessimists and instead asking which techs, where, why and who'd own it? First published in 1972, *Undercurrents* styled itself as 'the magazine of radical science and people's technology'* and soon

---

* The early editions of *Undercurrents* were produced as a collection of printed leaflets and articles, all collected in a bag. The idea being that an ordered contents page seemed to run counter to the philosophies of decentralisation underpinning the project, the bag would act as a 'common carrier' of the different items inside, and readers could pick and choose the order they read it in. As a result, complete early copies of *Undercurrents* are very rare.

became the in-house magazine for the alternative technology movement. With instructions for how to build your own wind turbine, and festivals where you could hang out in a field and play with technology outside of the structures of industrial production, it was part William Morris, part *Popular Mechanics*. Economist Ernst 'Fritz' Schumacher made a splash with his 1973 book *Small is Beautiful*, promoting the idea of 'appropriate technology' championing small-scale, decentralised and energy-efficient technologies compared with large-scale grids and production lines. He'd spent his career as a statistician for the Coal Board in the UK and was now promoting solar heating, small hydro and small-scale windmills; energy tech that was easier to control at a local level. Around the same time a whole Centre for Alternative Technology was built in a disused slate quarry in mid-Wales. You can still visit today, entering via their water-balanced funicular.

In north London, a group of workers at Lucas Aerospace responded to the threat of redundancy with a report suggesting they restructure the business to what they called 'socially useful projects' including wind turbines and heat pumps, rather than supplying military contracts. It was met with less than a warm reception from the management, but the so-called 'Lucas Plan' would go on to inspire others to question the ways decisions are made about technology (in particular who gets to make those decisions, for what end). Amory Lovins, a physicist who'd quit the University of Oxford when they wouldn't let him do a PhD on energy and was acting as the British representative for Friends of the Earth, wrote an influential article for *Foreign Affairs* in 1976 attacking the more 'brittle' systems like nuclear favoured by US energy strategists, and instead championed 'soft' paths of recycling and renewables. Still, although 'system change not climate change' would later become a refrain of the environmental movement, at the time, no one with much power seemed up for the disruption.

The oil crisis would help wind power make the leap from small-scale generation, largely useful in 'off-grid' contexts, to an energy source that might compete with fossil fuels, hydro and nuclear. The Danish government built on the ideas and networks Poul la Cour had set up at the turn of the century, leading to growth in the industry there and improvements in wind turbine design. Over in the US, William Heronemus, a former navy captain who'd specialised in the design of nuclear submarines before setting up a department of ocean engineering at the University of Massachusetts, had been worried about an energy crisis from the late 1960s. After flirting with nuclear he decided it was too expensive and instead started talking about 'grand scale renewables', stumbling across Palmer Putnam's book on his failure of big wind at Grandpa's Knob. Heronemus even suggested building big turbines out at sea, moored 200 miles off the coast of Cape Cod on special wind ships (Heronemus was a birdwatcher and figured this was better for wildlife). Where once Americans had mined the seas for whale oil, now they could reap the winds.

As worries around the oil crisis intensified, the National Science Foundation noticed Heronemus's work, and convened a workshop, inviting him along with veterans of the American wind industry like Putnam and Marcellus Jacobs, eventually pulling in money from NASA and the Department of Energy for a multi-million-dollar wind energy research project. Lockheed and General Electric saw business opportunities too, and although pretty much everyone dismissed the offshore idea for now, they agreed with Heronemus's idea that turbines could be grouped together as 'farms' and, like a large hydro plant, feed substantial amounts of energy into the grid. In 1980 the world's first wind farm opened in New Hampshire. There were problems with television interference, so they had to turn the blades off during prime-time viewing (before finally giving in and buying local people cable subscriptions), but on the whole it was a success. Cuts under Reagan got in the way, but his old state of California would step in with generous tax incentives, leading to a small 'wind rush' that Danish companies found themselves well placed to cash in on. With this new opportunity turbines became even more efficient, larger and more powerful, eventually leading to the first offshore wind farm in 1991 and the incredible sea giants of turbines we have today.

Solar made some progress, although it would take a while longer before it ever caught up with the dreamy-eyed hype surrounding it. The market for solar in space brought prices down when it came to more terrestrial applications, and by 1973 the Exxon-funded Solar Power Corporation was making silicon panels for $10 a watt. Still, that was way too expensive for mainstream use. Exxon's chairman Clifton Garvin made a big deal of how he heated his family swimming pool with solar power, but it wasn't really a challenge to his oil. The Solar Power Corporation tried selling panels to the coastguard, but its main application ended up back at Exxon, on offshore oil rigs. By the end of the 1970s, solar modules had been sold to pretty much all the major oil companies – not just Exxon, but Shell, Texaco, Chevron and more. It turned out it was useful in drilling for gas in Kansas too.

When Jimmy Carter became president in 1977, he supported a programme of solar research at the Solar Energy Research Institute (SERI) in Golden, Colorado. The founding director of SERI had been Paul Rappaport, an established expert on solar energy who'd spent two decades in the RCA labs and already had over a dozen patents to his name. A profile in the *New York Times* introduced the president's new 'solar advocate', saying he'd first rolled up his sleeves to look into solar energy back in 1960 and they'd stayed rolled up ever since. As Golden was also home to Coors beer, Rappaport quipped they'd be cooling the Coors with solar energy. Congress designated 3 May 1978 'Sun Day' with sunrise services, 'sunrise soup' of various orange and yellow fruits, songs and films about the Sun, and Sun salutations in yoga classes. Carter visited SERI, promising to put solar thermal panels on the roof of the White House, and announcing an extra $100 million for solar energy research. Sadly, it rained for Carter's visit, and SERI never really fulfilled its promise for renewables R&D. It never had quite enough funding, and had to be split between a long wish list of projects that included not only ways to make electricity or store heat from the Sun, but also explore tapping the heat in oceans, growing algae for fuel and wind power too (Rappaport argued wind came from the Sun, so might as well be on the list). There were criticisms of Rappaport too – he who was seen as rather chaotic and was asked to leave not long after the unveiling of the White House panels. His replacement was Denis Hayes, the activist who'd helped found Earth Day in 1970.

Heading into the 1980s, SERI had $131 million and a five-year plan that would give solar PV the bulk of its budget, aiming to bring down the costs year on year in the way computer chips had in previous years. But as with wind, the change at the White House would shift things. Hayes had been hopeful initially. Regan had spoken supportively on decentralised energy and there was a lot about off-grid solar that could appeal to American conservatives. But a nuclear advocate was appointed to head the Department of Energy and SERI's budget was slashed. This signalled to other investors that a bet on solar wouldn't be supported by the government, making it harder to gather momentum elsewhere too. On the morning of the summer solstice, 1981, Hayes resigned and went back to university to qualify as a lawyer, but not before he penned a scathing op-ed for the *New York Times*. 'Unless pro-solar Americans, the solid majority, put their senators and representatives on the line over the next few weeks, insisting that solar energy not be discriminated against,' arguing in its conclusion, 'this fall will see the Federal solar programme quietly eclipsed.' For all that people loved solar, no one had managed to give that support a political expression, not even Mr Earth Day, Denis Hayes. It would be a problem that would continue to plague renewables' advocates, indeed campaigners for climate action at large. People like Hayes want to see a shift away from fossil fuels, but they aren't quite sure how to go about asking. For years, the public had been told not to worry about energy, that it could be buried in wires and pipes out of sight and we just needed to flick a switch.

* * *

The 1970s also brought an expansion of science's understanding of greenhouse gases. Worries that supersonic planes might be injecting nitrogen oxides and other pollutants into the atmosphere had been a big topic at the 1971 Man's Impact on Climate conference in Stockholm. In 1973, Mario Molina and Sherwood Rowland, chemists at the University of California, Irvine, were surprised to find the incredible warming potential for chlorofluorocarbons (CFCs), a group of compounds used in the production of aerosols and refrigerants. These had been thought to be environmentally 'safe' because they were very stable, not reacting with animals and plants, but this same stability made them dangerous as they could linger in the atmosphere for centuries, drifting up into the stratosphere where they could react with ultraviolet rays to destroy ozone. Journalists alerted the public, who called for a ban on CFCs. It didn't have a direct link to climate change but did show the fragility of the atmosphere. Moreover, it inspired Veerabhadran Ramanathan at NASA to look further at CFCs and in 1975 he reported that they too were a potent greenhouse gas. Other scientists followed up with other gases that hadn't been looked at in detail – methane and nitrates, especially the nitrous oxide emitted in agriculture. It wasn't simply 'the carbon dioxide problem' any more.

Not all climate change scientists agreed that warming was the problem though. Indeed, one of the reasons that 1974 CIA report feels dated now is that it talks about both the idea of global warming or cooling. Remembering the 1971 Stockholm workshop a decade and a half later, meteorologist William Kellogg recalled that by the end of the three weeks he could see two opposing schools of thought laid out: 'The climate "coolers" and the climate "warmers", if you will.' This he put down to whether a scientist came from what he called the 'atmospheric particle or aerosol camp' (who thought dust from industry or agriculture would scatter sunlight back into space, causing cooling) or the 'carbon dioxide and infrared-absorbing gases camp' (closer to the approach to global warming accepted today). Although global temperatures had gradually crept up in the first half of the twentieth century, there had been some observable cooling post-war, feeding both scepticism about the impact of carbon dioxide and inspiring journalists and scientists alike to speculate about the return of a new ice age. At least for some, Callendar's concern over the return of the deadly glaciers was back. The delegates called an evening meeting to try to thrash out a consensus, but there were just too many differences of opinion. Moreover, there was a reluctance to make any predictions at all about the future, to 'stick out one's neck' in the face of what was still seen as a reasonably new and uncertain science. The ghost of FitzRoy was haunting them perhaps (or maybe it was the ghost of Galton, standing ready to stick the boot in).

One of several researchers to have picked up Dangaard's new history hidden in ice cores was Wallace Broecker. He'd been one of the scientists applying radiocarbon to the ocean back in the 1950s, and in 1965 had sat with Revelle and Keeling on the carbon dioxide subcommittee for the PSAC report on the environment. A dyslexic, he always worked with pencil and paper, often drawing graphs by hand too – 'I understand the data better this way,' he would say. He'd been looking at the long records of climate buried in the deep-sea mud, and in 1970 had introduced the prospect of abrupt climate change – large ice sheets might take tens of thousands of years to grow, but they could melt rapidly too. In 1975 he addressed the ongoing fight between the coolers and the warmers using Dangaard's ice core studies to argue that any recent cooling was just one of a series of natural climatic fluctuations. This cooling had masked the warming effect of carbon dioxide, but as the cooler period ends, he warned, the impact of the greenhouse effect will be all the more noticeable: 'This compensation cannot long continue both because of the rapid growth of the CO_2 effect and because the natural cooling will almost certainly soon bottom out. We may be in for a climatic surprise.' The paper's title 'Are we on the brink of pronounced global warming?' is generally seen as the first use of the term 'global warming',* although it was far from the end of that debate.

In 1976, a young climate modeller named Stephen Schneider decided it was time for someone in the climate science community to stick their neck out and make a public splash. As a grad student at Columbia University, Schneider wanted to find a research project that could make 'a difference' to the world. He'd hang out at the NASA Goddard Institute for Space Studies and it was there he stumbled on a talk on climate models. He was inspired: 'How exciting it was that you could actually simulate something as crazy as the Earth, and then pollute the model, and figure out what might happen – and have some influence on policy in a positive way,' he later recalled. He met Kellogg and managed to wrangle an invitation

---

*Broecker hated the idea that he might be remembered for this and offered $200 as a prize to anyone who could find an earlier use of the phrase 'global warming'.

to act as a rapporteur for the 1971 workshop in Stockholm. Initially working on how the climate might cool through aerosols, he'd been convinced by team warming and, after years of headlines about droughts and famine, figured the time was right for a popular science book on the danger climate change could cause, so he wrote *The Genesis Strategy*. Although he wanted to avoid positioning himself alongside either what he called the 'prophets of doom' on one side or the 'Pollyannas' on the other, he felt it was important to impart the gravity of climate change and catch people's attention.

And it got attention, with a jacket endorsement from Carl Sagan, reviews in the *Washington Post* and *New York Times*, and an invitation to Johnny Carson's *The Tonight Show*. This rankled some of the old guard, who felt this just wasn't the way to do science. Schneider's book drew an especially scathing attack from Helmut Landsberg, the scientist we met at the start of Chapter Eight, sprinkling soot on ice in 1940. He had moved on to direct the Weather Bureau's office of climatology (via work with Rossby and the Air Force) and was now a well-respected professor at the University of Maryland.* Landsberg reviewed the book for the American Geophysical Union, calling it a 'wide-ranging potpourri of science, nature, and politics' which 'is multidisciplinary, as promised, but it is also very undisciplined'. Complaining Schneider never really got to grips with the basic science, Landsberg wrote that he'd relied too much on newspapers and magazines rather than science: 'Conjectures are elevated to facts. Inadequate information is clothed in the mantle of scientific authority.' Landsberg disliked what he saw as an activist spirit in Schneider, believing that climate scientists should stay out of the public spotlight, especially when it came to the uncertainties of climate modelling. He'd only endanger the credibility of climatologists, he worried; much better to stay collecting data to iron out as many uncertainties as possible, only guardedly briefing politicians behind closed doors when absolutely needed. With first-class scientific bitching, Landsberg concludes his review by noting

---

* This spat is outlined in detail in Joshua Howe's accessible history of climate change, *Behind the Curve* (2014). Stephen Schneider's own book, *Science as a Contact Sport* (2009) is also worth a read.

that Schneider advocates that scientists run for public office and perhaps he'd better try that, but if he does want to be a serious scientist 'one might suggest that he spend less time going to the large number of meetings and workshops that he seems to frequent' and join a scientific library.

Their disagreement was partly down to style of science communication and a difference of view over computer modelling, but there was a generational issue here too. Schneider was a part of the new, rebellious generation who'd been inspired by Earth Day and wasn't scared to take science to the streets. Landsberg had spent a career working carefully with government and the military, generally behind closed doors, and was scared about what public involvement might do to this relationship. What's more, the cultural norms of scientific behaviour that expect a 'good' scientist to be guarded and avoid anything that even smells the slightest of drama are deeply embedded even if, like any deeply embedded cultural norm, they can skew the science.* Landsberg was far from the only established meteorologist bristling at all this new attention given to climate change. Some felt uneasy about the drama, others didn't trust the new technologies, disciplines and approaches being used. Many had also spent many years rebutting hype and conspiracy theories about cloud seeding. You could see how that might harden you to the new fears of fossil fuel warming. In the UK, the head of the Met Office John Mason called concern about climate change a 'bandwagon' and set about trying to 'debunk alarmist United States views'. In 1977 he gave a public talk at the Royal Society of Arts stressing there are always fluctuations in climate and the recent droughts were not unprecedented. He agreed that if we were to continue to burn fossil fuels at the rate we are, then we might have 1°C warming, which he thought was 'significant', sometime in the next 50–100 years, but on the whole the atmosphere is a system that

---

* Concern they might be criticised for over-dramatising the issue is sometimes given by members of the scientific community as a reason to be even more guarded but, as Keynyn Brysse and others reported in 2013, it appears that if there is a bias, it's towards under-predicting the rate and extent of climate change – 'erring on the side of least drama'.

will take what we throw at it.* Plus, he figured, we'd all move over to nuclear anyway. Writing up the talk for *Nature*, John Gribbin described the overall message as 'don't panic'. With a line possibly suggested by his editor, he reassured there was no need to listen to 'the prophets of doom'.

★ ★ ★

Change was coming though. As Nathaniel Rich argues in his gripping (2019) book *Losing Earth*, it would be a combination of an establishment scientist and an activist that would kick it off.† A lobbyist at the DC offices of Friends of the Earth, Rafe Pomerance, was reading a reasonably obscure 1978 Environmental Protection Agency (EPA) report on coal. It mentioned the 'greenhouse effect', noting fossil fuels could have significant and damaging impacts on the atmosphere in the next few decades. He hadn't heard of this greenhouse effect. He asked around the office and someone handed him a recent newspaper article by a geophysicist called Gordon MacDonald. MacDonald was a high-ranking American scientist who had worked on weather modification in the 1960s as an advisor to President Johnson and was one of the scientists flown in to help deal with the 1969 Santa Barbara oil spill. He'd written an essay back in 1968 called 'How to wreck the environment', imagining a future where we've resolved threats of nuclear war but people have weaponised the weather. Since then he'd watched people do this, not deliberately as a means of war, but more carelessly, simply by burning fossil fuels. More importantly, MacDonald was also something called a Jason.

---

* According to Jon Agar's 2015 paper on the UK's early response to climate change, when veteran science advisor Solly Zuckerman heard of Mason's scepticism, he remarked: 'Everyone knew that Dr Mason did not believe in long-term planning or indeed looking ahead at all.' Which is quite the burn when directed at a nation's chief weather forecaster.

† Rich manages to make what could be a rather boring period in the story of the climate crisis (his argument is partly that not enough happened) read like a political thriller. His interviews provide vital context to understand this period and it's a blistering read; a book I find myself regularly recommending.

The Jasons were a secret group of elite scientists brought together after the shock of *Sputnik* to give the government advice, outside of the public eye. Or at least they were secret until they were outed, to some public controversy, by the *New York Times* in the 1971 Pentagon Papers. However, they continued to meet every summer and prepare briefings for the government.[*] Whether it was because MacDonald was concerned by Jimmy Carter's plans to research high-carbon synthetic fuels as a way out of the oil crisis, or the Department of Energy (DoE) was as worried about changing weather as the CIA and asked the Jasons to look into it (reports differ), the group met to discuss carbon dioxide and climate change over both summers of 1977 and 1978, issuing a report co-authored by MacDonald the following year.

You might imagine there was some culture clash between this Friends of the Earth lobbyist Pomerance and secret military scientist MacDonald, but they made for a powerful combo. Pomerance would later remember MacDonald loved to eat at French restaurants, and they had a special place they'd go together for lunch. 'He was always very patient with me' he told a 2020 special Q&A for the American Academy of Arts and Sciences, 'explaining the different pieces of the climate problem.' They got a meeting with Frank Press, the president's science advisor, who brought along the entire senior staff of the Office of Science and Technology. After MacDonald outlined his case – telling the story of Tyndall, Arrhenius, Callendar, Revelle and finally Keeling's ever creeping curve – Press said he'd ask Jule Charney to look into it. If Charney said a climate apocalypse was coming, the president would act.

Charney summoned a team of scientists and officials, along with their families, to a large mansion at Woods Hole, in the southwest corner of Cape Cod. Their question was how sensitive the atmosphere was to carbon dioxide – is it a crushed butterfly level of sensitive or do you need a rhino's worth of carbon dioxide before you'll see much

---

[*] For more on the Jasons, see Ann Finkbeiner's *Secret History of Science's Postwar Elite* (2007).

warming? The Jasons' report had built a model that showed if you doubled the concentration of carbon dioxide in the atmosphere from what it had been before the Industrial Revolution, you'd get an increase of average surface temperature of 2.4°C, with rises as high as 10°C or 12°C at the poles. This wasn't especially new, and the doubling was rather arbitrary, but it provided an example that you could use to talk to politicians – and the Jasons knew politicians better than most scientists. Charney's brief was to assemble atmospheric scientists to check the Jasons' report, and he also invited two leading climate modellers to present the results of their more detailed, richer models; James Hansen at the Goddard Institute for Space Studies and Syukuro Manabe from the Geophysical Fluid Dynamics Lab in Princeton (the successor to the computer project Charney had set up with Harry Wexler and John von Neumann all those years ago).

The scientific part of proceedings was held in the carriage house of the mansion, the scientists on a rectangle of desks in the middle with political observers around the side. They dryly reviewed principles of atmospheric science and dialled in Hansen and Manabe. The models were different, providing two different warnings. After one of the team, Akio Arakawa, a pioneer of atmospheric modelling, sat up late in his room with a load of printouts, they decided to split the difference; if you double the amount of climate change in the atmosphere you'll have about 3°C warming, with an error of 1.5°C either side (that is, if you're lucky it might stay at 1.5°C, but it could be over 4°C too). The final report was unequivocal: 'We have tried but have been unable to find any overlooked or underestimated physical effects.'. They had high confidence that the Earth would warm by about 3°C in the next century, plus or minus 50 per cent (that is, we'd seen warming between 1.5°C or 4°C). *Science* magazine declared 'Gloomsday predictions have no fault'.

★ ★ ★

Back at Scripps, Keeling had managed to keep his curve going through the late 1970s, just about. A new opportunity came in 1977 when the DoE formed a study group on the Global Effects of Carbon Dioxide.

The group was based in an offshoot of the Atomic Energy Commission, and partly driven simply by an interest in nuclear power. The group was chaired by Alvin Weinberg, and looking back in the late 1990s, Keeling would say Weinberg didn't hide his motive. Nuclear power had some vocal critics by this point, but the group wanted to suggest 'the burning of fossil fuels might be more dangerous to mankind than any perceived side-effects of nuclear energy. It was time to find out.' The study group recommended the DoE sponsor a Carbon Dioxide Effects Research Programme and an Office of Carbon Dioxide Research was also set up, with scientists invited to help draft a research programme, which included $200,000 a year for carbon dioxide monitoring (i.e. Keeling).

However, the work in the DoE was seen as too bureaucratic, running study groups and workshops about research rather than doing it. 'If anything has been meetinged to death it's CO_2,' one administrator remarked. 'If conferences could solve problems, it should be solved by now.' By the end of the 1970s, the DoE work had got stuck, with the carbon dioxide monitoring work assigned to the National Oceanic and Atmospheric Administration, which wasn't exactly Keeling's biggest fan. The National Science Foundation (NSF) was terminating its support too. At the start of 1981 it seemed as if Keeling had run out of possible sponsors. A reprieve came at the start of 1981, when the DoE again agreed to pick up the bill, but it wasn't to last long. This was also the year Ronald Reagan became president. His administration was eager to subdue what they saw as 'alarmist' environmentalism. The DoE's climate research made a juicy target for cuts, it was easy for the White House to ask why the NSF wasn't funding it anyway?

Help, however, came from a new senator, Al Gore. In the mid 1960s, he'd taken an undergraduate course at Havard on population taught by Roger Revelle. Revelle had shared the first few years of Keeling's data with his students and it had startled Gore. Ever since, Gore had watched the Mauna Loa data come through. He held several hearings on climate issues in the early 1980s, which helped throw a public spotlight on the threat to the carbon dioxide monitoring programme. Embarrassed by media attention, the Reagan administration backed off and a small carbon dioxide monitoring budget was kept at the DoE.[*]

* * *

Meanwhile, Jersey Standard, the biggest tentacle in J. D. Rockefeller's old SO octopus, changed its name to Exxon and was starting to wonder if climate change might finally be about to arrive on the political agenda, messing with its business model. Maybe it was the CIA report. Or the reference in the Kissinger speech. Or Schneider's appearance on *The Tonight Show*. Or maybe it was just that the year 2000, that point after which scientists warned things were going to start to hurt, didn't seem quite so far off. In the summer of 1977, James Black, one of the top science advisors at Exxon, made a presentation on the greenhouse effect to the company's most senior staff. This was a reasonably big deal, executives at that level would only want to know about science that would affect the bottom line. The same year, they hired Edward David Jr to head up their research labs. He'd spent two decades at Bell Labs, a leader in the sort of expansive basic science Van Bush advocated, and had also learnt about climate change while working as an advisor to Nixon. Under David, Exxon started to build a small research project on carbon dioxide.

At least it was small by Exxon measures. At $1 million a year, it was a good chunk of cash, just not all that much compared with the $300 million a year it spent on research at large. In December 1978, Henry Shaw, the scientist leading Exxon's carbon dioxide research, wrote in a letter to David that Exxon 'must develop a credible scientific team', one that can critically evaluate science that comes in on the topic, and 'be able to carry bad news, if any, to the corporation'. Exxon fitted out one of its largest supertankers with custom-made

---

[*] Cuts under President George Bush Jr in 2013 threatened it again, and the Curve, by now run by Dave's son, Ralf Keeling, appealed to the public, launched a crowdfunder and went online to increase awareness of the work. As one science journalist quipped: 'The end of the world gets a Twitter account.' This raised $20,000, just enough to help keep staff on during some very lean months but, moreover, drew enough publicly for larger funding to come in.

instruments to do ocean research and recruited Broecker and his colleague Taro Takahashi to work with them. Exxon wanted to be taken seriously as a credible player, so wanted leading scientists, and was willing to ensure Broecker and Takahashi had scientific freedom.[*] Another one of its projects included running radiocarbon analysis on 100 bottles of vintage French wine. When *Inside Climate News* went through the files in 2015 they found *New York Times* wine reviews of classic Bordeaux vintages, including one for a $300 bottle from 1945. That's how the oil barons do climate research, with vintage fine wines and massive supertankers.[†]

In October 1982, David told a global warming conference financed by Exxon: 'Few people doubt that the world has entered an energy transition away from dependence upon fossil fuels and toward some mix of renewable resources that will not pose problems of CO_2 accumulation.' The only question, he said, was how fast this would happen. Riffing off Roger Revelle's 'experiment' line from the 1950s, giving it a brighter spin, he said: 'I'm generally upbeat about the chances of coming through this most adventurous of all human experiments with the ecosystem.' Maybe he really saw Exxon as about to lead the way on innovation to zero carbon fuels, with his R&D lab at the centre of it. Or maybe the enormity of the challenge hadn't really sunk in. Or possibly he was just enjoying the vintage wine. It wasn't to last long though; despite increasing the overall R&D budget, by the mid 1980s the carbon dioxide work had been seriously curtailed.

★ ★ ★

[*] This work would be used by Takahashi in 1990 for a paper showing land-based carbon sinks, like forests, absorb more than the ocean. He'd use it again in 2009, for work concluding the oceans absorb only 20 per cent of the carbon dioxide emitted from human activities; work that earnt him a Champion of the Earth prize from the United Nations.

[†] This story and many more on the oil industry's climate change science was unearthed in a 2015 series by *Inside Climate News* – 'Exxon: The road not taken' and can be read in full on its website.

In June 1980, the Energy Security Act included a call for a multi-year, $1 million study called Changing Climate to explore some of the unanswered questions in the Charney report. It also invited two dozen or so experts to a meeting that October, to be held at the Don CeSar Hotel in St Petersburg, Florida. Known locally as the 'Pink Palace', the hotel looks a bit like an overgrown sandcastle painted with candyfloss. Built in the Jazz Age, in its heyday it attracted the likes of F. Scott Fitzgerald and Al Capone and is said to be haunted by the ghost of its founder. The meeting was just before Halloween, and it must have felt quite surreal as a setting to turn the maths of climate change into policy recommendations. As Nathaniel Rich describes, the invitation list included Henry Shaw from Exxon, David Slade from the DoE, Rafe Pomerance from Friends of the Earth as well as a range of other experts, engineers, economists, EPA staffers and local politicians, all asked to explore proposals for what to do about climate change. A copy of the Charney report was left on each of their chairs, which was fine, but it only told some of the problem, not what to do about it and no one knew where to start. John Perry, a meteorologist who'd helped on the Charney report, suggested they try the post-mortem trick and work backwards. Imagine they were all sitting in the early-twenty-first century, the other side of those big 'by the year 2000' statements, and we'd sorted it. How did we do that? They all agreed some sort of international treaty was needed, but were not sure about what or how it would be achieved. Pomerance started to ache with frustration. He wanted to be out building a movement demanding action, not this rather inert brains trust at a pale pink Florida hotel.

A few days later, Reagan was elected. He appointed lawyer James G. Watt to run the department of the interior. Watt had headed a legal firm that fought to open public lands for drilling and mining, and already had a reputation for opposing conservation projects, as a matter of policy and of faith. He once famously described environmentalism as 'a left-wing cult dedicated to bringing down the type of government I believe in'. The head of the National Coal Association pronounced himself 'deliriously happy' at the appointment, and corporate lobbyists started joking: 'How much power does it take to stop a million environmentalists? One Watt.' He didn't, as people initially worried, simply close the EPA, but he did appoint Anne

Gorsuch, a famously tough and well-manicured anti-regulation activist who cut the Agency by 22 per cent, boasting she'd shrunk the book of clean water regulations from six inches thick to a mere half inch. Pomerance and his colleagues in the environmental movement were going to be busy. They'd make a certain amount of lemonade from the problem – membership of the Sierra Club more than doubled in the first two years Watt was in office – but it didn't exactly leave much time for picking up that lingering and still quite abstract problem of climate change. It'd still be a while before Pomerance would see a public movement for climate action.

Just before the Florida meeting and the November 1980 election, the National Academy of Sciences (NAS) had set up a new Carbon Dioxide Assessment Committee to do a larger follow up to the Charney report. The chair was Bill Nierenberg, who'd followed in Revelle's footsteps and become director of Scripps. One of the generation of scientists who'd been through both the war and the post-war science boom, he was quite at home working with the military, he was even a Jason. He'd been a fierce defender of the Vietnam War, which had put him apart from some of his colleagues, and he was still bitter about some of the left-wing protests on campus at the end of the 1960s, and the push-back against military-sponsored science they'd inspired. He also hated the environmentalist movement, which he saw as a band of luddites, especially on the issue of nuclear power. In many ways, he must have seemed like the perfect person to lead a review that would report back to the new President Reagan.

Nierenberg decided to build his report around a mix of economics and science. In theory, this should have been brilliant; the sort of multidisciplinary approach to the problem Margaret Mead had tried to kick off with her conference in 1975. But when it came to publication, the two sides didn't cohere. The writers hadn't worked together, but rather been sent off to be scientists in one corner and economists in another. It's been described as a report of two quite different views – five chapters by scientists that agreed global warming was a major problem, and then two more by economists that focused on the uncertainty that still existed about the physical impacts, especially beyond the year 2000, and even greater uncertainty still about how this would play out economically. What's more, it was the economists' take on things that got to frame the report as the first and last chapters, and whose analysis dominated the overall message. Back in 1963 the Conservation Foundation had cautioned about a wait-and-see approach and yet, three decades later, that was pretty much what Nierenberg seemed to be advocating. There is no particular solution to the problem, he argued at the start of the report, but we can't avoid it either so, 'We simply must learn to deal more effectively with their twists and turns as they unfold.'

For their 2010 book on climate scepticism, *Merchants of Doubt*,[*] Naomi Oreskes and Erik Conway dug out the peer review notes in the NAS archives. One of the peer reviews was Weinberg – and he was less than impressed. It might be better to say he was appalled by the stance Nierenberg had taken. At one point the report had suggested people will probably adapt, largely by moving. People had migrated because of climate change in the past, it argued, and they'll manage again: 'It is extraordinary how adaptable people can be,' the report muses. Weinberg was scathing: 'Does the Committee really believe the United States or Western Europe or Canada would accept the huge influx of refugees from poor countries that have suffered a drastic shift in rainfall pattern?' Oreskes and Conway did some digging into the reviews and note Weinberg's wasn't the only negative one (although the others were slightly more polite). Puzzled as to why these criticisms weren't responded to, a senior scientist later explained to them: 'Academy review was much more lax in those days.'

In the end, the Commission's report was launched in October 1983, with a formal gala with cocktails and dinner at the Academy's cathedral-like Great Hall. As Nathaniel Rich notes, Peabody Coal, General Motors and Exxon were all on the invite list, whereas Pomerance from Friends of the Earth had to sneak in via the press conference. The White House had briefed the Academy from the get-go, making it clear it didn't approve of speculative, alarmist or 'wolf-crying' scenarios; that it thought technology would find the answer and it didn't expect to do anything other than fund research

---

[*] Justifiably a classic within months of its first publication, this tells the historical story of climate scepticism and manages to weave in what could be rather turgid philosophy of science with great clarity. Highly recommended.

and see what happened. The Academy knew these people would be in charge for the next few more years, and possibly figured the best idea was to give them the most scientific version they could find of what the White House wanted. Or possibly it simply was what Nierenberg believed. Either way, looking back, it's hard not to see it as a pretty big misstep.

The report's introduction stated up front: 'Our stance is conservative: we believe there is reason for caution, not panic.' At the press conference, Revelle told reporters they were flashing an amber light, not a red one. And so, the *Wall Street Journal* reported: 'A panel of top scientists has some advice for people worried about the much-publicised warming of the Earth's climate: You can cope.'

CHAPTER TWELVE

Already Happening Now

In 1988 there was yet another heatwave in the US. It wasn't just hot. It was weirdly hot. Again. In late June, the climate modeller James Hansen from the Goddard Institute for Space Studies had been invited to testify to another senate hearing. As the media choked in the oppressive DC heat, they were told the man from NASA was going to make a 'major statement' about the latest temperature records.

The hearing was introduced by J. Bennett Johnston, a senator from Louisiana and chairman of the committee on energy and natural resources. 'The greenhouse effect has ripened beyond theory now. We know it is fact,' he told his audience. 'What we don't know is how quickly it will come upon us as an emergency fact, how quickly it will ripen from just simply a matter of deep concern to a matter of severe emergency.' Senator Dale Bumpers of Arkansas then introduced Hansen, promising his statement 'ought to be cause for headlines in every newspaper in America tomorrow morning'. He meant this rhetorically – that it ought to cause a stir, but probably wouldn't. Hansen took the microphone. His testimony was clear. The temperature was the highest on record, the rate of warming in the past 25 years was also the highest on record – the four warmest years all having been in the 1980s – and even though it was barely halfway through the year, it had been so hot it'd take 'remarkable and improbable cooling' for 1988 not to be declared the new warmest year on the record. 'Global warming has reached a level such that we can ascribe with a high degree of confidence a cause and effect relationship between the greenhouse effect and observed warming,' he stated, underlining: 'It is already happening now.'

Hansen had made a note to himself in the cab on the way there to say 'it is time to stop waffling', that the evidence is strong 'the greenhouse effect is here and is affecting our climate now'. After his testimony he picked up these notes, finding the phrase unused.

'Darn,' he thought, annoyed at himself, thinking he'd missed an opportunity. Phillip Shabecoff of the *New York Times* caught him on his way out and asked if there was a particular temperature point when we'd be able to confirm humans were changing the climate with greenhouse gases. Hansen rebuffed the premise of the question – there wasn't some magic number where we should all sit up – and instead pulled out the line about needing to stop waffling and accept the greenhouse effect is here and now. It made the front page. 'Global warming has begun, expert tells Senate' ran the headline, right in the centre, up top under the masthead, along with a graph of undulating temperatures from the 1880s, the tread very clearly going up. The story was continued on page 14, along with a photo of Hansen looking resolute.

The *New York Times* wasn't the only place covering the story. The resulting coverage was the first time climate science really made a news splash. It'd had the odd bit of coverage – Stephen Schneider on *The Tonight Show*, for example, or the minor-chord dramatic bit at the end of Frank Capra's show on meteorology – but otherwise it had just rolled on in the background. Hansen was invited to appear on a television news programme the following Sunday, and made a special set of large dice for the occasion – one standard die to represent 'normal' weather up until 1980, the other loaded to demonstrate how global warming weighted the climate to more extreme weather events like drought or storms. That night, back home, his wife told him she had breast cancer, she'd put off telling him until after the Senate testimony. Understandably distraught, struggling to string sentences together, Hansen diverted media requests to other scientists, like Schneider, for the time being. But it was hardly the end of the interest he managed to ignite.

* * *

The mix of years of wondering about the weird weather, this deeply hot summer and Hansen's to-the-point 'it is already happening' approach prompted a change in public discourse; global warming started to be talked about as a mainstream issue. As Spencer Weart notes in his 2003 book *The Discovery of Global Warming*, in September

1988 a poll found that 58 per cent of Americans recalled having heard or read about the greenhouse effect, a big jump from the 38 per cent that had said the same in 1981 and a reasonably high level of public awareness for any area of science. Most of these citizens recognised 'greenhouse effect' meant the threat of global warming, and most thought they would live to experience climate changes. Moreover, other polls showed that a majority of Americans thought the greenhouse effect was 'very serious' or 'extremely serious', and they worried 'a fair amount' or even 'a great deal' about global warming.

It wasn't the only environmental issue hitting the news at the time – acid rain, deforestation, that 'hole' in the ozone layer* and endangered species had all made headlines in the 1980s. Membership of environmental groups had grown in most Western countries throughout the decade and they were now a clear political force to contend with. Over in Hollywood, Ted Turner was planning a new cartoon about children from around the world cooperating to battle pollution, *Captain Planet*, signing up Whoopi Goldberg, Meg Ryan, Sting and other luminaries to do the voices. For 1989, *Time* didn't do a person of the year, but ran with 'the Endangered Earth' as Planet of the Year instead. 'This wondrous globe has endured for some 4.5 billion years,' it explained, 'but its future is clouded by man's reckless ways: overpopulation, pollution, waste of resources and wanton destruction of natural habitats.'

UK Prime Minister Margaret Thatcher also came out for the cause, with a couple of major speeches anthropologist Steve Rayner quipped were 'the Iron Maiden's transfiguration to Green Goddess'. Thatcher had initially been sceptical the greenhouse effect was a problem, exclaiming to her science advisor in 1980: 'Are you telling me I should

---

* Antarctic scientist Joe Farman recalls in his Oral History of British Science interview that no one ever owned up to naming it 'a hole', just that NASA did a press release to that effect and the phrase ended up in the *Washington Post*. There's arguably some rhetorical power in the term 'hole'. It feels like something you should actively try to fix (as opposed to a graze, which might heal on its own). Possibly if the press officers at NASA had grasped for a different bit of the English language to describe the phenomenon, the issue would have played out differently.

worry about the weather?' But by the end of the decade she'd come around. In an address to the Royal Society a few months after Hansen's testimony in 1988, she riffed off Revelle's old experiment line: 'For generations, we have assumed that the efforts of mankind would leave the fundamental equilibrium of the world's systems and atmosphere stable. But it is possible that with all these enormous changes (population, agricultural, use of fossil fuels) concentrated into such a short period of time, we have unwittingly begun a massive experiment with the system of this planet itself.' Just over a year later she was speaking at the UN, calling for an international convention on climate change. Here she takes Stewart Brand's space-based image of the Earth and makes it her own, replacing his San Francisco LSD trip with a bit of British Victoriana: 'What if Charles Darwin had been able, not just to climb a foothill, but to soar through the heavens in one of the orbiting space shuttles?' she asks. 'What would he have learnt as he surveyed our planet from that altitude? From a moon's eye view of that strange and beautiful anomaly in our solar system that is the Earth?'

She also made action on climate change her own too, outlining how a right-wing politician might approach the issue. 'We must resist,' she argued, 'the simplistic tendency to blame modern multinational industry for the damage which is being done to the environment.' She strongly distanced herself from left-wing uses of the climate challenge, and instead invited the audience to find salvation in – what else – the market: 'It is industry which will develop safe alternative chemicals for refrigerators and air-conditioning. It is industry which will devise biodegradable plastics. It is industry which will find the means to treat pollutants and make nuclear waste safe – and many companies as you know already have massive research programmes.' Moreover, 'as people's consciousness of environmental needs rises, they are turning increasingly to ozone-friendly and other environmentally safe products. The market itself acts as a corrective, the new products sell and those which caused environmental damage are disappearing from the shelves.'[*]

---

[*] People often put Thatcher's climate moment down to her scientific training. I suspect it's more because she got the politics. She knew that to tackle climate change seriously would require a massive economic shift and didn't want this to provide an opportunity for the left.

Thatcher wasn't the only right-winger momentarily enamoured with a bit of climate rhetoric. On the campaign trail for the 1988 election, George Bush Sr declared: 'Those who think we are powerless to do anything about the greenhouse effect forget about the White House effect.' During the election the topic of climate change made the front page not just of the newspapers, or magazines like *Time* and *Newsweek*, but *Sports Illustrated* too. Two weeks after polling day, former presidents Carter and Ford met with Bush, handing him a copy of 'American agenda', a bipartisan report on issues facing the country, with climate change a major national priority. A gas lobbyist told *Oil & Gas Journal*: 'A lot of people on the Hill see the greenhouse effect as the issue of the 1990s.' He might have been touting for work, but it was also a fair summary.

★ ★ ★

Americans had dominated climate change research in the 1960s and 1970s, but they were far from the only player. Bill Nierenberg's NAS report in 1983 might have drawn a curtain over one stage of American leadership on climate change, but there were others ready to pick up where it'd left off. Back in 1979 there'd been a World Climate Conference held in Geneva, which had launched a new research programme run by the World Meteorological Organization and the International Council of Scientific Unions. Inspired by the assessment Jule Charney had led in 1979, an attempt was made at an international version. This was limited to a week's work in Villach in Austria, and the conclusions didn't add much more detail, but it was the start of something bigger.

On the train ride back from Villach, several of the scientists involved started to chat about how they might more effectively build an international assessment of climate change science for policymakers. In 1982, Swedish meteorologist Bert Bolin took advantage of a visit to Stockholm from Mostafa Tolba – the Egyptian biologist who was heading up the UN Environment Programme – and pitched the idea to him. Together, they set up an international meeting of climate scientists in Tbilisi (then in the USSR) in early 1985. Under constant surveillance from the KGB – which Bolin felt was 'hardly surprising' considering the Soviet Union's carbon emissions – this opened with a powerful speech from Tolba, warning about future disasters if we didn't act on climate change. The speech worried Bolin slightly. He wasn't willing to paint quite such a scary picture of the future while the science was, for him, still uncertain. Typical of many others of his time and profession, he also worried about scientists getting involved in politics beyond what he saw as simply providing knowledge.

Bolin was also one of the scientists involved in drafting the 'Our common future' report, for a commission on environment and development chaired by the prime minister of Norway Gro Harlem Brundtland. Presented to the UN General Assembly in New York in the autumn of 1987, the report painted an image of a rapidly changing world and promoted an idea of 'sustainable development'; that is, meeting the needs of the present (especially those of the world's poor) without compromising the lives of future generations. The report noted many improvements in the world. Infant mortality was falling and human life expectancy increasing. More people could read and write. But at the same time, as global population increased, there were more hungry people in the world than ever before. What's more, the gap between rich and poor was expanding, and new environmental trends – among them, the warming of the atmosphere caused by the burning of fossil fuels – threatened the lives of many species, including the human species. It was decided a new UN Conference of Environment and Development should be held on the twentieth anniversary of the Stockholm event – an 'Earth Summit' – in 1992 and the location would be Rio de Janeiro.

Thatcher wasn't the only person talking about the need for an international agreement on climate change. Indeed, she mentioned it in her speech to the UN in 1989 because by that point people expected the Earth Summit to include the signing of something on climate change. A large World Conference on the Changing Atmosphere was held in Toronto in September 1988, just days after Hansen's front-page splash, bringing politicians as well as scientists together to discuss the issue. There, Brundtland stood next to Canadian Prime Minister Brian Mulroney to propose a new global 'law of the air', an international treaty to stabilise the Earth's climate, along with a new international research programme on renewable energy sources and greater scientific research into climate problems. The fuss around Hansen's testimony worried Bolin; he could see the public on climate change becoming quite, as he put it, 'chaotic'. Scientists would need to organise to brief politicians for any law of the air and they'd need to do it well. Bolin had been inspired by the way science had fed into the Montreal Protocol on substances such as CFCs that could deplete the ozone layer, in particular the way a group of scientists had been brought together by UNEP, led by NASA's Bob Watson. Bolin knew the issue of global warming was a lot bigger, and a lot stickier, but could something like the Montreal Protocol be built for greenhouse gases?

In the end, only 28 countries responded to the call for scientists to attend a meeting in Geneva in November 1988 to form a new Intergovernmental Panel on Climate Change (IPCC); climate change still wasn't high enough on the political agenda. Tolba was in charge of preparations, asking Bolin to serve as chairman, and suggested the project be run in three working groups – one on the physical basis of climate change, another on the impacts and a third on possible response strategies (or, to put it another way: is it happening, how bad will it be and what can we do about it?). Malta pushed hard for the IPCC's first report in time for the UN General Assembly in 1990, so it might properly set the scene for the negotiations that would lead up to the Earth Summit in 1992. Throughout 1989, 170 scientists worked through 12 workshops to build the report.

Jeremy Leggett – a scientist who'd jumped ship from oil tanker to *Rainbow Warrior*, moving from Imperial College's Royal School of Mines to a role at Greenpeace – remembers sitting in a hotel in Berkshire in May 1990 as about 100 scientists worked over three days to put finishing touches to the report. At the back of the room were 11 scientists from fossil fuel companies, including two from Exxon, and one each from BP and Shell. Like Leggett and other NGO representatives, they were allowed to take part as observers, but this was loosely defined, and they'd make suggestions on wording. Brian Flannery from the International Petroleum Industries Environmental Conservation Association (and Exxon) commented that uncertainties should be stressed in relation to the 60–80 per cent cuts to carbon dioxide emissions mentioned in the text, but no one agreed. Leggett left the meeting with mixed feelings, thinking the scientists had pulled their punches somewhat, not quite able to bring themselves to spell out the worst cases graphically. Still, he knew that the world had been given a sobering warning about global warming that would be hard to ignore. The chair of the event, the Met Office's John Houghton, briefed Thatcher who went on to address the press, warning 'people will be crying out not for oil wells but for water' as drought hits and deserts extend. The *Daily Express* declared under the headline 'Race to save our world' that Britain was leading a new crusade against the greenhouse effect.

★ ★ ★

Where were the activists in all of this? Where was that big public movement for action on climate change Rafe Pomerance had ached for, sitting in a stuffy science policy conference at the Florida hotel in the summer of 1980? There was a huge environmental movement growing, both in mainstream NGOs and more radical groups like Earth First!, which deliberately eschewed what they saw as increasingly 'corporate' systems utilised by even the more radical organisations such as Greenpeace or Friends of the Earth. Climate change could be found embedded in many of the environmental campaigns of the time, although it wasn't always obvious. Save-the-forest and block-the-road protests are about greenhouse gases after all, even if they tended to foreground other issues. Greenpeace even rocked up to a World Climate Conference held in Geneva in November 1990 with a giant balloon emblazoned with 'Cut CO_2 now'.* Still, it wasn't really until the 2000s that we saw the emergence of climate-specific groups and climate dominating the larger NGOs' portfolios.

If anything, the first really active climate campaigners were the climate sceptics. We've already seen a fair amount of what we might call climate scepticism, in terms of scientists looking at early research

---

* According to Leggett, the local police had been reasonably lenient with the climate protestors at this conference. He argues it suited the Swiss government at the time to look as if they were being pushed by the citizenry for greater climate action. His book on his career in the 1990s, *The Carbon War* (1999), is full of great details like this, and an essential source for this book.

and raising a quizzical eyebrow. For the most part, this was par for the course of good science. When people looked at new lab research at the start of the twentieth century and thought it meant they should dismiss Svante Arrhenius's warnings about coal emissions, that was good science even if it would also turn out to be incorrect. When the members of the Royal Meteorological Society ignored Guy Callendar in the 1930s, they possibly should have been less dismissive, but at the time it was understandable they reacted the way they did. When in 1977, head of the Met Office John Mason pushed back against what he saw as a climate change 'bandwagon', he was arguably dragging his heels a bit, but the odd old fogey like that was still all pretty much science as normal. However, starting around the late 1980s, just as the consensus about the greenhouse effect was really hardening, there was a deliberate, organised effort to amplify this natural doubt, extend it, and use it to dismiss and distract from warnings to take action on climate change. And that wasn't science, even if on occasion it used scientists – that was PR.

To understand the shape and style of the climate sceptic movement, we wind back a bit to the 1950s, and the link between cancer and tobacco. In Chapter Ten, we saw a 1959 edition of *New Scientist* that featured an article by a scientist from Shell assuring readers no matter how many tonnes of fossil fuels we burn, it's nothing compared with the magnitude of nature's carbon cycles. In the same edition, there's a news piece on how the tobacco industry had started to increase investment in scientific research following the link between smoking and cancer. Although the writer expresses some cynicism about the motivations behind this investment, they also welcome news that tobacco researchers have managed to cut the amount of arsenic in cigarettes and are excited to hear more about work on the possible benefits of smoking too. We now know this was a deliberate tactic to use new scientific research to distract and sometimes actively confound negative coverage. Links between smoking and cancer had been discussed for decades, but new experiments in 1950 caused a fresh press sensation. When in December 1952 *Reader's Digest* – then the most widely read publication in the world – printed the headline 'Cancer by the carton', the tobacco industry was thrown into crisis. The presidents of the four major tobacco companies in the US convened at the Plaza Hotel in New York to meet John Hill of Hill and Knowlton, one of America's largest PR companies.

John Hill's great PR innovation was to fight science with science. This didn't mean creating phoney science. That could only get you so far – you'd fund real, decent science, but in a way that'd confuse and muddy the message. He created the Tobacco Industry Research Committee and instructed his clients to invest in scientific research that could pay up the uncertainties, making those who caution for a more preventive approach look Cassandra-like.* This played into science's inherent scepticism, the sense that you are on a continual quest for knowledge, and should be ready to question and doubt everything, but amplified to make a political point. As these new research papers were reported, they also played into journalism's commitment to balanced fair play. They were turning the Enlightenment values of knowledge, scepticism and free speech in on themselves. They made use of, for example, Dr Clarence Cook Little, a renowned geneticist (a eugenicist to be more accurate) who was strongly committed to the idea that cancer, like everything else in his mind, was down to genetics. He'd emphasise the role of genetics over environmental factors like tobacco smoke; it was standard scientific chatter for the time but, distilled into a news report pitting one against the other, it could be made to weaken the case against tobacco. They similarly hired Wilhelm Hueper, an expert on asbestos who'd been frustrated by the ways in which tobacco smoke was itself used to dismiss links between cancer and asbestos. They'd use targeted research to ask questions – all reasonable, but also distracting – such as how much is the increase in cancer rates down to people not dying of other things? Why does Britain have a lung cancer rate higher than the US? Why aren't we also seeing cancers of the throat and tongue on the rise? As historians Naomi Oreskes and Erik Conway stress in

---

* We're talking about tobacco because it was their PR playbook the oil industry would later take on. But there were other links, not least because petrol stations were key retailers of cigarettes. One of the first places the tobacco industry went when they wanted to study the effects of smoking was the oil industry, who already had labs to help them tackle air pollution issues. The result was the cigarette filter (not much use to health, but a great marketing ploy, and today, one of the most ubiquitous bits of plastic pollution).

Merchants of Doubt, none of these questions were illegitimate, but they were all disingenuous.

The other important bit of context behind the use and abuse of climate scepticism was the strong cultural – sometimes generational – divides that had built up among parts of late-twentieth-century American science. By the early 1980s there was a small but powerful group of scientists who had spent their career enmeshed in the Cold War and now felt ostracised by campus campaigns against military science. They'd fallen out with colleagues over the Vietnam War and come to dislike the environmental movement intensely too. Many also maintained Cold War concerns about the rising influence of the left, suspicious of communists hiding everywhere. They loathed *Limits to Growth*; they hated Paul Erlich's *The Population Bomb*; they still were annoyed by Rachel Carson's stuff on DDT even though it was two decades ago. They had been annoyed at that Stephen Schneider guy on *The Tonight Show*, but they quite liked John Maddox's tirade against the doomists and they loved Reagan's new Strategic Defense Initiative idea. Many of them had been government advisors back when their knowledge was vital to the Cold War. Just as Gordon MacDonald's status as a Jason opened doors for his concern about global warming when he lobbied Carter's scientific advisor in the late 1970s, those on the other side could also get themselves listened to.

In 1984, three of these angry old men – Bill Nierenberg (former director of Scripps, who we've already met), Frederick Seitz (former President of the National Academy of Sciences), and Robert Jastrow (founder of NASA's Goddard Institute for Space Studies) – formed something called the George C. Marshall Institute. The idea was to act as a counterweight to the Union of Concerned Scientists, which they saw as having a clandestine leftist influence over many other scientific institutions. Initially they'd been focused on the Strategic Defense Initiative but, as that died down, they shifted to environmental issues, publishing a book in 1990, *Global Warming: What Does the Science Tell Us?* It wasn't out-and-out 'denial' of climate change or the greenhouse effect. They simply argued the Sun had caused the slight warming of the past century, and when its natural variation calmed down again that would balance out any future greenhouse warming. Like tobacco-industry-sponsored booklets before it, this promised to set the record straight. The real truths. In an era where experts were proliferating you can understand how it caught hold. It was already so much more souped up than the web of knowledge Van Bush had wanted to weave together with his memex in 1945.

In particular, it appealed to White House staff who Nierenberg took time to brief personally. They saw it as eminent scientists speaking their language and the colour started to drain from Bush's call to beat the greenhouse effect with the White House effect. 'I was impressed,' one White House staffer told *Science* magazine. 'Everyone has read it. Everyone takes it seriously.' Schneider quipped that White House chief of staff John Sununu was 'holding the report up like a cross to a vampire, fending off greenhouse warming.' In the UK, Downing Street advisors started sharing it to counter the briefings from climate scientists Thatcher was pulling in. To be fair on the people who picked up this book, Nierenberg, Seitz and Jastrow looked legit. Jastrow had founded the NASA institute where Hansen worked and Nierenberg had taken up 'grandfather' of climate change research Roger Revelle's old job at Scripps. What's more, they were talking a language that suggested mainstream science had been corrupted by the left, giving the George C. Marshall Institute, despite its rather fringe views and lack of peer review, an extra special aroma of truthiness. Other scientists supported it too. These scientists tended not to be climate change experts and of a similar cultural and political bent to Nierenberg et al, but that was either not obvious or wilfully ignored by the George C. Marshall Institute's new fans. Schneider spent months exchanging angry correspondence with Nierenberg, and Nierenberg's claims were also explicitly rebuffed in the IPPC report the following May. If anything though, these complaints just encouraged Nierenberg, who took his book on tour.

★ ★ ★

The George C. Marshall trio were hardly the only characters playing these sorts of games. Remembered today as the 'granddaddy of fake science', as a child in the 1930s, Fred Singer had fled Nazi Germany to England on the Kindertransport programme before being reunited with his family in Ohio. He'd gone on to do a PhD in physics at

Princeton and have a reasonably illustrious career in science and engineering, including being the first director of the National Weather Satellite service. In the early 1970s, he was what might be described as an environmentalist, arguing for the preservation of nature and better use of resources, but had gradually started to worry that perhaps environmental action wasn't worth the economic cost. In the early 1980s he'd started casting doubt about the dangers of acid rain, supported by the Reagan White House and affiliated to the Heritage Foundation, following it with complaints over what he called the 'ozone scare'. His problems with the mainstream scientific community arguably went further back, harbouring a grudge that he had been insufficiently credited for his work on satellites in the International Geophysical Year. In 1990, he founded something called the Science & Environmental Policy Project which, like the George C. Marshall Institute, argued the need to correct what he saw as a corruption of science by left-wingers and 'politically correct' views, and started campaigning aggressively on global warming.

Singer was an active climate contrarian for several decades; if you're interested there are plenty of climate investigation sites online that will list his various projects. Here, I'll stick to just the one story of when, in the early 1990s, he spun a line that 'granddaddy' of climate change research Revelle had changed his mind on global warming. By this point Revelle was in his 80s. He was still talking about climate change though and had moved on from the early days where he'd seen it all as an exciting experiment, now he advocated action. Still, he remained the sort of scientist who would stress the uncertainties and was careful to avoid overly dramatic claims. He'd worked on Nierenberg's NAS report in the early 1980s, standing by him at the report launch in 1983, telling reporters they were flashing an amber light of warning on global warming, not a red one. Maybe he never saw the political naivety of this sort of statement. Or perhaps he still just had hope that we'd tackle it, the same hope that meant, back in the 1950s, he thought all this burning would be over in a few decades. In an address to the American Association for the Advancement of Science in February 1990 he stated 'there is a good but by no means certain chance' that we'd see significant warming in the following century, and it was quite likely that he was just saying 'no means certain' as a way to advocate for the actions we could still take. But Singer pounced and asked Revelle to co-author an article on the topic. Revelle rather naively said yes. He then became ill, which limited his ability to fight over drafts. There is evidence to suggest Revelle was bothered by what Singer was suggesting, but when an article was published in *Cosmos* in the summer of 1992, arguing that there was no likelihood of significant warming, it was co-authored by Singer and Revelle (alongside electrical engineer Chauncey Starr). By this point, Al Gore was running for president and the idea that his mentor on global warming had changed his mind was a story that could be put to political work. Revelle had died in the summer of 1991 so couldn't speak for himself. His daughter wrote to the *Washington Post*, stressing her father had supported immediate steps to mitigate climate change, many of which were far more radical than Gore or other politicians had advocated. Revelle's colleagues also fought to clear his name (one even ending up in a legal fight with Singer) but a fair bit of damage had been done.

Characters like Singer and Nierenberg might have followed the same pattern they'd started, getting a few more meetings with the White House than they perhaps deserved and annoying former colleagues, but little else. But, at the start of the 1990s, the oil industry decided to go big on climate change scepticism as a tactic to slow the growing momentum for action. The oil industry had let the issue of carbon dioxide emissions sit for a while in the mid 1980s. BP had started researching polar ice melt, worried about its $11 billion investment perched on top of the Alaskan permafrost, but no one was talking about the sort of large-scale project to get involved in climate change research Exxon had flirted with at the end of the 1970s. In August 1988, just weeks after Hansen's testimony, Exxon's public affairs department issued an internal memo laying out what it felt should be the company's position and it was straight from John Hill's playbook for tobacco PR: 'Emphasise the uncertainty.'

Duane LeVine, Exxon's manager of science and strategy development, gave a primer to the company's board of directors, making it clear there was now a generally agreed consensus that global warming was happening and we could see significant changes by the middle of the twenty-first century 'with generally negative consequences'. Greenhouse gases were the problem, with carbon dioxide from fossil fuels the biggest perpetrator. But the danger they saw wasn't melting ice caps, it was closer than that; it was the prospect of politicians doing something about climate change and putting limits on the burning of fossil fuels. CFCs had been regulated in the Montreal Protocol; were fossil fuel emissions next? As LeVine put it: 'Arguments that we can't tolerate delay and must act now can lead to irreversible and costly Draconian steps.' Sticking with the old amber-light position of Nierenberg's wait-and-see recommendations in 1984 was, by far, the oil industry's preferred position in on it; how much longer could they delay emissions' cuts? The American Petroleum Institute, too, was in on it. After a series of internal briefings, the chief execs of the major companies agreed to set aside funds – only $100,000 for now, but it'd grow – to work on climate policy, establishing the Global Climate Coalition. It sounds like a green group doesn't it? They knew the best way to keep everyone sitting on the amber light was to get involved in the scientific and policy debate; it was there they would be there they'd been best placed to push the uncertainties and question regulations. They'd done this before with air pollution and their PR companies had had success delaying action on tobacco too. Before long, groups like this started to proliferate: the Information Council on the Environment, the Cooler Heads Coalition, the Global Climate Information Project.

★ ★ ★

Back in the White House, Bush's chief of staff John Sununu found a lot he liked in the George C. Marshall Institute's report. Sununu had a PhD in engineering from MIT and had long fostered a personal theory that commie-smelling anti-growth campaigners were using science as a cover. The worst, he thought, was the Club of Rome and its *Limits to Growth*, but he didn't trust Hansen and the climate modellers either. He banned the use of the terms 'climate change' and 'global warming' from any White House staffers who didn't know the details (and he had enough scientific knowledge to run rings around someone who simply took Hansen on trust).

In April 1990, an internal White House memo on talking to the press about climate change was leaked to environmentalists. It confirmed suspicions that the oil industry was working from the doubt-mongers playbook and that the White House was following, so passed it on to the *New York Times*. The memo had included the advice that there was no point arguing global warming wasn't happening, as it'd just loose that debate 'in the eyes of the public', but 'a better approach is to raise the many uncertainties that need to be better understood on this issue'. Still, there were people pushing for greater action on climate change from within the Bush administration too. The EPA administrator William K. Reilly had worked on Nixon's environmental policy before going on to be president of the Conservation Foundation and then, when it merged with WWF, that larger group too. He was keen the US didn't cede leadership on climate to Europeans – it's been suggested Thatcher's UN speech was partly because she wanted to get in before the French did and possibly Reilly was attuned to this – and wasn't about to let the whole thing fall to the delayers without a fight.

A key ministerial-level conference on climate change was set for November 1989 in Noordwijk in the Netherlands. Policymakers had met alongside scientists at earlier climate meetings in the 1980s, but this would be the first major international political event on climate change. Held near the end of the year, as climate conferences would continue to be, it was freezing cold. Discussions went on late into the night. When a Swedish minister came out briefly, a Sierra Club delegate asked what was going on. His reply: 'Your government is fucking things up.' Reilly would later tell Nathaniel Rich that Sununu was critical to the US 'fucking things up'. Sununu himself, however, told Rich that the other countries had been looking for the US to play this role. They didn't want to sign something they saw as endangering their economies and really hadn't much idea about how they'd go about reducing emissions either. The IPCC was just a face-saving exercise. 'That was the dirty little secret at the time,' Sununu told Rich.

The negotiations progressed on to the second World Climate Conference in Geneva a year later. The first of these, back in 1979, had chiefly been a scientific affair. This time, it would combine science and policymaking, with scientists meeting to discuss the IPCC report that had come out a few months before and produce a scientists' declaration off the back of it, while governments negotiated a ministerial one, the framework for a treaty to be signed at Rio. The scientists' declaration made it clear there was a consensus that global warming could be expected in the twenty-first century and 'countries are urged to take immediate actions to control the risks'. Some countries needed more than 'urging' by scientists though. After rumours that the US, USSR and Saudis were colluding to slow progress, the ministerial declaration ended up with a compromise that avoided committing to cut or even freeze emissions. Frustrated, a group of island nations formed a new bloc in the UN that had a particular interest in climate, the Alliance of Small Island States. But they weren't invited to the closed-door session that agreed the final text. That was for those larger states whose interest on climate change cantered more around what carbon cuts might do to their economies.

The discussions continued to rumble on, with the aim to have a text of a UN Framework Convention on Climate Change (the UNFCCC) ready for heads of state to sign at the Earth Summit in June 1992. A set of last-ditch talks were scheduled for May 1992, in New York, just weeks before the summit. Earlier that month, an Oxford University study for the UN reported that if global temperatures continued to creep up as they were, one in eight people could face famine within 50 years. Another, from the World Health Organization, warned of climate change increasing the spread of tropical diseases. Sununu did not want America to sign, but Reilly was pushing the other side, and eventually Bush decided the US did have to be part of the climate convention, but only if it was watered down enough that they didn't have to commit to any cuts right away. Everyone knew America's participation in the Earth Summit was reluctant and they certainly weren't going to lead the world to carbon cuts. The *New York Times* memorably dubbed Bush 'the Darth Vader of the Rio meeting'.

Photo banks of the Earth Summit tend to lead with the 100ft art deco statue of Christ the Redeemer in the Tijuca Forest National Park overlooking the city of Rio de Janeiro, his arms outstretched as if to welcome visitors. As Leggett described the event, however, down in the city itself, the streets were full of armed police, the areas where the delegates were staying 'sanitised', local children nowhere to be seen. Al Gore was buzzing around though, promoting his new book, *Earth in the Balance*. It was quite the media event: 10,000 press were accredited, with a further 30,000 to support them – more journalists than the Olympic Games. Prime minister of Tuvalu Bikenibeu Paeniu gave a press conference, describing how his people saw their land being washed away by the creeping seas, how drought and cyclones were becoming more and more of a problem. The first journalist's question brought it right back to America though: 'How do you feel about President Bush then?' Paeniu, a deeply religious man, emphasised that the president of the United States like everyone else had a moral obligation to 'face reality' and do what was right. 'There is someone up there who will judge us,' he added. 'We will all be judged. Even President Bush.' The world's press hadn't always paid much attention to Tuvalu, but that line hit headlines.

Boutros Boutros-Ghali, then the secretary general of the UN, declared: 'Ultimately if we do nothing, then the storm will break on the heads of future generations. For them it will be too late.' And in one of the more memorable moments of the event, 12-year-old Severn Cullis-Suzuki took the podium. With friends back home in Canada she'd founded Eco, the Environmental Children's Organization, and they'd raised money to, as she put it, 'travel 5,000 miles to tell you adults you must change your ways'. She was 'fighting for my future', she told them. 'Losing my future is not like losing an election, or a few points on the stock market.' Talking about the environmental worries she and others in her generation held, she asked the delegates: 'Did you have to worry of these things when you were my age?' Reminding the delegates that they had no particular technofix for any of these problems, she pleaded with world leaders to put the brakes on pollution: 'If you don't know how to fix it, please stop breaking it.' Still, if anything, all this emphasis on doing it for the kids let people think it was still a while off, even if the prospect of a degree or so global warming in the twenty-first century was inching ever closer.

One by one the heads of state signed the UNFCCC. It felt toothless to many of the climate experts watching, but it did give the world something to work from. The climate negotiations in the run up to Rio had hung on two key points. What would be the targets for emissions cuts and what was the timetable for action? In the end they agreed nothing specific, simply the loose agreement to 'stabilise greenhouse gas concentrations in the atmosphere at a level that would prevent dangerous anthropogenic interference with the climate system'. The other was more moral: how would these cuts be weighted and, in particular, should developing countries be exempt? Poorer countries, quite reasonably, wanted richer countries like the US or those in the EU to take the bulk of the burden. In the end they agreed on a principle of 'common but differentiated responsibilities', with responsibility largely on the shoulders of developed countries.

Most importantly, perhaps, this framework wouldn't be a static bit of paper, but rather the start of a longer process. A whole new UN body, also called the UNFCCC, was set up to facilitate the continuation of the discussion with an annual Conference of the Parties, also known as COP ('parties' being the countries signed up to the convention). It was nowhere near what was needed, but in principle left the door open for future rapid action. Of course, it also left the door open to more prevaricating. After several more years of negotiations, at the third COP, held in Kyoto at the end of 1997, they were (supposedly) ready for a treaty.

* * *

Kyoto's conference centre is a low-rise concrete building a little north of the city, sitting among leafy hills and lakes dotted with fish. It's an incredibly peaceful spot, but the big UN climate show would puncture that somewhat. On the first day, 1,500 government delegates from over 160 countries registered. There were a further 3,500 journalists and slightly more than that number of 'observers' (NGOs, business reps). Greenpeace alone had brought 45 staff members. Things had moved on from days when Rafe Pomerance and a few others were the only NGOers working on climate change. What's more, they were all connected to teams back home. It was the early days of the internet (or 'information superhighway' as Al Gore called it back then), but it was present in a way it just wasn't for Rio in 1992. News was starting to move faster. By the end of the first week, 30 delegates had to be hospitalised with exhaustion and dehydration.

There was a sense of hope pervading the event. John Gummer, who a few months before had been the British environment secretary (the Conservatives having just lost to Tony Blair's 'New' Labour), wrote in the *Washington Post*: 'The entire scheme seems inspired by a misplaced optimism that "something will turn up", as if global warming was a bad dream.' Even before the conference had started, the US Senate declared by a vote of 95–0 that it would reject any treaty that exempted developing countries. As the conference got going, the *New York Times* published a major new opinion survey: 65 per cent of Americans felt the US should cut its own emissions immediately, whatever other nations did. This was despite a $13 million advertising campaign pushing the idea that action on climate change would ruin the economy. Not for the first or last time (or in the US alone), the public at large were seemingly more willing for action than their policymakers.

At the start of the second week, the ministers arrived, thousands more press behind them. The Costa Rican president José María Figueres spoke on behalf of many other developing nations: 'My friends from the North, the ball is in your court.' President of Nauru, Kinza Clodumar went further: 'The wilful destruction of entire countries and cultures with foreknowledge would represent an unspeakable crime against humanity. No nation has the right to place its own, misconstrued national interest before the physical and cultural survival of whole countries.' Turning to Gore directly, to thunderous applause, he declared: 'Let us take action, effective action, prompt action, here in Kyoto, without reservation, without delay, for now and for ever.' The discussions quickly became tangled up in that tricky issue of fairness, a point those wanting to delay anyone finding an agreement made much out of. In the end they agreed another watered-down compromise: poorer countries would be exempt for the time being, and although richer countries would have to curb their emissions, they'd have another decade or so to get ready. Tony Juniper from Friends of the Earth started reading out passages from Gore's 1992 book. Others repeated the comments Gore himself had made about Bush's turn at Rio: 'It is about far more than hopping on a plane for a quick photo opportunity' and 'pretending to be doing something when actually nothing is being done'.

★ ★ ★

And then, there we were, at the year 2000. That weird, science-fiction sounding date that people had been talking about for so long; a point it was seemingly hard to really imagine much beyond, especially when it came to global warming. The world had a party, woke up the next morning relieved the Y2K bug hadn't hit, muttered an attempt at a joke about the absence of jetpacks and went about their lives, pretty much the same as they had in 1999. Back in 1971 Keeling had wondered if the generation of people then at college, by this point solidly middle aged, might demand serious action on climate change. That hadn't really happened, at least at the size required. So, now what?

The sixth COP negotiations were held in the Hague just after the 2000 US election and provided a key moment for global networks of climate activists to vent some frustrations over the continued decades of delay, as well as network for further action. Campaigners interested in climate change had been meeting alongside climate negotiations since Noordwijk in 1989. It had remained a relatively small subset of environmental or global justice campaigning in the 1990s, but things were about to change. A coalition of groups worked together under the banner of Rising Tide; they found space in a former-tax-office-turned-squat, putting on a Climate Justice Conference and running a memorial for Ken Saro-Wiwa on the fifth anniversary of his execution. An electric blue samba band played while a troupe of 'Greenwash Guerrillas' burst into a climate tech event, splattering it with green paint before moving on to paint-bomb Shell's HQ. The US lead negotiator Frank Loy got a custard pie in the face during a press conference. Outdoors, a semi-official protest from Friends of the Earth (part funded by the Dutch government) gathered several thousand activists to build a sandbag dyke around the conference centre. The talks ended without agreement, the UK Deputy Prime Minister John Prescott storming out in anger. Activists dressed as mourners, all in black, carried a ceremonial 'body' of the talks, wrapped in the event's flag, doused it in fuel and set it alight, before joining a street party back at the squat with that electric blue samba band, along with bagpipes, fire jugglers and a homemade light-sensitive synthesiser.

A new IPPC report in 2001 included new analysis of paleoclimatology data with a graph that looked a little like a hockey stick lying on its side, the temperatures running reasonably steadily before whooshing up in the last century or so. The sceptics went for it, hitting climate scientist Michael Mann especially hard. This was in part because by now climate scepticism had grown from just a few crotchety right-wing scientists at the George C. Marshall Institute to an active and frequently online community of people who'd seemingly taken on climate scepticism as a sort of hobby.

One of the other reasons the delayers went hard on Mann was, undoubtedly, because they were feeling the heat (or rather worried everyone else was). There'd been yet more 'hottest ever' years, including a heatwave in 2003 that had hit even those rich Western Europeans – usually insulated from climate impacts – hard. There had been 15,000 deaths in France alone, with temporary mortuaries set up in refrigerated lorries. BP started the expensive but short-lived process of rebranding as 'Beyond Petroleum', with billboards professing that BP 'believes in alternative energy. Like solar and cappuccino.' BP, along with other major companies, left the Global Climate Coalition, officially embarrassed by the association. In 2002, a leaked memo from Republican consultant Fred Luntz suggested the delayers felt there was still a lot to play for: 'Voters believe that there is no consensus about global warming within the scientific community. Should the public come to believe that the scientific issues are settled, their views about global warming will change accordingly. Therefore, you need to continue to make the lack of scientific certainty a primary issue in the debate.'

The annual UN climate talks continued, but didn't seem to get anyone much further along the line. As *Guardian* environment correspondent Suzanne Goldenberg describes in a brilliant 2015 essay, as the cycle of COPs came around, year after year, they fell into a repeating pattern of business seminars, complaints about terrible food and lack of sleep. There was the same parade of arguments and counter-attacks, and pretty much the same science year-on-year too, just more solidly making the case. It started to pick up traditions: veterans sharing advice to newbies on survival skills, a 'Fossil of the Day' awards from protesters to highlight those fighting the hardest against climate action. Yvo de Boer, who led the UNFCCC from 2006 to 2010, compared attending each COP to watching the soap *As the World Turns*: 'Every episode is really exciting, but if you don't watch for three years, you haven't missed anything.' For some delegates, it felt like a bit of a circus, more an excuse for a trip to see old friends than actually build change. Speaking at the 2006 talks in Nairobi, Kenyan environmentalist Sharon Looremetta gained a standing ovation when she stood up to declare: 'Fine, we can have Western countries coming, but some came here with their own agenda, to protect themselves and their economies; others came here as climate tourists who wanted to see Africa, take snaps of the wildlife, the poor, dying African children and women.' Nearly a decade on from Kyoto, understandably people were wondering whether the process was helpful.

Al Gore had fought and conceded the election back in 2000 and George Bush Jr moved into the White House. Bush would go on to pull the US out of the Kyoto agreement and then pull both his own country and several others into a protracted war in Iraq, which many believed to be largely about oil. Gore, meanwhile, retreated for a while, coming back in 2006 with the film *An Inconvenient Truth*, based on a presentation on climate change he'd been developing since the early 1990s. In 2007, the IPCC delivered its fourth report. With climate change riding high in public consciousness and coming off the back of a review on the economics of climate change the year before (the 'Stern Review' after its lead author, Nicholas Stern), it received slightly more media attention than its previous efforts. Later that year, the IPCC and Al Gore were jointly awarded the Nobel Peace Prize.

The British offices of Friends of the Earth tapped into the growing interest in climate change with the Big Ask campaign, calling for a climate change law with tactics that involved a 'carbon speed dating' event in Westminster, a concert featuring Radiohead's Thom Yorke and 100,000 people sending postcards to their MP. In 2008, they got their ask too; the UK adopted its world-leading Climate Change Act and started to build the political infrastructure needed to enact its part of the Kyoto agreement. New climate-specific campaign groups also emerged, including 350.org; an org so painfully noughties it had a dot in the title. Reflecting its avowedly 'science-based' approach, the 350 in 350.org is a reference to 350 parts per million (ppm), the amount of carbon dioxide in the atmosphere that James Hansen dubbed 'safe' (although we passed 350ppm before some of 350.org's current staff members were born). One of the founders May Boeve describes the founding of 350: they were coming to the end of their undergraduate degrees not long after the turn of the millennium, and planning ways to play a role in helping the climate movement up its game. There's a sweet and very millennial story of how they made a map overlaying coal reserves, wind energy potential and microbreweries (the latter being a proxy for somewhere they wanted to live), and settled on Billings, Montana, as a place to build a base. Then science writer and activist Bill McKibben came down, having walked for five days across the state of Vermont. He asked them if they'd be up for building a national version of their project and 350.org was born.

On the more grassroots end of things, a series of 'climate camps' emerged, initially in the UK, but then later in other parts of Europe, as well as New Zealand and Canada. The first Camp for Climate Action was a week-long protest by coal-fired power station Drax in North Yorkshire in August 2006. It had roots in earlier direct-action environmental protests, especially the anti-roads and anti-capitalism protests of the 1990s. Crucially they weren't just a protest, they were a camp. Campers would spend time with each other, participating in daily consensus-based meetings, sharing cooking and cleaning, and the building of compost toilets. As such, the camps became a huge space for learning and development, as well as relationship building for a new generation of climate activists. The way the camp was built and run – solar-powered, with horizontal management – was, in many respects, just as important as what they were protesting against. When the time came to pack up and go home, even though they hadn't shut down Drax, the campers felt elated; they felt they had been part of a model for a new way of living. As the camps continued over the following few summers, they picked up major media attention. After controversies around the cost of policing and accusations of risk of violence, activists at the 2008 camp by the Kingsnorth coal power station in Kent marched behind a banner declaring: 'We are armed only with peer review', copies of the latest IPCC report stuck to their hands.

* * *

At the start of 2009, it felt like finally, the world might be ready to take action. George Bush Jr was moving out of the White House and Barack Obama was moving in. The sceptics were looking increasingly ridiculous, increasingly isolated, increasingly desperate. Mann worried people were getting complacent and emailed colleagues at the University of East Anglia, warning them to 'expect lots more attacks'. He was right too. On 17 November 2009, RealClimate – a blog set up by Mann and other climate scientists a few years before – was hacked, temporarily disabling editing access. When the editors got back in, they found a draft post: 'We feel that climate science is, in the current situation, too important to be kept under wraps. We hereby release a random sleeting of correspondence, code and documents. Hopefully it will give some insight into the science and the people behind it. This is a limited time offer, download now.'

The download was thousands of emails and other documents taken from a server at the University of East Anglia. The hackers hadn't managed to publish through RealClimate, but plenty of other parts of the internet were keen to promote it, presenting this secret cache of scientist-to-scientist correspondence as a sort of 'smoking gun', evidence that climate change was just a hoax all along and really the Earth was cooling. Much was made about a line from an email from UEA's Phil Jones about a 'trick' Mann had done with some data. He meant a technique, but it was taken out of context. Senator James Inhofe built an igloo on the National Mall in Washington DC, with a sign inviting passing traffic to 'honk if you heart global warming'. *The Telegraph* tried to sell it as the greatest scandal in modern science, claiming it was 'the final nail in the coffin of anthropogenic global warming', but despite the initial hype, the fuss soon dissipated when people who weren't already committed climate sceptics realised there really wasn't much to see there (if anything, the emails were incredibly dull, as most emails are). In January 2010 a committee from the House of Commons Science and Technology Select Committee found no evidence of scientific misconduct or any reason to challenge the consensus over anthropogenic global warming. Still, it caused immense distress to the scientists involved. Jones even contemplated suicide.[*]

This was all just the warm-up act for the Copenhagen talks, COP15, which were to be a bumper set of UN climate talks. Kyoto had failed, but that was then; this was the new, climate-conscious late noughties and things were different. This was the big one, the great new climate treaty everyone had been waiting for. It was rather late, granted, but not too late. A new campaign group TckTckTck emphasised the sense of time running out as a clock tick-tocked. Bradley Whitford – who had recently finished playing White House staffer Josh Lyman in *The West Wing* – invited readers of the *Huffington Post* to become ambassadors of something called 'Hopenhagen'. But it wasn't long before all the hype and dreams of Hopenhagen dissolved into fear, anger and recriminations of 'Brokenhagen'. The fact it had been hyped up as the conference to save the world made the comedown all the more difficult.

Looking back on the event a few years later, Suzanne Goldenberg recalled: 'It is hard even now to fully grasp the degree of dysfunction that took hold of the conference. Some delegates still say, only half-jokingly, that the experience left them with PTSD.' The Danish hosts had planned for 15,000 delegates but 35,000 turned up, leading to hours of queues. It was unusually cold too, with a rare pre-Christmas snowfall. Frustrations were running high. At one point in a queue of VIP limos waiting to exit, Nicolas Sarközy turned to Hillary Clinton, rolled his eyes like a teenager and sighed, 'I want to die.' Perhaps

---

[*] When I started writing this book, I figured Climategate would be a chunk of the story. But as we passed the 10-year anniversary in the winter of 2019, although it was clearly personally very difficult for many people involved to live through, it just didn't seem too significant in the larger picture. If you want to know more, see Fred Pearce's book *Climategate* (2010).

unsurprisingly, files later leaked by Edward Snowden suggested the US and UK were both spying on delegates to give their teams an advantage.

China had recently taken the title as the world's biggest polluter from the US. Still, in line with the principles agreed in Kyoto (running back to Stockholm in 1972) most countries agreed that the US, Europe and Japan should lead on emissions. The legacy of their historical greenhouse gas emissions was already warming the earth, just as the legacy of their years of economic dominance gave them a political advantage. Obama knew he needed action from China to sell a deal back home. He flew in on the last day – fresh from his Nobel Peace Prize win – and promptly gatecrashed a meeting of leading developing countries, the so-called 'BRICS' of Brazil, Russia, India, China and South Africa. After 45 minutes of negotiation, Obama and Chinese premier Wen Jiabao brokered a loose accord that 'recognised' the scientific case for keeping temperature rises to no more than 2°C (still highly dangerous) and did not contain commitments to emissions' reductions to achieve that goal. Around 11pm that evening, Obama addressed a press conference, announcing this, before jumping on a plane home, hoping to get ahead of a blizzard that'd been forecast on the east coast of the US. For most of the delegates, the president's TV appearance was the first they'd heard of any 'agreement'.

India's environment minister Jairam Ramesh was furious: 'What is this? How can an agreement between four heads of state and the US be taken as a decision of 193 countries?' Venezuelan delegate Claudia Salerno banged her hand on the table until it was covered in blood. 'Do I have to bleed to grab your attention?' she cried out, raising her bloody fist. 'International agreements cannot be imposed by a small exclusive group. You are endorsing a *coup d'état* against the United Nations.' Obama spun it as, 'We've come a long way but we have much further to go', which might have sounded encouraging if it wasn't exactly what George Bush Sr had said in Rio nearly two decades before. He also said: 'We are in this together', which for many was clearly not the case. Not only had he worked a deal with, in Salerno's words, 'a small exclusive group' and then scarpered, anyone who'd been paying even the slightest bit of attention knew the inequalities of the global warming issue all too well. Climate change affects some parts of the world much worse than others and some countries have caused more of it than others. For Lumumba Di-Aping, the leader of the G77 group of the world's 130 poorest countries, the Obama–Jiabao deal was 'a suicide pact, an incineration pact in order to maintain the economic dominance of a few countries'. Or, as the Greenpeace press release put it: 'The city of Copenhagen is a crime scene tonight, with the guilty men and women fleeing to the airport.'

Conclusion: End Point?

At some point you have to draw a line between history and 'recent events'. Attempting a history of anything less than 10 years old is a mug's game. You don't have the distance; it's not been digested. Plus, it makes me feel old. So, I'm going to stop at those broken, Hopenhagen, Copenhagen talks.

I should still offer a quick run-through of some of what happened next. There's been much more of that extreme weather scientists in the 1960s and 1970s warned us would start to bite after the year 2000, and we're calling it out as climate change more often too. We're also getting more used to identifying something we now call 'fake news', as those games of spin and astroturfing the doubt-mongers played in the 1990s have become more and more intense in an era of social media, making the truths of the issue even harder to navigate.

After Copenhagen, the job of running the UNFCCC was given to Costa Rican diplomat Christiana Figueres, sister of the president who back in 1997 had looked to the global North at Kyoto and said 'the ball is in your court'. In her first few days in office, a journalist asked if Figueres thought a climate change agreement was possible and she responded, to the despair of her press officers: 'Not in my lifetime.' But she knew that wasn't good enough. Not only had we needed an agreement at Copenhagen – not the broken talks we ended up with – but we needed that agreement decades before. The world couldn't wait. She started to realise that if they were going to change things, they'd have to get out of the post-Copenhagen despondency and start beating a more optimistic drum for change. Building a new climate agreement would be, she figured, partly a matter of changing the tone: 'Impossible is not a fact, it's an attitude. And I decided, right then and there, that I was going to change my attitude, and I was going to help the world change its attitude on climate change.'

Things were shifting. Wind and solar had come of age, the drop in prices exceeding even the most generous of estimates, and Figueres and her PR team started to talk about renewables as 'unstoppable'. The sceptics still had power – terrifying amounts of it in places – but they also looked increasingly more desperate. A sizeable chunk of the world's population had now been carrying worries about climate change for decades and wanted to finally see action. When he retired from NASA in 2013, James Hansen told journalists he was looking forward to getting arrested for demanding climate action: 'At my age, I am not worried about having an arrest record.' Obama had said in a speech at a special UN climate conference held in New York in 2014, quoting Governor Jay Inslee: 'We are the first generation to feel the impact of climate change and the last generation that can do something about it.' Increasingly, people around the world really felt that – very strongly in places.

When the next big climate talks came around, in Paris in 2015, the world was slightly different. The city was badly bruised after brutal attacks a couple of weeks before. Protests were banned and everyone was on edge. There was still some of the carnivalesque campaigning that had become a routine part of these talks, but it was quieter, limited, surrounded by policemen holding giant guns who often looked more scared than any of the protesters. It was just before Christmas, and the shops and streets were twinkling with those spectacular electric lights Paris had pioneered back in the late-nineteenth century. The trees on the Champs-Élysées were dressed up to look like sparkling champagne glasses and children peeked up to the window displays at the Galeries Lafayette to follow a glittering fairy tale of a robot travelling through outer space. Such festivities, along with a cohort of activists from Australia who for some reason had come dressed as angels complete with large outstretched wings, made for a sometimes unsettling juxtaposition to all the public displays of grief about the attacks and fears about whether the UN could finally agree a decent climate deal. The weather was also creepily warm.

In the end, the Paris talks weren't the disaster of Copenhagen. If anything, they surprised even the optimists by how well they went (although it's worth stressing that by this point climate optimism had set an exceedingly low bar and many climate activists were still left angry by the outcome). Delegates agreed to not only try to keep global warming to less than 2°C above pre-industrial levels, but aspire to 1.5°C, commissioning the IPCC to conduct a report on the feasibility of such a goal. Still, arguably, the Paris Agreement was only achieved because it still left it up to countries to make their own plans for carbon cutting, letting them start low and grow in the future, hardly radical action.

In yet another blisteringly hot summer, this time in 2018, Swedish teenager Greta Thunberg – who in a gift to history writers is a relative of Svante Arrhenius – started skipping school to sit outside the parliament building, holding a handwritten sign: *Skolstrejk för klimatet* ('school strike for climate'). She was fed up and wasn't going to let temperatures continue to go up quietly. Students around the world followed her. The IPCC's special report on 1.5°C was published in October 2018 it seemed to spark a new anger. When Thunberg told the 2019 World Economic Forum in Davos: 'I want you to act as you would in a crisis. I want you to act as if our house is on fire. Because it is.' It was met with the usual eye-rolls about alarmism from some quarters, but it resonated with a lot of people too. A new space to be angry about climate change had opened up, with stronger, more vehement roars for action that were harder for politicians to ignore. Oil companies could see their carefully developed 'social licence to operate' rapidly eroding and started issuing statements about how serious they were about helping us all get to something called 'net zero' emissions (the 'net' is the slippery bit, it means you can cause emissions as long as you find a way to soak them up or bury them, which is both easier said than done and offers quite a lot of space to hide within). How much of this is just a smoke-and-mirrors PR game and how much serious transition is yet to be seen. My gut tells me some of the players are deadly serious, some are not. My gut also tells me that even where they are serious, it's not nearly enough.

★ ★ ★

Why did it take so bloody long? Why is it still taking so bloody long? In his conclusion to *Losing Earth*, Nathaniel Rich takes aim at the campaigners' call that 'Exxon knew'. Of course Exxon knew. We all did, and we let it happen. A cohort of AT&T-sponsored cartoon characters and Prince Philip were both openly discussing the issue on prime-time television back in the 1950s. The dangers of carbon dioxide weren't some sort of secret and yet, as Spencer Weart pointed out, less than 40 per cent of Americans in 1981 recalled having heard or read about the greenhouse effect. What's more, for the people who heard about the issue, it had generally been presented as something far off, or that would be sorted soon enough. From Frank Capra giving his *Unchained Goddess* a cosy Hollywood ending in the 1950s to Roger Revelle telling a 1983 press conference that 'we're flashing an amber light, not a red one', any members of the public watching were routinely reassured it'd turn out ok.

I'm not sure a blame game is useful and if it was it'd be the people who deliberately peddled doubt who would be first in line. But in terms of simply trying to better understand what happened, I think it's worth looking at the cultures of scientific work that have gradually developed over centuries, and could do with a bit of an update. This isn't the same as blaming individual climate scientists themselves. Scientists working on climate change have been put in an incredibly difficult position, but all too often other scientists have aggravated this problem, or at least the dominant working cultures of science haven't exactly helped things. Jobbing climate scientists should have been given time, expert support and a decent budget to think about the multiple challenges and transformations that have to happen when you take a contentious bit of science out of the scientific community and put it in the public sphere. And yet, if anything, many of these scientists have been ridiculed by their colleagues for speaking to media or, perish the thought, showing emotion. It didn't help, for example, that editor of *Nature* John Maddox spent a lot of the 1970s laughing at what he called the 'prophets of doom'. As historian Keynyn Brysse has argued, the tendency for climate scientists to err on the side of least drama is a bias; the sort of bias scientists should be extracting themselves from like any other bias. And yet a sense that it was somehow better science to appear undramatic was allowed to hide, to fester, while anyone saying something that smelled even the slightest bit of caution was shouted out as an activist. When the National Academy of Science was preparing its major report on climate change in the early 1980s, they were briefed by the new Reagan White House that the new administration didn't approve of speculative, alarmist or 'wolf-crying' scenarios, and instead of challenging the ideology inherent in such a statement it simply acquiesced. Hardly speaking truth to power.

But if aspects of the scientific community need to look at themselves, so do parts of the environmental movement. Sure, they had plenty of other battles to fight. It's also true that even the larger environmental NGOs are still tiny compared with what they're up against. Still, if even a bit of the time devoted to fights over population in the 1970s had been spent on climate change instead, the movement might have been in a better place to do more in the 1980s. I sometimes wonder if one of the reasons environmental groups dragged their heels a bit on climate change was that they saw it as a topic pro-nuclear campaigners talked about, not people like them. The mid-century conservation movement had good links with many aspects of science – ecology, zoology, botany, etc. – but climate change, on the whole, came from geophysics with stronger links to the military. There was a bit of a culture clash. You still sometimes hear people in the conservation end of the green movement talk disdainfully about 'the climate change lot'. And if scientists can be accused of occasionally pulling punches in the past 50 years, so can professional campaigners. Are environmental NGOs really happy to settle for 2°C warming and the number of people that would kill? Is the green movement still a group of rich white people for rich white people, or is it pushing for something more radical?.

Still, I should stress that it's the doubt-mongers and delayers who are the key problem. The scientists and activists have been the small number of people sticking their necks out trying to do something about this issue, while all too many people simply let it pass them by. Plus, it's hard to know the right thing to do or say. Back in 1953 when the *New York Times* magazine covered the growing science of global warming, the journalist pointed out it was understandable the scientists weren't sure of causes yet – it was the first climatic change they'd had a chance to examine first hand, after all, so they were learning as they were going. This continued to be the case throughout the twentieth century. Arguably, we are all still learning as we're going – living through this big experiment, seeing what works – even if we've had a bit of a chance to try a few things out. Moreover, everything would have been much easier to get right if scientists and campaigners had anything like the resources the oil industry had amassed by the end of the twentieth century. Exxon was buying up vintage wine to chase carbon isotopes back when it was doing climate change research, meanwhile scientists are fighting over whether to hire a postdoc or run an outreach programme.

★ ★ ★

This has been a book about the past, but one of the key lessons I took from researching it is that we're not very good at thinking about the future. There's the ongoing problem that it's hard to get politicians to think beyond a four- or five-year term, and the tendency for media commentators to draw on extremes of utopia and dystopia. But bigger than either of those annoyances, promises of future technologies all too often obscure what should be done about the technologies of here and now. It's become a well-worn trick from the delayer camp; push images of an emerging tech to hide the problems of existing technological systems. Fossil fuel companies' ads are full of slight-of-hand images of glowing algae, sparkling solar farms or football floodlights powered by the movement of the players themselves – you'd hardly think they actually spent their time burning oil. It also happens in less malign ways too, for example how a genuine belief in the promise of nuclear power meant geophysicists in the mid-twentieth century brushed off worries about global warming. Too often, we also imagine a world beyond fossil fuels as something that lives in some sort of futureworld, forgetting that futures catch up with us and we need to get building that world now (in fact we needed to do it decades ago). As a result, electric cars are still seen as a bit weird and futuristic even though they've held that role for well over 150 years. Arguably this problem haunts our ability to see climate change impacts too. We keep thinking of global warming as a future problem, even when it quite pertinently hasn't been that for a long time. We're not great at appreciating that what was 'once upon a time' has not only arrived but we passed it a while back. We're living our ancestors' dystopias and it is time we woke up to that.

Another big underlying problem was the way we ended up doing industrialisation, not just a commitment to fossil fuels or letting profits trump caution, but the distance we allowed to grow between consumers and producers. We let ourselves sit ignorant next to the amazing infrastructure of gas and electricity; the complex networks of production and trade and resource extraction that built the world we're sitting on. In some ways that's great, we can flip a switch and not worry where it comes from. But it's also one of the reasons we've struggled to act. Most of us are pretty clueless about how we built this world in the first place, and so struggle to work out where to start rebuilding it. This cluelessness is far from just a problem for energy. It's a price we pay for modernity; there's so much stuff to know we have to live our lives in a lot of ignorance. On the flip side, we get to draw on a range of areas of expertise without needing to have it all ourselves. It's the same problem Van Bush was hitting up against when he dreamt up the idea of the memex. It's a challenge climate scientists themselves know very well. One of the things Bill Nierenberg wrote in his 1983 National Academy of Sciences report which I wholeheartedly agree with is: 'The CO_2 issue is so diverse in its intellectual components that no individual may be considered an expert on the entire problem.' In many respects that reflects something wonderful; we are able to see climate change because we can pool expertise; but it brings a lot of challenges too.

We have to manage specialism in society more effectively. We need to celebrate and cultivate generalists more and invest in people whose job it is to help communicate and share specialist knowledge. We need to remember that science isn't done when a paper is published. If anything, that's just the start of a load more work of taking it out in the wider world. We also need to be more inventive about ways in which everyday people can be connected to technological and industrial systems that might otherwise be hidden. We don't all have to build our own solar farms. But if a few more of us did, it'd be easier for all of us to have a say, to ask questions when we're not sure about a development and celebrate what's going right. Back in 1980 Denis Hayes complained that unless 'pro-solar' Americans stood up for the tech, it would be cut. That's a problem I've seen repeated more recently too. People love solar in theory, but for most of us there's no particular way of expressing that, which makes it easy for politicians to cut. Similarly, if you don't love oil, gas or coal it can be hard to know where to start unravelling their grip. It could be different. In fact, in some parts of the world, it is. One of the legacies of Poul la Cour's approach to wind energy in Denmark is that, even in the twenty-first century, a third of the population has links with a co-operatively owned wind project. A more democratic relationship with energy systems (indeed, technologies and industries in general) would be good for a host of reasons, but as we face the need for extremely rapid change to tackle climate change, it's more vital than ever.

We need to be more mindful consumers of technology and make better choices. As Ruth Schwartz Cowan puts it in her classic essay on the fridge's hum, technological history is 'littered with the remains of abandoned machines', not just junked cars but 'the rising hulks of aborted ideas'; patents that were never exploited, test models that couldn't be scaled up at affordable prices, devices that made it to market, but for whatever reason didn't sell well. Tracing through the remnants of some of these in nineteenth-century women's magazines, Cowan wonders at some of the technological roads not travelled, and whether we might have been better off otherwise. 'What resident of a drought-prone area would not be grateful for a toilet that does not use water? How many energy-conscious housewives would be unwilling to try out a fireless cooker?' It echoes a point Ida B. Wells and her co-authors made in their critique of the Chicago World's Fair back in 1893: whose future are you really selling, who is it made by, for whom, what is it celebrating and what is it ignoring? It's a critique not just of those who came before us, but one for us to keep in mind as we strategise our way out of the climate crisis too. If we want to live well with the benefits of technology, we need to have better conversations about it; for the environment, but for so much more too – for equality, for fun.

All this matters because we still have choices. Nothing was inevitable about our addiction to fossil fuels, and yet it happened. The stories of gas fridges, *fin-de-siècle* electric cabs, and a Ford Model T that could flex between alcohol and gasoline reflect the many technological options people in the past had on offer. The story of the climate crisis has always been a choose your own adventure and continues to be, even if we have fewer choices now than we did a few decades ago. This has been a story of possibilities and it will continue to be full of possibility even if our space for action is diminishing fast.

★ ★ ★

The story of the climate crisis is one that still hasn't found its conclusion. If I was at my day job as a campaigner, this would be the point I'd pull on a Captain Planet costume and try to spin some inspirational line like 'the next chapter is up to you'.

But this is a history book, not campaign copy. Is it too late for a happy ending? For a lot of people, yes. One of toughest things about campaigning on climate change is that you can't really win, because in many ways we've already lost. The more optimistic spin on this is that because climate change happens by degree, there is still always something to fight for. Because 1°C is better than 1.5°C, 1.5°C is better than 2°C, and that will continue to be the case for some time to come. Climate change simply isn't a pass or fail issue. It's not something you win or lose. As the ever-articulate NASA climate scientist Kate Marvel puts it, 'Climate change isn't a cliff we fall off, but a slope we slide down.' That slope is getting very steep, but we haven't fallen off yet.

I don't know what the situation will be by the time you read this. I don't know how the next set of climate talks will go, or the one after that, how each crucial election will play out, where the next storm will hit, when or how hard. I do know it's all going to be difficult though. We've inherited a position of having to radically change the world's economy to have even a snowball's chance in hell of staying anywhere near 2°C, let alone 1.5°C. What's more, we have to make these enormous changes while also tackling challenging climate change impacts. And we have a rapidly vanishing snippet of time in which to do all this. It would have been a lot easier if we'd started earlier (whether this 'earlier' was the 1970s, 1980s, 1990s or 2000s, any of it would be easier than now). A few countries have passed some OK-ish laws promising action in line with the Paris Agreement and plenty of people are taking action. For a big chunk of the time I wrote this book, the electricity my laptop runs off – that is, the British national grid – has been powered without anyone needing to burn coal. That's historic; not a word, having spent so many months researching and writing this book, I'd use lightly. But it's no way fast enough. No way near fast enough.

One way to look at it is that we're lucky action has happened at the speed it has, or that we even know climate change is happening at all. As Spencer Weart notes, until Keeling, pretty much everyone who studied global warming took it up as a side issue, 'a chance for a few publications, a detour from work that aimed elsewhere and to which they soon returned'. If they hadn't taken those detours, our awareness of climate change could so easily have been delayed even more. There are plenty of points in the story where we can shout 'we could have had more time!' But equally there are plenty where we might never have had the opportunities we did.

What modern climate scientists do is amazing. The knowledge they can pull out of the skies, seas and soils is just incredible. The joy of writing this book has been tracing through their history, from Eunice Foote putting a few cylinders of gas on her sunny windowsill, to Willi Dansgaard playing around with bubbles caught in ice or Mostafa Tolba setting up the first working groups of the IPCC. It really wasn't all that long ago that our ancestors simply looked at air and thought it was just that, air, not an array of different chemicals, some of which might make you high, explode or, over several centuries of burning coal and oil, could have a warming effect on the Earth. When climate fear starts to grip, it's worth remembering that we have all this incredible knowledge and that offers us a chance to act. We could, all too easily, be sitting around thinking 'the weather's a bit weird today. Again.'

We've inherited an almighty mess, but we've also inherited a lot of tools that could, if we choose wisely and make the most of them, help us and others survive. If you're looking for something optimistic to end this story, that's it. We owe it not just to future generations who'll inherit whatever we do to this Earth, but also to the memory of those who came before, who gave us something to work with.

Key Sources

Introduction

Abram, Nerilie J. et al. 2016. 'Early onset of industrial-era warming across the oceans and continents'. *Nature* 536:411–8.

Allen, Myles. 2019. 'Why protesters should be wary of "12 years to climate breakdown" rhetoric'. *The Conversation* 18 April.

Dunne, Daisy. 2020. 'Q&A: Could climate change and biodiversity loss raise the risk of pandemics'. *Carbon Brief* 15 May.

Foote, Eunice. 1856. 'Circumstances affecting the heat of the Sun's rays'. *The American Journal of Science and Arts* November 22:382–3.

Gore, Timothy. 2020. 'Confronting carbon inequality'. *Oxfam* 21 September.

Jackson, Roland. 2019. 'Eunice Foote, John Tyndall and a question of priority'. *Notes and Records of the Royal Society* 74(1):105–18.

Lanchester, John. 2019. 'How the little ice age changed history'. *New Yorker* 25 March.

Loch, Alexander et al. 2019. 'Earth system impacts of the European arrival and great dying in the Americas after 1492'. *Quaternary Science Reviews* 207:13–36.

Maslin, Mark & Lewis, Simon. 2018. *The Human Planet: How We Created the Anthropocene*. Penguin.

McNeill, Leila. 2016. 'This lady scientist defined the greenhouse effect but didn't get the credit, because sexism'. *Smithsonian Magazine* 5 December.

Ortiz, Joseph D. & Jackson, Roland. 2020. 'Understanding Eunice Foote's 1856 experiments: heat absorption by atmospheric gases'. *Notes and Records of the Royal Society* 26 August.

Parker, Geoffrey. 2017. *Global Crisis: War, Climate Change and Catastrophe in the Seventeenth Century*, revised edition. Yale University Press.

Schwartz, John. 2020. 'Overlooked no more: Eunice Foote, climate scientist lost to history'. *New York Times*, series of obituaries on previously unreported people 21 April.

Steadman, Ian. 2019. 'The forgotten woman who unlocked the greenhouse effect'. *How We Get to Next* 15 March.

Weart, Spencer R. 2003. *The Discovery of Global Warming*. Harvard University Press.

Chapter One: A Steam-Powered Greenhouse

Auerbach, Jeffrey. 1999. *The Great Exhibition of 1851: A Nation on Display*. Yale University Press.

Brindle, Steven. 2005. *Brunel: The Man Who Built the World*. Weidenfeld & Nicolson.

Burton, Anthony. 2000. *Richard Trevithick: Giant of Steam*. Aurum Press.

Dodson, John et al. 2014. 'Use of coal in the Bronze Age in China'. *The Holocene* 24(5):525–30.

Doe, Helen. 2017. *The First Atlantic Liner: Brunel's Great Western Steamship*. Amberley Publishing.

Freese, Barbara. 2003. *Coal: A Human History*. Perseus.

Grace, Richard J. 2014. *Opium and Empire: The Lives and Careers of William Jardine and James Matheson*. McGill-Queen's University Press.

Greenhalgh, Paul. 1988. *Ephemeral Vistas: The Expositions Universelles, Great Exhibitions and World's Fairs, 1851–1939*. Manchester University Press.

Khilnani, Sunil. 2016. *Incarnations: India in 50 Lives* 327–39. Allen Lane.

Malm, Andreas. 2016. *Fossil Capital: The Rise of Steam Power and the Roots of Global Warming*. Verso.

Marsden, Ben. 2002. *Watt's Perfect Engine: Steam and the Age of Invention*. Icon.

Marsden, Ben & Smith, Crosbie. 2005. *Engineering Empires: A Cultural History of Technology in Nineteenth-Century Britain*. Palgrave Macmillan.

McKendrick, Neil. 1982. *The Birth of a Consumer Society: The Commercialization of Eighteenth-Century England* 100–45. Europa Publications.

Mullem, Stephen & Newman, Simon. 2018. 'Slavery, abolition and the University of Glasgow', University of Glasgow History of Slavery Steering Committee.

Philip, Cynthia Owen. 1985. *Robert Fulton: A Biography*. Franklin Watts.

Rhodes, Richard. 2018. *Energy: A Human History*. Simon & Schuster.

Ross, David. 2010. *George and Robert Stephenson: A Passion for Success*. The History Press.

Sale, Kirkpatrick. 2001. *The Fire of His Genius: Robert Fulton and the American Dream*. The Free Press.

Uglow, Jenny. 2002. *The Lunar Men: The Friends Who Made the Future*. Faber and Faber.

Wolmar, Christian. 2007. *Fire and Steam: A New History of the Railways in Britain*. Atlantic Books.

Wolmar, Christian. 2009. *Blood, Iron and Gold: How the Railways Transformed the World*. Atlantic Books.

Chapter Two: Discovering Our Hothouse Earth

Arrhenius, Svante. 1896. 'On the influence of carbonic acid in the air upon the temperature of the ground'. *Philosophical Magazine* 41(251):237–76.

Barry, R. G. 1978. 'HB de Saussure: The first mountain meteorologist'. *Bulletin of the American Meteorological Society* 59(6):702–5.

Barton, Ruth. 1990 '"An influential set of chaps": The X-Club and Royal Society Politics 1864–85'. *The British Journal for the History of Science* 23(1):53–81.

Carozzi, Albert V. 1966. 'Agassiz's amazing geological speculation: the ice age'. *Studies in Romanticism* 5(2):57–83.

Chang, Hasok. 2004. *Inventing Temperature: Measurement and Scientific Progress.* Oxford University Press.

Crosland, Maurice. 1984. 'A practical perspective on Joseph Priestley as a pneumatic chemist'. *British Journal for the History of Science* 16(3)223–38.

Degroot, Dagomar. 2018. *The Frigid Golden Age: Climate Change, the Little Ice Age, and the Dutch Republic, 1560–1720.* Cambridge University Press.

de Beer, Gavin & Hay, Max H. 1955. 'The first ascent of Mont Blanc'. *Notes and Records of the Royal Society* 11(2):236–55.

Dixon, Joshua. 1801. *The Literary Life of William Brownrigg, M. D. to which are added an account of the coal mines near Whitehaven, and observations on the means of preventing epidemic fevers.* Longman & Rees.

Dorgon, Lauren R. 2002. 'Committee renames local Agassiz school'. *The Harvard Crimson* 22 May.

Fleming, James Rodger. 1998. *Historical Perspectives on Climate Change.* Oxford University Press.

Fleming, James Rodger. 1999. 'Joseph Fourier, the "greenhouse effect" and the quest for universal theory of terrestrial temperatures'. *Endeavour* 23(2):72–5.

Fleming, James R. 2000. 'T. C. Chamberlin, climate change, and cosmogony'. *Studies in History and Philosophy of Modern Physics* 31(3) 293–308.

Grattan-Guinness, Ivor & Ravetz, Jerome R. 1972. *Joseph Fourier, 1768–1830.* MIT Press.

Hartocollis, Anemona. 2019. 'Who should own photos of slaves? The descendants, not Harvard, a lawsuit says'. *New York Times* 20 March.

Herivel, John. 1975. *Joseph Fourier: The Man and the Physicist.* Clarendon Press.

Jackson, Joe. 2005. *A World on Fire: A Heretic, An Aristocrat, and the Race to Discover Oxygen.* Viking.

Jackson, Roland. 2018. *The Ascent of John Tyndall.* Oxford University Press.

Jackson, Roland. 2019. 'Eunice Foote, John Tyndall and a question of priority'. *Notes and Records of the Royal Society* 74(1):105–18.

Johnson, Steven. 2009. *The Invention of Air: A Story of Science, Faith, Revolution, and the Birth of America.* Penguin.

Kunzig, Robert & Broecker, Wallace. 2008. *Fixing Climate: The Story of Climate Science and How to Stop Global Warming.* Profile Books.

Morris, Robert J. 1972. 'Lavoisier and the caloric theory'. *The British Journal for the History of Science* 6(1):1–38.

Pawley, Emily. 2008. 'Powerful effervescence'. *Distillations, Science History Institute* 1 June.

Rodhe, Henning & Charlson, Robert. 1998. *The Legacy of Svante Arrhenius Understanding the Greenhouse Effect*. Royal Swedish Academy of Sciences.

Seitz, Frederick. 2005. 'Henry Cavendish: the catalyst for the chemical revolution'. *Notes and Records of the Royal Society of London* 59(2):175–99.

Schaffer, Simon. 1984. 'Priestley questions: a historiographic survey'. *History of Science* 22(2):151–83.

Sokolow, Jayme A. 1980. 'Count Rumford and late Enlightenment science, technology, and reform'. *The Eighteenth Century* 21(1):67–86.

Thomas, John Meurig. 1999. 'Sir Benjamin Thomson, Count Rumford and the Royal Institution'. *Notes and Records of the Royal Society of London* 53(1):11–25.

Townsend, Chris. 2016. 'Year without a summer'. *Paris Review* 25 October.

Tyndall, John. 1861. 'The Bakerian lecture – on the absorption and radiation of heat by gases and vapours, and on the physical connexion of radiation, absorption, and conduction'. *Philosophical Transactions of the Royal Society* 151:1–37.

Wulf, Andrea. 2015. *The Invention of Nature: The Adventures of Alexander von Humboldt, the Lost Hero of Science*. John Murray.

Chapter Three: From Whale to Shale

Dolin, Eric Jay. 2007. *Leviathan: The History of Whaling in America*. W. H. Norton.

Garfield, Simon. 2000. *Mauve: How One Man Invented a Colour that Changed the World*. Faber and Faber.

Hoare, Philip. 2008. *Leviathan or, the Whale*. Fourth Estate.

Holmes, Richard. 2008. *The Age of Wonder: How the Romantic Generation Discovered the Beauty and Terror of Science*. Harper Collins.

Holmes, Richard. 2013. *Falling Upwards: How We Took to the Air*. William Collins.

Jay, Mike. 2009. *The Atmosphere of Heaven: The Unnatural Experiments of Dr Beddoes and his Sons of Genius*. Yale University Press.

Lucier, Paul. 2008. *Scientists and Swindlers: Consulting on Coal and Oil in America, 1820–1890*. Johns Hopkins University Press.

McGlade, Christophe & Ekins, Paul. 2015. 'The geographical distribution of fossil fuels unused when limiting global warming to 2°C'. *Nature* 517:187–90.

Metzger, Peter. 1970. 'Project Gasbuggy and Catch-85★'. *New York Times* 22 February 329:355–8.

Mills, Mary. 2009. 'Golden Lane Brewery'. *Mary's Gas Book blog* 6 August.

Pagel, Walter. 1970–80. *Dictionary of Scientific Biography* 6:253–9.

Rhodes, Richard. 2018. *Energy: A Human History*. Simon & Schuster.

Roberts, Jacob. 2014. 'Whales in space'. *Distillations Magazine* 11 January.

Schivelbusch, Wolfgang. 1988. *Disenchanted Night: Industrialization of Light in the Nineteenth Century*. University of California Press.

Stone, Maddie. 2019. 'How much is a whale worth'. *National Geographic* 24 September.

Tomory, Leslie. 2012. *Progressive Enlightenment: The Origins of the Gaslight Industry, 1780–1820*. MIT Press.

Werrett, Simon. 2007. 'From the grand whim to the gasworks: philosophical fireworks in Georgian England', Roberts, Lissa et al (eds). *The Mindful Hand: Inquiry and Invention from the Late Renaissance to Early Industrialisation* 325–48.

Wise, Damon. 2017. 'How we made Stanley Kubrick's *Full Metal Jacket*' *Guardian* 1 August.

Yergin, Daniel. 2008. *The Prize: The Epic Quest for Oil, Money and Power*. Simon & Schuster.

York, Richard. 2017. 'Why petroleum did not save the whales'. *Socius* 1:1–13.

1879. 'Heinrich Geissler'. *Scientific American* 5 April, 209.

1968. 'Science: Nevada's big blast'. *Time* 3 May, 1968, 50.

Chapter Four: The Weather Watchers

Anderson, Katherine. 2005. *Predicting the Weather: Victorians and the Science of Meteorology*. University of Chicago Press.

Burton, Jim. 1986. 'Robert FitzRoy and the Early History of the Meteorological Office'. *The British Journal for the History of Science*. 19(2):147–76.

Callendar, Guy. 1938. 'The artificial production of carbon dioxide and its influence on temperature'. *Quarterly Journal of the Royal Meteorological Society* 64(275):223–40.

Chase, Thorington. 1912. 'Should the atmosphere be conserved?'. *Scientific American* 106(9):195.

Coen, Deborah R. 2011. 'Imperial climatographies from Tyrol to Turkestan'. *Osiris, Klima* 26(1):45–65.

Coen, Deborah R. 2016. 'Big is a thing of the past: climate change and methodology in the history of ideas'. *Journal of the History of Ideas* 77(2):305–21.

Coen, Deborah R. 2018. *Climate in Motion: Science, Empire and the Problem of Scale*. University of Chicago Press.

Coxe, E. J. D. 1912. 'Conserving the atmosphere'. *Scientific American* 107(4):75.

Dry, Sarah. 2019. *Waters of the World*. Scribe.

Fleming, James Rodger. 1998. *Historical Perspectives on Climate Change*. Oxford University Press.

Fleming, James Rodger. 2007. *The Callendar Effect: The Life and Work of Guy Stewart Callendar (1898–1964)*. The American Meteorological Society.

Hamblyn, Richard. 2001. *The Invention of Clouds: How an Amateur Meteorologist Forged the Language of the Skies*. Picador.

Holmes, Richard. 2013. *Falling Upwards: How We Took to the Air*. William Collins.

Molena, Francis. 1912. 'Remarkable weather of 1911'. *Popular Mechanics Magazine* March 1912 339–42.

Monmonier, Mark. 1999. *Air Apparent: How Meteorologists Learned to Map, Predict, and Dramatize Weather*. University of Chicago Press.

Moore, Peter. 2015. *The Weather Experiment: The Pioneers who Sought to See the Future*. Penguin.

Ratcliffe, R. A. S. 1978. 'The story of the Royal Meteorological Society'. *Weather* 33(7):261–68.

Walker, J. M. 1991. 'The Meteorological Societies of London'. *Weather* 48(11):364–72.

Walker, Malcolm. 2012. 'History of the Meteorological Office'. Cambridge University Press.

— . 1912. 'The increasing temperature of the world'. *Scientific American* 107(5):99.

Chapter Five: Electric Avenues

Baldwin, Neil. 2001. *Edison: Inventing the Century*, 2nd Edition. University of Chicago Press.

Ben-Chaim, Michael. 1990. 'Social mobility and scientific change: Stephen Gray's contribution to electrical research'. *The British Journal for the History of Science* 23(1):3–24.

Berton, Pierre. 2009. *Niagara: A History of the Falls*. SUNY Press.

Blyth, James. 1892. 'On the application of wind power to the production of electric currents'. *Proceedings of the Philosophical Society of Glasgow*, 25 January 1–2.

Boase, George Clement. 1885–1900. 'Willoughby Smith', *Dictionary of National Biography* 53. Smith, Elder & Co.

Brox, Jane. 2010. *Brilliant: The Evolution of Artificial Light*. Houghton Mifflin Harcort.

Cantor, Geoffrey, Gooding, David & James, Frank A. J. L. 1991. *Faraday*. Macmillan.

Cohen, I. B. 1952. 'The two hundredth anniversary of Benjamin Franklin's two lightning experiments and the introduction of the lightning rod'. *Proceedings of the American Philosophical Society* 96(3):331–66.

Desmond, Kevin. 2015. *Gustave Trouvé: French Electrical Genius (1839–1903)*, foreword by John Devitt. McFarland & Company.

Goodrick-Clarke, Nicholas. 2002. *Black Sun: Aryan Cults, Esoteric Nazism, and the Politics of Identity*. NYU Press.

Greenhalgh, Paul. 1988. *Ephemeral Vistas: The Expositions Universelles, Great Exhibitions and World's Fairs, 1851–1939*. Manchester University Press.

Grummitt, Julia. 2017. 'Princeton and slavery: the scientist's assistant'. *Princeton Alumni Weekly* 8 November.

Heald, Henretta. 2010. *William Armstrong: Magician of the North*. McNidder & Grace.

Holst, Bodil, & Zimmer, N. 2002. 'The discovery of the electric shock'. *Science* 298(5602):2327–8.

Hughes, Thomas P. 1985. 'Edison and electric light', in MacKenzie, Donald & Wajcman, Judy (eds). *The Social Shaping of Technology: How the refrigerator got its hum*. Open University Press 39–52.

Isreal, Paul. 1998. *Edison: A Life of Invention*. John Wiley.

James, Frank A. J. L. 2010. *Faraday: A Very Short Introduction*. Oxford University Press.

Jonnes, Jill. 2003. *Empires of Light: Edison, Tesla, Westinghouse and the Race to Electrify the World*. Random House.

Moran, Richard. 2002. *Executioner's Current: Thomas Edison, George Westinghouse, and the Invention of the Electric Chair*. Knopf.

Morus, Iwan Rhys. 2009. 'Radicals, Romantics and electrical showmen: placing galvanism at the end of the English Enlightenment'. *Notes and Records of the Royal Society of London* 63(3):263–75.

Morus, Iwan Rhys. 2011. *Shocking Bodies: Life, Death and Electricity in Victorian England*. The History Press.

Morus, Iwan Rhys. 2019. *Nikola Tesla and the Electrical Future*. Icon.

Nye, David E. 1990. *Electrifying America: Social Meanings of a New Technology*. MIT Press.

Nye, David. 2018. *American Illuminations: Urban Lighting, 1800–1920*. MIT Press.

Paddon, Anna R. & Turner, Sally. 1995. 'African Americans and the world's Columbian exposition'. *Illinois Historical Journal* 88(1):19–36.

Pepper, John Henry. 1869. 'Some experiments with the great induction coil at the Royal Polytechnic'. *Proceedings of the Royal Society of London* 18:65–72.

Price, Trevor J. 2005. 'James Blyth – Britain's first modern wind power pioneer'. *Wind Engineering* 29(3):191–200.

Reed, Christopher Robert. 2002. *All the World is Here!: The Black Presence at White City*. Indiana University Press.

Rinehart, Melissa. 2012. 'To Hell with the Wigs! Native American representation and resistance at the World's Columbian Exposition'. *American Indian Quarterly* 36(4):403–42.

Roche, Maurice. 2002. *Mega-events and Modernity: Olympics and Expos in the Growth of Global Culture.* Routledge.

Rydell, Robert W. 1985. *All the World's a Fair: Visions of Empire at American International Expositions, 1876–1916.* University of Chicago Press.

Schaffer, Simon. 1997. 'Experimenters' techniques, Dyers' hands, and the electric planetarium'. *Isis* 88(3):456–83.

Schivelbusch, Wolfgang. 1988. *Disenchanted Night: Industrialization of Light in the Nineteenth Century.* University of California Press.

Secord, James A. 2002. 'Portraits of science: quick and magical shaper of science'. *Science* 297(5587):1648–9.

Smith, Willoughby. 1873. 'Effect of light on selenium during the passage of an electric current'. *Nature* 7:303.

Weeden, Brenda. 2008. *The Education of the Eye: History of the Royal Polytechnic Institution 1838–1881.* University of Westminster Press.

Wells, Ida B. (ed). 1893. *The Reason Why the Colored American is Not in the World's Columbian Exposition* (reprinted by University of Illinois Press, edited by Robert W. Rydell, 1999).

Wolmar, Christian. 2009. *The Subterranean Railway: How the London Underground was Built and How it Changed the City Forever.* Atlantic Books.

1883. 'An electrically-moved tramcar'. *The Times* 13 March, 4.

1890. 'Mr Brush's windmill dynamo'. *Scientific American.* 20 December, 389.

1892. 'The Crystal Palace Electrical Exhibition'. *Nature.* 14 January, 45(1159):261.

1902. 'Bovril is Liquid Life', from an advertisement in Abel Heywood & Son's *Influenza: Its cause, cure and prevention*, republished by Wellcome Collection website.

1936. 'Stephen Gray: the first Copley Medallist'. *Nature* 22 February, 137(3460):299–300.

2017. 'James Blyth and the world's first wind-powered generator'. British Library science blog, 17 August.

Chapter Six: Tree Huggers

Allen, Garland. 2013. 'Culling the herd: eugenics and the conservation movement in the United States, 1900–1940'. *Journal of the History of Biology* 46(1):31–72.

Baker, David. 1984. 'A serious time: forest satyagraha in Madhya Pradesh, 1930'. *The Indian Economic and Social History Review* 21(1):71–90.

Bandyopadhyay, Jayanta. 1999. 'Chipko movement: of floated myths and flouted realities'. *Economic and Political Weekly* 34(15):880–2.

Cantor, Geoffrey. 2012. 'Science, providence, and progress at the Great Exhibition'. *Isis* 103(3):439–59.

Chapman, Sasha. 2017. 'The woman who gave us the science of normal life'. *Nautilus* 30 March, 46.

Cronon, William. 1995. 'The trouble with wilderness; or, getting back to the wrong nature', In William Cronon (ed). *Uncommon Ground: Rethinking the Human Place in Nature.* W.W. Norton & Co 69–90.

Egan, Kristen R. 2011. 'Conservation and cleanliness: racial and environmental purity in Ellen Richards and Charlotte Perkins Gilman'. *Women's Studies Quarterly* 39(3–4):77–92.

Hale, Piers J. 2003. 'Labor and the human relationship with nature: the naturalization of politics in the work of Thomas Henry Huxley, Herbert George Wells, and William Morris'. *Journal of the History of Biology.* 36(2):249–84.

Hans, Namit. 2016. 'Khejri, the tree that inspired Chipko movement, is dying a slow death'. *The Indian Express* 4 December.

Hartnett, Kevin. 2013. 'Was Dickens's *A Christmas Carol* borrowed from Lowell's mill girls?'. *The Boston Globe* 15 December.

Hickman, Leo. 2011. 'The 1847 lecture that predicted human-induced climate change'. *Guardian* Environment Blog 20 June.

Holmes, Richard. 2008. *The Age of Wonder: How the Romantic Generation Discovered the Beauty and Terror of Science.* Harper Collins.

Jackson, Roland. 2018. *The Ascent of John Tyndall.* Oxford University Press.

Jevons, William Stanley. 1866. *The Coal Question; An Inquiry Concerning the Progress of the Nation, and the Probable Exhaustion of Our Coal Mines,* 2nd edition. Macmillan.

Johnson, Benjamin Heber. 2017. *Escaping the Dark, Gray City: Fear and Hope in Progressive-era Conservation.* Yale University Press.

Katz, Esther. 1995. 'The editor as public authority: interpreting Margaret Sanger'. *The Public Historian* 17(1):41–5.

Krausman, Paul & Mahoney, Shane P. 2015. 'How the Boone and Crockett Club (B&C) shaped North American conservation'. *International Journal of Environmental Studies* 72(5):746–55.

Nobles, Gregory. 2020. 'The myth of John James Audubon'. *Audubon Magazine* 31 July.

Malm, Andreas. 2016. *Fossil Capital: The Rise of Steam Power and the Roots of Global Warming.* Verso.

Marsden, Ben & Smith, Crosbie. 2005. *Engineering Empires: A Cultural History of Technology in Nineteenth-Century Britain.* Palgrave Macmillan.

Mitchell, Timothy. 2011. *Carbon Democracy: Political Power in the Age of Oil.* Verso, London.

Monk, Brentin. 2017. 'The green movement is talking about racism? It's about time'. *Outside Magazine* 27 February.

Montrie, Chad. 2004. '"I think less of the factory than of my native dell"': labor, nature, and the Lowell 'mill girls'. *Environmental History* 9(2): 275–95.

Montrie, Chad. 2011. *A People's History of Environmentalism in the United States.* Continuum.

Mosley, Stephen. 2001. *The Chimney of the World: A History of Smoke Pollution in Victorian and Edwardian Manchester.* White Horse Press.

Phillips, H. A. 1882. 'Pollution in the atmosphere'. *Nature* 27:127.

Purdy, Jedediah. 2015. 'Environmentalism's racist history'. *The New Yorker* 13 August.

Reid, T. D. W. & Reid, Naomi. 1979. 'The 1842 Plug Riot in Stockport'. *International Review of Social History* 24(1):55–79.

Richards, Ellen H. 1910. *Euthenics, the Science of Controllable Environment.* Whitcomb & Barrows.

Sanger, Margaret. 1926. 'The function of sterilization', extract of an address delivery to the Institute of Euthenics at Vassar College, 5 August, republished by the *NYU Margaret Sanger Project.*

Schivelbusch, Wolfgang. 1988. *Disenchanted Night: Industrialization of Light in the Nineteenth Century.* University of California Press.

Schulz, Kathryn. 2015. 'The moral judgements of Henry David Thoreau'. *The New Yorker* 12 October.

Shanahan, Mike. 2010. 'A challenge: to anyone who ever used the phrase '"tree-hugger"'. Under the Banyan Tree blog, 5 September.

Shiva, Vandana & Bandyopadhyay, Jayanta. 1986. 'The evolution, structure, and impact of the Chipko movement'. *Mountain Research and Development* 6(2):133–42.

Spence, Mike. 1996. 'Dispossessing the wilderness: Yosemite Indians and the National Park ideal, 1864–1930'. *Pacific Historical Review* 65(1): 27–59.

Spence, Mike. 1999. *Dispossessing the Wilderness: Indian Removal and the Making of the National Parks.* Oxford University Press.

Spiro, Jonathan Peter. 2009. *Defending the Master Race: Conservation, Eugenics, and the Legacy of Madison Grant.* University of Vermont Press.

Stillwell, Devon. 2012. 'Eugenics Visualised: The Exhibit of the Third International Congress of Eugenics, 1932'. *Bulletin of the History of Medicine* 86(2):206–36.

Stradling, David. 1999. *Smokestacks and Progressives: Environmentalists, Engineers and Air Quality in America, 1881–1951.* John Hopkins University Press.

Strandling, David & Thorsheim, Peter. 1999. 'The smoke of great cities: British and American efforts to control air pollution, 1860–1914'. *Environmental History* 4(1):6–31.

1842. 'The Riots in the Manufacturing Districts'. *The Times* 17 August, 4–6.

1880. 'Life and talk in London'. *New York Times* 16 February 1–2.

1881. 'Smoke-abatement exhibition'. *Nature* 25:121–2.

1883. 'The atmosphere'. *New York Times,* 6 January, 4.

1926. 'Vassar girls to study home-making as career'. *New York Times* 23 May, 220

1937. 'Madison Grant, 71, zoologist, is dead'. *New York Times* 31 May, 15.

Chapter Seven: The Rise, Fall and Rise of Big Oil

Cummins, Ian & Beasant, John. 2005. *Shell Shock: The Secrets and Spin of an Oil Giant.* Mainstream.

Edgerton, David. 2013. *England and the Aeroplane,* revised edition. Penguin.

Gandhi, Lakshmi. 2013. 'A history of snake oil salesmen'. NPR Code Switch 26 August.

Gilson, Dave. 2011. 'Octopi Wall Street!'. *Mother Jones* 6 October.

Gore, Timothy. 2020. 'Confronting carbon inequality'. *Oxfam* 21 September.

Mitchell, Timothy. 2011. *Carbon Democracy: Political Power in the Age of Oil.* Verso.

Sampson, Anthony. 1980. *The Seven Sisters: The Great Oil Companies and the World They Shaped,* third edition. Hodder & Stoughton.

Weinberg, Steve. 2008. *Taking on the Trust: The Epic Battle of Ida Tarbell and John D. Rockefeller.* W. W. Norton.

Yergin, Daniel. 2008. *The Prize: The Epic Quest for Oil, Money and Power,* new edition with epilogue. Simon & Schuster.

Chapter Eight: Big Science

Agar, Jon. 2012. *Science in the Twentieth Century and Beyond.* Polity.

Baker, Zeke. 2017. 'Climate state: science-state struggles and the formation of climate science in the US from the 1930s to 1960s'. *Social Studies of Science* 47(6):861–87.

Berkner, Lloyd V. 1954. 'International scientific action: the International Geophysical Year 1957–58'. *Science* 119(3096):569–75.

Blair, William R. 1909. 'The exploration of the upper air by means of kites and balloons'. *Proceedings of the American Philosophical Society* 48(191)8–33.

Born, Maria. 2011. 'Concentrating on CO_2: The Scandinavian and Arctic measurements'. *Osiris, Klima* 26(1):165–79.

Budiansky, Stephen. 1984. 'Military and basic science offered more rapid growth'. *Nature* 307:491–2.

Bush, Vannevar. 1945. 'As we may think'. *The Atlantic* July, republished online.

Bush, Vannevar. 1945. 'Science, the endless frontier: a report to the president'. US Government Printing Office, Washington DC.

Capshew, James H. & Rader, Karen A. 1992. 'Big science: price to the present'. *Osiris* 7:2–25.

Day, Deborah. Undated. *Roger Randall Dougan Revelle Biography*. PDF downloadable from UC San Diego Library.

Doel, Ronald E. 2003. 'Constituting the post-war Earth sciences: the military's influence on the environmental sciences in the USA after 1945'. *Social Studies of Science* 33(5):635–66.

Eisenhower, Dwight D. 1955. 'The President's news conference'. 2 March; transcript at The American Presidency Project, UC Santa Barbara.

Eisenhower, Dwight D. 1957. 'Remarks in connection with the opening of the International Geophysical Year'. 30 June; transcript at The American Presidency Project, UC Santa Barbara

Engel, Leonard. 1953. 'The weather is really changing'. *New York Times*, 12 July 160:184–5.

Fleming, James. 2007. 'A 1954 color painting of weather systems as viewed from a future satellite'. *Bulletin of the American Meteorological Society* October 1525–7.

Fleming, James Rodger. 2012. *Fixing the Sky: The Checkered History of Weather and Climate Control*. Columbia University Press.

Fleming, James Rodger. 2016. *Inventing Atmospheric Science: Bjerknes, Rossby, Wexler, and the Foundations of Modern Meteorology*. MIT Press.

Freeman, Chris. 1999. 'The social function of science' in Swann, Brenda & Aprahamian, Francis (eds). *J. D. Bernal: A Life in Science and Politics*. Verso, 101–31.

Hamblin, Jacob Darwin. 2013. *Arming Mother Nature: The Birth of Catastrophic Environmentalism*. Oxford University Press.

Harper, Kristine C. 2003. 'Research from the boundary layer: civilian leadership, military funding and the development of numerical weather prediction (1946–55)'. *Social Studies of Science* 33(5):667–96.

Harper, Kristine C. 2008. *Weather by the Numbers: The Genesis of Modern Meteorology*. MIT Press.

Hollingham, Richard. 2014. 'V2: The Nazi rocket that launched the space age'. *BBC Future*, 8 September.

Howkins, Adrian. 2011. 'Melting empires? Climate change and politics in Antarctica since the International Geophysical Year'. *Osiris* 26(1):180–97.

Kaempffert, Waldemar. 1945. 'Julian Huxley pictures the more spectacular possibilities that lie in atomic power'. *New York Times* 9 December, 77.

Kimble, George H. T. 1950. 'The changing climate'. *Scientific American* 182(4):48–53.

Landsberg, H. 1940. 'The use of solar energy for the melting of ice'. *Bulletin of the American Meteorological Society* March 102–7.

Langmuir, Irving. 1950. 'Control of precipitation from cumulus clouds by various seeding techniques'. *Science* 12(2898)35–41.

Martin, D. C. 1957. 'The inauguration of the International Geophysical Year'. *Notes and Records of the Royal Society of London* 12(2):160–2.

Martin, D. C. 1959. 'Some achievements of the International Geophysical Year'. *Journal of the Royal Society of Arts* 107(5034):406–22.

McCray, W. Patrick. 2006. 'Amateur scientists, the International Geophysical Year, and the ambitions of Fred Whipple'. *Isis* 97(4):634–58.

Monmonier, Mark. 1999. *Air Apparent: How Meteorologists Learned to Map, Predict, and Dramatize Weather.* University of Chicago Press.

Painter, Thomas H. et al. 2013. 'End of the little ice age in the Alps forced by industrial black carbon'. *PNAS* 110(38):15216–21.

Pielke Jr, Roger. 2010. 'In retrospect: science – the endless frontier'. *Nature* 466:922–3.

Pielke Jr, Roger. 2020. 'A "sedative" for science policy'. *Issues in Science and Technology* 37(1).

Plass, Gilbert N. 1956. 'The carbon dioxide theory of climatic change'. *Tellus* 8(2):140–54.

Plass, Gilbert N., Fleming, James Rodger & Schmidt, Gavin. 2010 'American Scientist Classics: carbon dioxide and the climate'. Edited version of 1956 article with modern commentary. *American Scientist* 98(1):58–67.

Potter, Robert D. 1940. 'Can science defrost the Antarctic'. *San Bernardino Sun* 10 March 46:30, via the California Digital Newspaper Collection, University of California, Riverside.

Reinholds, Robert. 1974. 'Dr. Vannevar Bush is dead at 84' *New York Times* 20 June, 36.

Revelle, Roger & Suess, Hans E. 1957. 'Carbon dioxide exchange between atmosphere and ocean and the question of an increase of atmospheric CO_2 during the past decades'. *Tellus* 9(1):18–27.

Sapolsky, Harvey M. 1990. *Science and the Navy: The History of the Office of Naval Research.* Princeton University Press.

Shalett, Sidney. 1946. 'Electronics and weather figuring'. *New York Times* 11 January, 12.

Sörlin, Sverker. 2011. 'The anxieties of a science diplomat: field coproduction of climate knowledge and the rise and fall of Hans Ahlmann's "polar warming"'. *Osiris, Klima* 26 (1):66–88.

Steward, Fred. 1999. 'The social function of science' in Swann, Brenda & Aprahamian, Francis (eds). *J. D. Bernal: A Life in Science and Politics.* Verso, 37–77.

Rainger, Ronald. 2001. 'Constructing a landscape for postwar science: Roger Revelle, the Scripps Institution and the University of California, San Diego'. *Minerva* 39(3):327–52.

Rainger, Ronald. 2004. '"A wonderful oceanographic tool"': the atomic bomb, radioactivity and the development of American oceanography' in Rozwadowski, Helen M. and Van Keuren, David K. (eds). *The Machine in Neptune's Garden: Historical Perspectives on Technology and the Marine Environment.* Science History Publications 96–132.

Taylor, Geoffrey Ingram. 1962. 'Gilbert Thomas Walker: 1868–1958'. *Biographical Memoirs of Fellows of the Royal Society* 8 November 166–74.

Vaughan, William W. & Johnson Dale L. 1994. 'Meteorological satellite – the very early years, prior to launch of TIROS-1'. *Bulletin of the American Meteorological Society* 75(12):2295–2302.

Waterman, Alan T. 1956. 'The International Geophysical Year'. *American Scientist* 44(2):130–3.

Weinberg, Alvin M. 1961. 'Impact of large-scale science on the United States'. *Science* 134(3473):161–4.

Wexler, Harry. 1954. 'Observing the weather from a satellite vehicle'. *Journal of the British Interplanetary Society* 13:269–76, reprinted in Logsdon, John M. (ed). 1998. *Exploring the Unknown: Selected Documents in the History of the U.S. Civil Space Program Vol 3 Using Space* 178–83. NASA History Office.

Wood, Colonel F. B. & Wexler, Major Harry. 1945. 'A flight into the September, 1944, hurricane off Cape Henry, Virginia'. *Bulletin of the American Meteorological Society* 26 May 153–6.

Zachary, G. Pascal. 1997. *Endless Frontier: Vannevar Bush, engineer of the American century*. The Free Press.

Zworykin, V. K. 1945. 'Outline of weather proposal'. Republished with introduction from James Rodger Fleming in *History of Meteorology* 4:57–78.

1940. 'Coal dust speeds melting of ice by absorbing Sun's heat'. *Popular Mechanics* April 544.

1950. 'Getting Warmer?' *Time* 15 May, 76–7.

1950. 'Weather or Not'. *Time* 28 August, 52–6.

1953. 'Invisible Blanket'. *Time* 25 May, 82–5.

1953. 'Growing blanket of carbon dioxide raises Earth's temperature'. *Popular Mechanics* August 119.

1956. 'One big greenhouse'. *Time* 28 May, 59.

1957. 'Geophysical Year on television'. *The Times* 12 June, 10.

1957. *The Restless Sphere*, BBC Television, first broadcast 30 June, archive recording at BBC Archive.

1962. 'Dr Wexler dies, a meteorologist'. *New York Times* 12 August, 80.

2019. 'Obituary notice: Gustaf Arrhenius, 1922–2019'. *Scripps News* 13 February.

Chapter Nine: A Carousel of Progress

Ansell, Nick. 2019. 'Daddy long-legs – a weird and wonderful railway'. Railway Museum blog, 14 January.

Asmus, Peter. 2001. *Reaping the Wind: How Mechanical Wizards, Visionaries, and Profiteers Helped Shape Our Energy Future*. Island Press.

Boon, Rachel. 2014. 'Alexander Parkes: living in a material world'. Science Museum Blog 14 January.

Brady, Hillary. 2016. 'Walt Disney's Progressland'. Smithsonian Institution Archives blog 21 July.

Cowan, Ruth Schwartz. 1985. 'The Industrial Revolution in the home' and 'How the refrigerator got its hum', in MacKenzie, Donald & Wajcman, Judy (eds). *The Social Shaping of Technology: How the refrigerator got its hum*. Open University Press 181–218.

Cummins, Ian & Beasant, John. 2005. *Shell Shock: The Secrets and Spin of an Oil Giant*. Mainstream.

Demeter, Michelle. 2019 'Advancing an optimistic technological narrative in an age of skepticism: General Electric and Walt Disney's Progressland at the 1964–1965 New York World's Fair', in Molella, Arthur P. (ed). *World's Fairs in the Cold War: Science, Technology, and the Culture of Progress*. Pittsburgh Press 87–95.

Fleischer, Matthew. 2020. 'Opinion: Want to tear down insidious monuments to racism and segregation? Bulldoze LA freeways'. *Los Angeles Times* 24 June.

Hermann, Matthias. 1998. 'Signs of hubris: the shaping of wind technology styles in Germany, Denmark, and the United States, 1940–1990'. *Technology and Culture* 39(40):641–70.

Joerges, Bernward. 1999. 'Do politics have artefacts?' *Social Studies of Science* 29(3):411–31.

Kirsch, David A. 2000. *The Electric Vehicle and the Burden of History*. Rutgers.

Madrigal, Alexis. 2011. *Powering the Dream: The History and Promise of Green Technology*. Perseus Books.

Mitchell, Timothy. 2011. *Carbon Democracy: Political Power in the Age of Oil*. Verso.

Norton, Peter D. 2007. 'Street rivals: jaywalking and the invention of the motor age street'. *Technology and Culture* 48(2):331–59.

Norton, Peter D. 2008. *Fighting Traffic: The Dawn of the Motor Age in the American City*. MIT Press.

Nye, David E. 1990. *Electrifying America: Social Meanings of a New Technology*. MIT Press.

Nye, David E. 2001. *Consuming Power: A Social History of American Energies*. MIT Press.

Owens, Brandon N. 2019. *The Wind Power Story: A Century of Innovation that Reshaped the Global Energy Landscape*. Wiley-IEEE Press.

Perkins, Frank C. 1909. 'A new solar power plant'. *Scientific American* 98(6):97.

Perlin, John. 1999. *From Space to Earth: The Story of Solar Electricity*. Aatec Publications.

Rhodes, Richard. 2018. *Energy: A Human History*. Simon & Schuster.

Robbins, Michael. 2000. 'The early years of electric traction: Invention, development, exploitation'. *The Journal of Transport History* 21(1):92–101.

Rogers, Heather. 2005. *Gone Tomorrow: The Hidden Life of Garbage*. The New Press.

Safire, William. 2009. 'The Cold War's hot kitchen'. *New York Times* 23 July.

Sampson, Anthony. 1980. *The Seven Sisters: The Great Oil Companies and the World They Shaped*, 3rd edition. Hodder & Stoughton.

Shuman, Frank. 1914. 'Feasibility of utilizing power from the Sun'. *Scientific American* 110(9):179.

Smith, John Caulfield. 1946. 'The house that Sun built'. *Macleans* 15 January 19:31.

Steyn, Phia. 2009. 'Oil exploration in colonial Nigeria, c.1903–58'. *The Journal of Imperial and Commonwealth History* 37(2):249–74.

Winner, Langdon. 1985. 'Do artefacts have politics', in MacKenzie, Donald & Wajcman, Judy (eds). *The Social Shaping of Technology: How the refrigerator got its hum*. Open University Press 26–38.

Yergin, Daniel. 2008. *The Prize: The Epic Quest for Oil, Money and Power*, new edition with epilogue. Simon & Schuster.

1916. 'American inventor uses Egypt's Sun for power'. *New York Times Magazine* 2 July, 15.

1939. 'Sir Henri Deterding: The international oil industry', *The Times* 6 February, 14.

1954. 'Vast power of the Sun is tapped by battery using sand ingredient'. *New York Times*, 26 April, 1:11.

2016. 'Science under President Trump'. *Nature* 9 November.

Chapter Ten: Growing Concern

Brannon, H. R. Jr et al. 1957. 'Radiocarbon evidence on the dilution of atmospheric and oceanic carbon by carbon from fossil fuels'. *Eos, Transactions American Geophysical Union* 38(5):643–50.

Brewer, Sam Pope. 1971. 'Study says man alters climate'. *New York Times*, 23 September, 22.

Capra, Frank. 1958. *The Unchained Goddess*, first broadcast 12 February. NW Ayer.

Chang, Kenneth. 2008. 'Edward N. Lorenz, a meteorologist and a father of chaos theory, dies at 90'. *New York Times* 17 April.

Dansgaard, Willi et al. 1969. 'One thousand centuries of climatic record from Camp Century on the Greenland ice sheet'. *Science* 166(3903):377–81.

Dansgaard, Willi. 2005. *Frozen Annals: Greenland Ice Cap Research*. Neils Bohr Institute, downloadable at Copenhagen University website.

Darling, Frank Fraser. 1969. 'Global changes – actual and possible', Reith Lectures 1969: Wilderness and Plenty, *BBC Radio*, 30 November, available via BBC Sounds archive.

Death, Carl. 2015. 'Disrupting global governance: protest at environmental conferences from 1972 to 2012'. *Global Governance* 21(4):579–98.

Dry, Sarah. 2019. *Waters of the World*. Scribe.

Edwards, Paul N. 2010. *A Vast Machine: Computer Models, Climate Data, and the Politics of Global Warming*. MIT Press.

Eichhorn, Noel. 1963. *Implications of Rising Carbon Dioxide Content of the Atmosphere. A Statement of Trends and Implications of Carbon Dioxide Research Reviewed at a Conference of Scientists*. Conservation Foundation.

Emmanuel, Kerry. 2008. 'Edward N. Lorenz (1917–2008)'. *Science* 320(5879):1025.

Emmelin, Lars. 1972. 'The Stockholm Conferences'. *Ambio* 1(4):135–40.

Franta, Benjamin. 2018. 'On its 100th birthday in 1959, Edward Teller warned the oil industry about global warming'. *Guardian*'s Climate Consensus – the 97% blog, 1 January.

Gilbert, James. 2008. *Redeeming Culture: American Religion in an Age of Science*. University of Chicago Press.

Gillis, Justin. 2010. 'A scientist, his work and a climate reckoning', *New York Times* 12 December.

Hart, David M. & Victor, David G. 1993. 'Scientific elites and the making of US policy for climate change research, 1957–74'. *Social Studies of Science* 23(4):643–80.

Howe, Joshua P. 2014. *Behind the Curve: Science and the Politics of Global Warming*. University of Washington Press.

Huxley, Julian. 1946. *UNESCO: Its purpose and its philosophy*. Preparatory Commission of UNESCO.

Johnson, Lyndon B. 1965. State of the Union, 4 January.

Johnson, Lyndon B. 1965. Natural Beauty Message, 8 February.

Johnston, Sean F. 2018. 'Alvin Weinberg and the promotion of the technological fix'. *Technology and Culture* 59(3):620–51.

Jones, Charles A. 1958. 'A Review of the Air Pollution Research Program of the Smoke and Fumes Committee of the American Petroleum Institute'. *Journal of the Air Pollution Control Association* 8(3):268–72.

Kaysen, Carl. 1972 .'The computer that printed out W★O★L★F★'. *Foreign Affairs* July, 660–8.

Keeling, Charles D. 1960. 'The concentration and isotopic abundances of carbon dioxide in the atmosphere'. *Tellus* 12(2):200–3.

Keeling, Charles D. 1970. 'Is carbon dioxide from fossil fuel changing man's environment?' *Proceedings of the American Philosophical Society* 114(1):10–7.

Keeling, Charles D. 1998. 'Rewards and penalties of monitoring the Earth'. *Annual Review of Environment and Resources* 23:25–82.

Kellaway, Kate. 2010. 'How the *Observer* brought the WWF into being'. *Observer* 7 November.

Kirk, Andrew G. 2007.*Counterculture Green: The Whole Earth Catalog and American Environmentalism*. University of Kanas Press.

Kroll, Gary. 2001.'The "Silent Springs" of Rachel Carson: Mass media and the origins of modern environmentalism'. *Public Understanding of Science* 10:403–20.

Langway, Chester C. Jr. 2011. 'Willi Dansgaard (1922–2011)'. *Arctic* 64(3):385–7.

Maddox, John. 1971. 'The Doomsday Syndrome'. *Nature* 233:15–6.

Maddox, John. 1972. *The Doomsday Syndrome*. Macmillan.

Malone, Thomas F., Goldberg, Edward D. & Munk, Walter H. 1998. *Roger Randall Dougan Revelle, 1909–1991, A Biographical Memoir*. National Academy of Sciences.

Matthews, M. A. 1959. 'The Earth's carbon cycle'. *New Scientist* 8 October 664–6.

Masood, Ehsan. 2015. 'Maurice Strong (1929–2015)'. *Nature* 528:480.

Maugh, Thomas H. II. 2008. 'His computer riddle led to chaos theory'. *Los Angeles Times* 18 April.

Morello, Lauren. 2014. 'Philanthropists aid Keeling curve'. *Nature* News Blog, 5 September.

Murphy, Priscilla Coit. 2005. *What a Book Can Do: The Publication and Reception of Silent Spring*. University of Massachusetts Press.

Palmer, T. N. 2009. 'Edward Norton Lorenz: 23 May 1917–16 April 2008'. *Biographical Memoirs of Fellows of the Royal Society* 55:140–55.

Plass, Gilbert N. 1956. 'Carbon dioxide and the climate'. *American Scientist* 44(3): 302–16.

Plass, Gilbert N. 1959. 'Carbon dioxide and climate'. *Scientific American* 201(1):41–7, also republished online.

Robertson, Thomas. 2012. *The Malthusian Moment: Global Population Growth and the Birth of American Environmentalism*. Rutgers University Press.

Robinson, E. & Robbins, R. C. 1968. *Sources, abundance, and fate of gaseous atmospheric pollutants*. Menlo Park, CA: Stanford Research Institute.

Rome, Adam. 2003. 'Give Earth a chance: The environmental movement and the sixties'. *The Journal of American History* 90(2):525–54.

Rome, Adam. 2010. 'The genius of Earth Day'. *Environmental History* 15(2):194–205.

Strong, Maurice. 1972. 'The Stockholm Conference: where science and politics meet'. *Ambio* 1(3):73–8.

Sullivan, Walter. 1961. 'Air found gaining in carbon dioxide'. *New York Times* 11 September, 29.

Weart, Spencer R. 1997. 'Global warming, Cold War, and the evolution of research plans'. *Historical Studies in the Physical and Biological Sciences* 27(2):319–56.

Weart, Spencer R. 2003. *The Discovery of Global Warming*. Harvard University Press.

Weart, Spencer R. 2007. 'Money for Keeling: monitoring CO_2 levels'. *Historical Studies in the Physical and Biological Sciences* 37(2):435–52.

Weinberg, Alvin. 1967. 'Can technology replace social engineering?'. *The American Behavioral Scientist* May, 7–10.

Weindling, Paul. 2012. 'Julian Huxley and the continuity of eugenics in twentieth-century Britain'. *Journal of Modern European History* 10(4):480–99.

Wexler, Henry. 1957. 'Meteorology in the International Geophysical Year'.
    The Scientific Monthly 84(3):141–5.

Winter, Anna. 2018. 'The complicated legacy of Stewart Brand's Whole
    Earth Catalog'. *New Yorker* 16 November.

1965. *Restoring the Quality of Our Environment*, a report of the Environmental
    Pollution Panel of the President's Science Advisory Committee. The
    White House.

1972. 'The case against hysteria'. *Nature* 235:63–4.

1972. 'Environment conference will offer some sideshows'. *New York Times*
    5 June, 24.

2016. *Smoke and Fumes*. Center for International Environmental Law,
    www.smokeandfumes.org.

2020. *Climate Files*, Climate Investigations Center, www.climatefiles.com.

Chapter Eleven: Crisis Point

Agar, Jon. 2015. '"Future Forecast – changeable and probably getting worse":
    the UK's early response to anthropomorphic climate change'. *Twentieth
    Century British History* 26(4):602–28.

Banerjee, Neela, Strong, Lisa & Hasemyer, David. 2017. 'Exxon's own
    research confirmed fossil fuels' role in global warming decades ago'. *Inside
    Climate News* 16 September.

Barry, R. G. 2015. 'The shaping of climate science: half a century in personal
    perspective'. *History of Geo- and Space Sciences* 6:87–105.

Broecker, Wallace. 1975. 'Climatic change: are we on the brink of a
    pronounced global warming?'. *Science* 189(4201):460–3.

Brysse, Keynyn et al. 2013. 'Climate change prediction: Erring on the side of
    least drama?'. *Global Environmental Change* 23(1):327–37.

Corfee-Morlot, Jan, Maslin, Mark & Burgess, Jacquelin. 2007. 'Global
    warming in the public sphere'. *Philosophical Transactions of the Royal Society
    A: Mathematical, Physical and Engineering Sciences* 365(1860):2741–76.

Cowan, Edward. 1973. 'Energy volunteerism'. *New York Times* 9 November, 26.

Finkbeiner, Anne. 2006. *The Jasons: The Secret History of Science's Postwar Elite*.
    Viking.

Gribben, John. 1976. 'Man's influence not yet felt by climate'. *Nature* 264:608.

Haitch, Richard. 1977. 'The President's "solar advocate"'. *New York Times* 5 June.

Hayes, Denis. 1981. 'Washington decrees a solar eclipse'. *New York Times*
    12 August, 27.

Henderson, Gabriel. 2014. 'The dilemma of reticence: Helmut Landsberg,
    Stephen Schneider, and public communication of climate risk, 1971–1976'.
    History of Meteorology 6:53–78.

Howe, Joshua P. 2017. *Making Climate Change History: Primary Sources from
    Global Warming's Past*, foreword by Paul S. Sutter. University of
    Washington Press.

Jennings, Katie, Grandoni, Dino & Rust, Susanne. 2015. 'Special Report: How Exxon went from leader to skeptic on climate change research'. *LA Times* 23 October.

Kellogg, William W. 1987. 'Mankind's impact on climate: the evolution of an awareness'. *Climatic Change* 10:113–36.

Kissinger, Henry A. 1974. 'Address to the sixth special session of the United Nations General Assembly'. *International Organization* 28(3):573–83.

Lamb, Peter J. 2002. 'The climate revolution: a perspective'. *Climatic Change* 54:1–9.

Lovins, Amory B. 1976. 'Energy strategy: the road not taken?'. *Foreign Policy* 20 October, 65–96.

Mason, B. J. 1977. 'Man's influence on weather and climate'. *Journal of the Royal Society of Arts* 125(5247):150–65.

Munk, Walter, Oreskes, Naomi & Muller, Richard. 2004. 'Gordon James Fraser MacDonald July 30, 1929–May 14, 2002'. *Biographical Memoirs of the National Academies of Science, Engineering and Medicine* 84.

Nye, David E. 2001. *Consuming Power: A Social History of American Energies*. MIT Press.

Oreskes, Naomi, Conway, Erik M. & Shindell, Matthew. 2008. 'From Chicken Little to Dr. Pangloss: William Nierenberg, global warming, and the social deconstruction of scientific knowledge'. *Historical Studies in the Natural Sciences* 38(1):109–52.

Oreskes, Naomi & Conway, Erik M. 2010. *Merchants of Doubt: How a Handful of Scientists Obscured the Truth on Issues from Tobacco Smoke to Global Warming*. Bloomsbury.

Perlin, John. 1999. *From Space to Earth: The Story of Solar Electricity*. Aatec Publications.

Prochnau, Bill. 1983. 'The Watt controversy'. *Washington Post* 20 June.

Rich, Nathaniel. 2019. *Losing Earth: The Decade We Could Have Stopped Climate Change*. Farrar, Straus and Giroux.

Rosenblum, Nancy L. & Pomerance, Rafe. 2020. 'A Conversation', *Daedalus*, 1 October 2020, 149(4):163–179.

Russakoff, Dale. 1983. 'Watt's Adversaries Would Almost Hate to See Him Resign'. *Washington Post*, 7 October.

Schneider, Stephen. 2009. *Science as a Contact Sport: Inside the Battle to Save Earth's Climate*. National Geographic.

Smith, Adrian. 2014. *Socially Useful Production*, STEPS Working Paper 58. University of Sussex/Institute of Development Studies.

Sterba, James P. 1977. 'Problems from climate changes foreseen in a 1974 CIA Report'. *New York Times* 2 February, 10.

Stoddard, Woody. 2002. 'The life and work of Bill Heronemus, wind engineering pioneer'. *Wind Engineering* 26(5):335–41

Sullivan, Patricia. 2004. 'Anne Gorsuch Burford, 62, Dies', Washington Post, 22 July 2004.

Wade, Nicolas. 1979. 'CO_2 in climate: gloomsday predictions have no fault'. *Science* 23 November 206(4421)912–3.

Weart, Spencer R. 2003. *The Discovery of Global Warming.* Harvard University Press.

1973. 'Excerpts from Nixon message: developing our domestic energy resources'. *New York Times* 19 April, 53.

1974. 'A study of climatological research as it pertains to intelligence problems'. CIA Office of Research and Development. US Government Printing Office.

1983. *Changing Climate: Report of the Carbon Dioxide Assessment Committee.* The National Academies Press.

Chapter Twelve: Already Happening Now

Agar, Jon. 2019. *Science Policy Under Thatcher.* UCL Press.

Bolin, Bert. 2007. *A History of the Science and Politics of Climate Change: The Role of the Intergovernmental Panel on Climate Change.* Cambridge University Press.

Brown, Paul & Rocha, Jan. 1992. 'Earth summit: Rio opens with plea for proof of global brotherhood'. *Guardian* 4 June.

Burkman, Oliver. 2003. 'Memo exposes Bush's new green strategy'. *Guardian* 4 March.

Carter, Neil Thomas & Childs, Mike. 2018. 'Friends of the Earth as a policy entrepreneur: "The Big Ask" campaign for a UK climate change act'. *Environmental Politics* 27(6):994–1013.

Corfee-Morlot, Jan, Maslin, Mark & Burgess, Jacquelin. 2007. 'Global warming in the public sphere'. *Philosophical Transactions of the Royal Society A: Mathematical, Physical and Engineering Sciences* 365(1860): 2741–776.

Doherty, Brian. 1999. 'Paving the way: the rise of direct action against road-building and the changing character of British environmentalism'. *Political Studies* June, 275–91.

Farman, Joseph. 2010–11. *Joseph Farman: An Oral History of British Science.* British Library, interviewed by Dr Paul Merchant.

Frey, Darcy. 2002. 'How Green is BP?'. *New York Times*, 8 December 274:606–10.

Goldenberg, Suzanne. 2015. 'Paris climate summit: the climate circus comes to town'. *Guardian* 26 November.

Gore, Al. 1992. *Earth in the Balance: Forging a New Common Purpose.* Earthscan.

Greenhouse, Steven. 1992. 'A closer look: ecology, the economy and Bush'. *New York Times* 14 June, 124, 129.

Hansen, James. 2009. *Storms of my Grandchildren: The Truth About the Coming Climate Catastrophe and our Last Chance to Save Humanity.* Bloomsbury.

Hirst, David G. 2014. 'Balancing scientific credibility and political legitimacy: The IPCC's first assessment cycle, 1988–1990'. *History of Meteorology*, 79–94.

Howe, Joshua P. 2014. *Behind the Curve: Science and the Politics of Global Warming*. University of Washington Press.

Hufbauer, Carolyn Revelle. 1992. 'Global warming: what my father really said'. *Washington Post* 13 September.

Lahsen, Myanna. 2008. 'Experiences of modernity in the greenhouse: A cultural analysis of a physicist "trio" supporting the backlash against global warming'. *Global Environmental Change* 18:204–19.

Leggett, Jeremy. 1999. *The Carbon War: Global Warming and the End of the Oil Era*. Penguin.

Mann, Michael E. 2012. *The Hockey Stick and the Climate Wars: Dispatches from the Front Lines*. Columbia University Press.

North, Peter. 2011. 'The politics of climate activism in the UK: a social movement analysis'. *Environment and Planning A* 43:1581–98.

Oreskes, Naomi & Conway, Erik M. 2010. *Merchants of Doubt: How a Handful of Scientists Obscured the Truth on Issues from Tobacco Smoke to Global Warming*. Bloomsbury.

Pearce, Fred. 2010. *The Climate Files: The battle for the truth about global warming*. Guardian Books.

Rayner, Steve. 1989. 'Fiddling while the globe warms?' *Anthropology Today* 6(5):1–2.

Roberts, Leslie. 1989. 'Global warming: blaming the Sun'. *Science* 246(4933)992–993.

Schlemback, Raphael. 2011. 'How do radical climate movements negotiate their environmental and their social agendas? A study of debates within the Camp for Climate Action, UK'. *Critical Social Policy* 31(2):194–215.

Schwartz, John. 2020. 'S. Fred Singer, a leading climate change contrarian, dies at 95'. *New York Times* 11 April.

Shabecoff, Philip. 1988. 'Global warming has begun, expert tells Senate: sharp cuts in burning of fossil fuels is urged to battle shift in climate'. *New York Times* 24 June, 1, 14.

Shabecoff, Philip. 1988. 'Norway and Canada call for pact to protect atmosphere'. *New York Times* 28 June, 4.

Shabecoff, Philip. 1990. 'Bush denies putting off action on adverting global climate shift'. *New York Times* 19 April, 30.

Smith, Heather. 2014. 'How 350.org went from "strange kid" to head of the green class'. *Grist* 5 February.

Thatcher, Margaret. 1988. 'Speech to the Royal Society', 27 September. Margaret Thatcher Foundation Archive.

Thatcher, Margaret. 1989. 'Speech to United Nations General Assembly. Global Environment', 8 November. Margaret Thatcher Foundation Archive.

Vanderheiden, Steve. 2005. 'Eco-terrorism or justified resistance? Radical environmentalism and the "War on Terror"'. *Politics and Society* 33(3):425–47.

Vidal, John, Stratton, Allegra & Goldenberg, Suzanne. 2009. 'Low targets, goals dropped: Copenhagen ends in failure'. *Guardian* 19 December.

Weart, Spencer R. 2003. *The Discovery of Global Warming.* Harvard University Press.

2020. *Climate Files*, Climate Investigations Center, www.climatefiles.com.

End point?

Banerjee, Neela, Strong, Lisa & Hasemyer, David. 2017. 'Exxon's own research confirmed fossil fuels' role in global warming decades ago'. *Inside Climate News*, 16 September.

Brysse, Keynyn et al. 2013. 'Climate change prediction: Erring on the side of least drama?'. *Global Environmental Change* 23(1):327–37.

Cowan, Ruth Schwartz. 1985. 'The industrial revolution in the home' and 'How the refrigerator got its hum', in MacKenzie, Donald & Wajcman, Judy (eds). *The Social Shaping of Technology: How the refrigerator got its hum.* Open University Press, 181–218.

Gillis, Justin. 2013. 'Climate maverick to retire from NASA'. *New York Times* 1 April.

Greene, Brian. 2016. 'Impossible isn't a fact; it's an attitude: Christiana Figueres at TED2016'. *Ted Blog,* 17 February.

Hayes, Denis. 1981. 'Washington decrees a solar eclipse'. *New York Times* 12 August, 27.

King, Ed. 2016. 'Christiana Figueres: Protecting the vulnerable is my priority'. *Climate Change News* 8 June.

Marvel, Kate. 2018. 'Thinking about climate on a dark, dismal morning'. *Scientific American* 25 December.

Obama, Barack. 2014. 'Remarks by the President at UN Climate Change Summit', 23 September. Obama White House Archives.

Rich, Nathaniel. 2019. *Losing Earth: The Decade We Could Have Stopped Climate Change.* Farrar, Straus and Giroux.

Simcock, Niel, Willis, Rebecca & Capener, Peter. 2016. *Cultures of Community Energy.* British Academy.

Weart, Spencer R. 1997. 'Global warming, Cold War, and the evolution of research plans'. *Historical Studies in the Physical and Biological Sciences* 27(2):319–56.

Weart, Spencer R. 2003. *The Discovery of Global Warming.* Harvard University Press.

Wells, Ida B. (ed.) 1893. *The Reason Why the Colored American Is Not in the World's Columbian Exposition* (reprinted by University of Illinois Press, edited by Robert W. Rydell, 1999).

1983. *Changing Climate: Report of the Carbon Dioxide Assessment Committee.* The National Academies Press.

Acknowledgements

This book started as a walk by the river. Back in 2015, a group of us decided to build an alternative walking tour of London, on energy and climate change. We started at the old Bankside Power Station (now Tate Modern), making our way under the Blackfriars solar bridge and the old site of Boulton & Watt's Albion Mills, along the Southbank to Shell's offices before crossing the river to Embankment, ending up in Westminster, outside the old Department for Energy and Climate Change offices.

It was while developing the script for that tour that I caught a bug for reading, writing and talking about the history of the climate crisis, which later turned into an email newsletter and blog, some public talks, another set of walking tours and, finally, this book. I owe a huge amount to my fellow tour guides Chitanya Kumar, Sophie Neuberg and Max Wakefield for taking those early steps with me, as well as everyone who joined us on those tours, especially those who asked questions and shared their own stories.

I'd never have been able to develop the idea in a full book if it wasn't for my colleagues at the climate charity Possible. They allowed me space to write and were early guinea pigs for some of the content (as well as being ingenious and dedicated people who inspire me daily). Many thanks to Possiblists past and present, including Amy Cameron, Sarah Barfield Marks, Hannah Bland, Matt Bonner, Flossie Boyd, Liz Clark-Lim, Imogen Downing, Patrik Ewe, Skye Golding, Esther Griffin, Dan Jones, Neil Jones, Emma Kemp, Amita Kotecha, Carly McLachlan, Leo Murray, David Rouane, Max Wakefield (again) and Alethea Warrington.

I've also learnt so much about science, history, climate change and politics from colleagues, tutors, students, and co-conspirators at the Science Museum, UCL's Department of Science and Technology Studies, Imperial's Science Communication Group, the Science Question Time events, Sussex's Science Policy Research Unit, the *Guardian* science network, the Road to Paris blog, Storythings, City's journalism school and the Art Not Oil collective. This includes (but is in no way limited to) Brian Balmer, Jess Bland, Joe Cain, Uslan Cevet, Hasok Chang, Danny Chivers, Giskin Day, Mel Evans, Kieron Flanagan, Owen Gaffney, Anna Galkina, Chris Garrard, Sam Geall, Duncan Geere, Jane Gregory, Victoria Herridge, Emma Hughes, Kathleen Issac, Imran Khan, Matt Locke, Mariana Mazzacato, Felicity Mellor, Johannes Mengel, Erik Milstone, Paul Nightingale, Nathan Oxley, Robert Seaman, Rachel Souhami, Adrian Smith, Beck Smith, Kevin Smith, Jack Stilgoe, Connie St Louis, Jon Turney, James Wilsdon, Chris Wlasnik, Jess Worth and Denise Young.

Without my astute and encouraging agent, Donald Winchester, this story might well have remained a draft I was too scared to show anyone. It also was greatly improved by his advice and notes. It's often said your first editor is your agent, and Donald was a brilliant one. Thanks also to my editors at Bloomsbury, Jim Martin and Angelique Neumann, and Counterpoint, Jennifer Alton, and the staff at the various libraries I used in working on this, especially the British Library whose careful reopening of reading rooms under COVID-19 was so well handled.

I also want to thank the hundreds of researchers and writers who helped me write this book without their knowledge (some long dead). There are the characters in the book like Eunice Foote or Guy Callender but also the authors of secondary sources found in the sources section and footnotes. Due to the breadth of the story I wanted to tell, exacerbated by the challenges of research under lockdown, I'm indebted to many researchers from a huge range of specialisms whose work I've used to weave my larger story together. I make no pretence to have generated original historical research myself. This is rather an exercise in popularisation and storytelling. I've peppered my book with recommendations for further reading, and I do hope you follow them, to read in detail the stories I could only really nod to here (plus they are all much better writers than me).

Finally, thanks to the people who don't fit into the categories above, but helped me write this book in some way or another, whether it was reading a draft, suggesting something I should read up on, an invitation to speak in their space-themed shed or radio show, general encouragement or an insightful question that sparked something: The

Bells – Angie, Jim, Fifi and Maryam, Jon Agar, Katherine Allen, Carolyn Cobbold, Adam Cole, Steve Cross, Anthony Cummings, Ruth Garde, Lucy Gilliam, Genevieve Guenther, Vanessa Heggie, Christian Hunt, Roland Jackson, Mun Keat Looi, Kirsty Lothian, Justin Pickard, Dave Powell, Becky Purvis, Hauke Riesch, Lucy Santos, Jon Spooner, Ian Steadman, Natalie Steed, Simon Werrett and Ed Yong. And thanks to Ian Pocock too, even if he was, from the start, very clear he thought the whole idea was quite boring.

Index